AN INTRODUCTION TO OCEAN REMOTE SENSING

An Introduction to Ocean Remote Sensing describes the use of satellite data in the retrieval of oceanic physical and biological properties. It gives many examples of the kinds of data that can be acquired and describes their oceanographic application. It also describes the numerous national and international programs in satellite oceanography that have been initiated during the past two decades, and reviews current and future programs up to 2019. The book covers radiative transfer, ocean surface properties, satellite orbits, instruments and methods, visible remote sensing of biological properties, infrared sea surface temperature retrieval, passive microwave measurements, scatterometer wind retrieval, altimetry and SAR. New and proposed techniques, such as polarimetric passive microwave radiometers and SARs, interferometric radar altimetry and sea surface salinity retrieval are also discussed.

This textbook is designed for use in graduate and senior undergraduate courses in satellite oceanography, and will prepare the student and interested scientist to use satellite data in oceanographic research.

SEELYE MARTIN received his Ph.D. in engineering mechanics from Johns Hopkins University in 1967 then spent two years as a research associate in the Department of Meteorology at the Massachusetts Institute of Technology. In 1969 he took up a position in the School of Oceanography at the University of Washington where he is now a Professor of physical oceanography. He has taught courses on remote sensing of the oceans since 1987. Professor Martin has been involved with passive microwave, visible/infrared, and radar ice research since 1979 and was a member of the NASA Earth Observing System (EOS) committee from 1984 to 1989. Since this time, he has served on additional NASA committees and panels involving SAR and high latitude processes. He has made many trips to the Arctic for research on sea ice properties and oceanography and was a Visiting Scientist at the National Institute of Polar Research in Tokyo from 1993 to 1994.

AN INTRODUCTION TO OCEAN REMOTE SENSING

SEELYE MARTIN

School of Oceanography
University of Washington

 CAMBRIDGE
UNIVERSITY PRESS

PUBLISHED BY THE PRESS SYNDICATE OF THE UNIVERSITY OF CAMBRIDGE
The Pitt Building, Trumpington Street, Cambridge, United Kingdom

CAMBRIDGE UNIVERSITY PRESS
The Edinburgh Building, Cambridge, CB2 2RU, UK
40 West 20th Street, New York, NY 10011–4211, USA
477 Williamstown Road, Port Melbourne, VIC 3207, Australia
Ruiz de Alarcón 13, 28014 Madrid, Spain
Dock House, The Waterfront, Cape Town 8001, South Africa

http://www.cambridge.org

First published 2004

Printed in the United Kingdom at the University Press, Cambridge

Typeface Times 10/13 pt.   *System* LATEX 2_ε [TB]

A catalog record for this book is available from the British Library

Library of Congress Cataloging in Publication data
Martin, Seelye, 1940–
An introduction to ocean remote sensing / Seelye Martin.
p.   cm.
Includes bibliographical references and index.
ISBN 0 521 80280 6
1. Oceanography – Remote sensing.   I. Title.
GC10.4.R4M375   2004
551.46′028 – dc22   2003061380

ISBN 0 521 80280 6 hardback

To the memory of my mother
Lucy Gray Martin

April 19, 1915–June 13, 2002

Contents

vii

The plates will be found between pages 196 and 197

Preface

During the past two decades, there has been a dramatic growth in the number and variety of ocean observing satellites. This growth, combined with a similar growth in computational resources and surface receiving and distribution networks, has greatly increased our knowledge of the properties of the upper ocean, its surface, and the overlying atmosphere. During the same period, an increasing number of countries have either launched or contributed instruments to ocean satellites. In the early 1980s, only the United States and the Soviet Union had ocean observing programs. In 2000, the countries or groups of countries with ocean observing programs included Brazil, Canada, China, Europe, India, Japan, Republic of Korea, Republic of China, Russia, Ukraine, the United States and at least one private corporation.

The developments in the application of the electromagnetic spectrum to ocean observing combined with our understanding of the ocean surface and atmospheric properties have served as the basis for a large variety of innovative instrumentation. Many of the sensors that were experimental in the 1980s are now essential tools of oceanography. These include the use of narrow band optical sensors to estimate biological productivity and observe plankton-associated fluorescence; infrared measurements of sea surface temperature that approach the accuracy necessary to observe climate change; passive microwave sensors that provide global cloud-independent observations of sea surface temperature; and altimeters capable of measuring sea surface height to a 2-cm precision. Because of the increase in computational resources, these data sets are rapidly made available and are often posted on public websites.

Remote sensing involves primarily the use of the electromagnetic spectrum to observe the ocean, and secondarily the use of gravity observations to observe ocean currents and tides. Because remote sensing involves many disciplines, the book provides under one cover the necessary background in electromagnetic theory, atmospheric and seawater properties, physical and biological oceanography, physical properties of the sea surface and satellite orbital mechanics. The material discussed ranges from the reflective and emissive properties of clouds and foam to radar scattering properties of ocean waves, to the optical properties of plankton-associated pigments. It also includes many examples. The book describes the development of satellite oceanography from 1975 to 2004, and summarizes future activities through 2019. As prerequisites, the book requires only an introductory knowledge of electromagnetic theory and differential equations.

The text divides into five parts. Chapters 1, 2 and 3 provide an introduction to satellite systems, ocean surface properties and electromagnetic theory. Chapters 4–7 discuss remote sensing in the visible and infrared spectrum, including atmospheric properties, the ocean/atmosphere interface, the visible retrieval of ocean color and the infrared retrieval of sea surface temperature. Chapters 8 and 9 discuss the passive microwave, including antennas, instruments, atmospheric properties and the retrieval of ocean surface and atmospheric variables. Chapters 9–13 discuss the active microwave, including the use of a large variety of radars to retrieve wind speed and direction, sea surface height, and images of the ocean surface. Finally, for 2004 through 2019, Chapter 14 describes the approved and proposed satellite missions.

I wrote the first draft of this book during 1993–94, when I was a visiting scientist at the National Institute of Polar Research in Tokyo. I greatly appreciate the opportunities and hospitality offered to me by the Institute and thank Takao Hoshiai and Nobuo Ono for inviting me. I also thank the Japanese Ministry of Education, Science, Sports and Culture (Monbusho) and Dean Ross Heath of the College of Ocean and Fisheries Sciences, University of Washington for financial support.

The book benefited from my work with the National Aeronautics and Space Administration (NASA); I am particularly grateful to Dixon Butler and the other members of the EOS steering committee during my three years of service, and to the research support of Robert Thomas, Kim Partington and Waleed Abdalati at the Oceans and Ice Division at NASA Headquarters. I taught the subject matter of this book both by myself and jointly with Miles Logsdon. I thank Miles and all of the students, who always managed to focus on those points that I did not understand. In my teaching, I also benefited from the class notes of Dudley Chelton and those of James Mueller and Carlyle Wash, and the textbooks of Charles Elachi, George Maul, Ian Robinson and Robert Stewart.

At NASA Goddard Space Flight Center (GSFC), I thank William Barnes, Don Cavalieri, Josephino Comiso, Wayne Esaias, Antony Liu, Charles Koblinsky, Charles McClain, Claire Parkinson and Menghua Wang; at Brigham Young University, David Long; at the Jet Propulsion Laboratory (JPL), Robert Benada, Ben Holt, Ron Kwok, Lee Fu, Timothy Liu, Son Nghiem, William Patzert, Ernesto Rodriguez and Simon Yueh. Also at GSFC, I particularly thank Gene Feldman for his help with SeaWiFS. At NOAA, I thank Pablo Clemente-Colón, Maria Colton, Michael Mignogno, Stan Wilson and Michael Van Woert; at the European Space Agency (ESA), Mark Drinkwater; at Centre National d'Etudes Spatiales (CNES), Yann Kerr; at Sequoia Scientific, Curtis Mobley. At the Naval Research Laboratory (NRL), I thank Gene Poe and Karen St. Germain; at Oregon State University, Dudley Chelton and Michael Freilich; at Monterey Bay Aquarium Research Institute, Francisco Chavez. At Earth and Space Research, I thank Gary Lagerloef; at the National Polar-orbiting Operational Environmental Satellite System (NPOESS) Integrated Project Office, James Jolley; at Remote Sensing Systems, Chelle Gentemann, Tom Meissner and Frank Wentz. At University of Alaska, I thank Kevin Engle and Nettie LaBelle-Hamer; at University of California at

San Diego, Detlef Stammer and William Melville; at University of California at Irvine, Djamal Khelif and Jon Stairs. At Scott Polar Research Institute, I thank Peter Wadhams; at Southampton Oceanography Centre, Paolo Cipollini and Meric Srokosz; at Tokyo Mercantile Marine University, Manami Ide and Shogo Hayashi. At Florida State University, I thank James J. O'Brien, Mark Bourassa and Josh Grant; at University of Miami, Peter Minnett; at University of Maine, Mary Jane Perry and Brandon Sackmann. I thank Leonid Mitnik for information on the Russian and Ukrainian space programs, and Alcatel Space, Ball Aerospace and Technologies Corporation, Boeing Satellite Systems, Northrop Grumman Corporation, Orbimage and Raytheon for use of their figures and data. Other acknowledgements are in the text.

At the University of Washington (UW), I thank Arthur Nowell for his support, both as Director of the School of Oceanography and later as Dean. I also thank the present director, Bruce Frost. I further thank Neal Bogue, Robert Brown, Laurie Bryan, Robert Drucker, David English, Charles Eriksen, Rita Horner, Andrew Jessup, Kristina Katsaros, Evelyn Lessard, Ellen Lettvin, David Martin, Jérôme Patoux, William Plant, Penny Rowe, Drew Rothrock, Kittie Tucker and Dale Winebrenner. I acknowledge the resources and staff of the UW Library system, and thank the staff for their role in the rapid recovery of the Oceanography and Engineering Libraries from the damage caused by the February 28, 2001 Nisqually Earthquake.

I also thank Lin Robinovitch for her skill and assistance with figure preparation and thank NASA for allowing the use of carryover funds from NAGW-6894 for partial support of her work. I thank the School of Oceanography for additional support of figure preparation. I thank my cousin Ann Warren Turner for her perspective on large writing projects, and for recommending the book *Bird by Bird* by Anne Lamott, as a primer on organization and writing. I thank Mary Jane Perry for help with the biology and for her critical reading of an early draft of the biology chapter, Ted Straub for his reading of several chapters, and Mike Alfultis for his early and detailed critique of the entire book. Any errors are my own.

Many of the papers, reports and presentations cited in this book are taken from websites cited in the book and maintained by agencies such as NASA, ESA and the Japan Aerospace Exploration Agency (JAXA), formerly the Japanese National Space Development Agency (NASDA). Although these websites are current as of 2003, their addresses may change. In this case, the material can generally be found using search engines such as Google. At Cambridge University Press I thank Jayne Aldhouse, Susan Francis, Matt Lloyd, Margaret Patterson and Sally Thomas.

The generosity of my parents, William Ted Martin and Lucy Gray Martin, made it possible for me to finish the book. I thank my son and daughter, Carl William Coryell-Martin and Maria Elizabeth Coryell-Martin for putting up with all this, and my wife Julie Esther Coryell for her optimism that I might finish the book, for reading all of the chapters in draft, and for her support. Finally, I ask the reader to remember that each of the satellites, instruments and algorithms described in this book began as an idea generated by a single individual or a small committee.

Chemical symbols

Ar	Argon
CH_4	Methane
CO	Carbon monoxide
CO_2	Carbon dioxide
Fe	Iron
H_2O	Water
N_2	Nitrogen
N_2O	Nitrous oxide
O_2	Oxygen
O_3	Ozone
$H\alpha, H\beta, H\gamma$	Hydrogen lines in the Fraunhofer spectrum
Mg-I	Magnesium-Iodine line
O_2-A	Oxygen-A line

Mathematical symbols

Symbol	Unit	Definition
A	m^2	Area, or instrument aperture area
A_e	m^2	Effective antenna aperture area
A_{FOV}	area	Antenna half-power field-of-view
$A_i(400)$	m^{-1}	Reference absorption at 400 nm; i refers to particulates or CDOM
$a(\lambda)$	m^{-1}	Volume absorption coefficient
$\hat{a}(\lambda; \theta, \phi)$	—	Ratio of gray body to blackbody absorption; in VIR, the absorptance, in microwave, the absorptivity
a_{CDOM}	m^{-1}	CDOM absorption coefficient
a_w	m	Amplitude of ocean surface waves
$a_w(\lambda), a_p, a_\phi, a_T$	m^{-1}	Absorption coefficients for seawater, particulate, phytoplankton and total absorption
B	$W\ m^{-2}\ sr^{-1}$	Brightness, used for radiance in the passive microwave
$\mathbf{B}$	$tesla\ m^{-1}$	Magnetic field vector
B_f	$J\ m^{-2}\ sr^{-1}$	Frequency form of spectral brightness
$b(\lambda)$	m^{-1}	Volume scattering coefficient of seawater
$b_b(\lambda), b_{bw}(\lambda)$	m^{-1}	Backscatter coefficient of pure seawater
$b_{bT}(\lambda)$	m^{-1}	Total backscatter coefficient of seawater
$^\circ C$		Degrees Celsius
C_a	$mg\ Chl\text{-}a\ m^{-3}$	Chlorophyll concentration
C_W, C_I	—	Concentrations of open water and sea ice
c	$m\ s^{-1}$	Speed of light in vacuum
$c(\lambda)$	m^{-1}	Volume attenuation coefficient of seawater
D	cm, m	Aperture diameter of a lens or length of an antenna
$\hat{d}(\lambda)$	—	Normalized absorption depth
$d_a(\lambda)$	m	Absorption depth of radiation in seawater

Symbol	Unit	Definition
E	W m^{-2}	Irradiance, the incident flux density per unit area
$\mathbf{E}$	V m^{-1}	Electric field vector
$\hat{E}$	J	Energy of a photon
$\mathbf{E}_0$	V m^{-1}	Reference amplitude of an electric field vector
$E_d(\lambda, 0_+)$	W m^{-2}	Downwelled solar irradiance measured just above the ocean surface
$E_R(\chi, \psi)$	km	Height of reference ellipsoid above Earth's center of mass
$E_u(0_-)$	W m^{-2}	Upwelled solar irradiance just below the water surface
E_V, E_H	V m^{-1}	Vertically and horizontally polarized components of the electric field vector
$e(\lambda; \theta, \phi)$	—	Emissivity, which is the ratio of a gray body to blackbody radiance
e_0	—	Temperature- and salinity-dependent emissivity of a specular ocean surface
$F(\lambda, z)$	W m^{-2}	Solar irradiance at a height z in the atmosphere
F_n	—	Normalized power or radiation pattern
$F_S(\lambda)$	W m^{-2}	Solar irradiance at the top of the atmosphere
$F'_S(\lambda)$	W m^{-2}	$F_S(\lambda)$ attenuated by two passes through the ozone layer
f	s^{-1}	Coriolis parameter
f	Hz	Frequency
$f(x)$	V m^{-1}	Antenna illumination pattern
f_L	m	Focal length
f_N	s^{-1}	Nyquist sampling frequency
$f_P(T, \lambda)$	W m^{-3} sr^{-1}	Planck blackbody radiance
G	—	Antenna gain
G_0	—	Maximum antenna gain
G_R	—	Gradient ratio used in the derivation of sea ice concentration
g	m s^{-2}	Acceleration of gravity
H	km	Radial distance of a satellite from Earth's center of mass
$H_{1/3}$	m	Significant wave height
Hz	s^{-1}	Cycles-per-second
h	length	Height of satellite above ocean surface
h_S	length	Height of sea surface above Earth's center of mass
$\overline{h}_S$	length	Temporal mean of sea surface height

Symbol	Unit	Definition
h	J s	Planck constant, 6.625×10^{-34} J s
I	deg	Inclination, the angle between the Earth's rotation axis and the normal to the orbit plane
$I(r, \theta, \phi)$	W sr^{-1}	Radiant intensity
I_0	W sr^{-1}	Maximum radiant intensity
i		Imaginary part of complex number
J		Joules
K		Degrees Kelvin
k, k_{im}	m^{-1}	Real and imaginary part of the wavenumber
$\mathbf{k}$	m^{-1}	Vector wavenumber
k_B	J K^{-1}	Boltzmann constant, 1.38×10^{-23} J K^{-1}
k_w	m^{-1}	Wave number of ocean waves
L	mm	Columnar equivalent of non-raining cloud liquid water
$L(\lambda)$	μW cm^{-2} nm^{-1} sr^{-1} W m^{-3} sr^{-1}	Radiance (Alternative units of L)
$L_A(\lambda)$	μW cm^{-2} nm^{-1} sr^{-1}	Path radiance generated by aerosol atmospheric scattering
L_E	km	Equatorial separation between successive orbits
$L_f(\lambda)$	J m^{-2} sr^{-1}	Frequency form of spectral radiance
$L_R(\lambda)$	μW cm^{-2} nm^{-1} sr^{-1}	Path radiance generated by Rayleigh scattering
$L_S(\lambda)$	μW cm^{-2} nm^{-1} sr^{-1}	Solar radiance at the top of the atmosphere
$L_T(\lambda)$	μW cm^{-2} nm^{-1} sr^{-1}	Total radiance received at the satellite
$L_w(\lambda)$	μW cm^{-2} nm^{-1} sr^{-1}	Water-leaving radiance
$[L_w(\lambda)]_N$	μW cm^{-2} nm^{-1} sr^{-1}	Normalized water-leaving radiance
$L_\lambda(\lambda)$	μW cm^{-2} nm^{-1} sr^{-1}	Wavelength form of spectral radiance
l	m	Length of an imaging radar
M	W m^{-2}	Exitance, or emitted flux or power density
$N(\chi, \psi)$	m	Geoid undulation, or height of geoid relative to the reference ellipsoid E_R
Np, nepers	—	Units of atmospheric absorption used in microwave
$NE\Delta T$	K	Noise-equivalent-delta-temperature
$NE\Delta L$	μW cm^{-2} nm^{-1} sr^{-1}	Noise-equivalent-delta-radiance
$NE\Delta\sigma_0$	—	Noise-equivalent-delta-sigma-zero
n	—	Real part of the index of refraction
P	—	For radiometers, subscript indicates V or H polarization. For radars, subscript indicates VV or HH polarization

List of mathematical symbols

Symbol	Unit	Definition
$P(\theta)$	sr^{-1}	Atmospheric scattering phase function
P_R	—	Polarization ratio used in the derivation of sea ice concentration
$P_R(\theta)$	sr^{-1}	Rayleigh atmospheric scattering phase function
p	$\mathrm{kg\ m}^{-1}\ \mathrm{s}^{-2}$	Atmospheric pressure
Q	—	Coefficient used in description of the water-leaving radiance
$R(\lambda)$	—	Plane irradiance reflectance
$R(\lambda, 0_-)$	—	Irradiance reflectance evaluated just below the surface
R_0	km	Distance from radar to target
R_c	mm, μm	Radius of curvature of the sea surface
$R_F(\lambda)$	—	Irradiance reflectance of foam
R_R	$\mathrm{mm\ h}^{-1}$	Rain rate
$R_{rs}(\lambda)$	—	Remote sensing reflectance
r	length	Radius
$\mathbf{r}$	length	Vector radius (r, θ, ϕ)
$r(\theta)$	—	Unpolarized radiance reflectance
S	psu	Salinity
S_N	—	Signal-to-noise ratio
S_S	psu	Surface salinity
T	°C, K	Temperature
$\overline{T}$	°C, K	Mean temperature of the lower troposphere
$T(\theta)$	—	Interface transmittance
T_3, T_4, T_5	K	AVHRR brightness temperatures for bands 3, 4, 5
$T_{22}, T_{23}, T_{31}, T_{32}$	K	MODIS brightness temperatures for bands 22, 23, 31, 32
T_A	K	Antenna temperature
T_a	K	Air temperature
T_B	K	Brightness temperature
T_b	°C	Buoy or bulk temperature
T_{BV}, T_{BH}	K	Vertically and horizontally polarized components of brightness temperatures
T_{ext}	K	Extraterrestrial brightness temperature exclusive of the sun
T_{gal}	K	Brightness temperature of the Milky Way galaxy
T_S	°C, K	Ocean surface skin temperature
T_{sfc}	°C, K	Externally supplied surface temperature to algorithms

Symbol	Unit	Definition
T_{sol}	K	Solar contribution to the antenna brightness temperature
T_{sun}	K	Solar brightness temperature
T_{univ}	K	The 2.7 K universe background temperature
T_w	s	Period of ocean surface waves
t		Time
t	—	In the visible/infrared, the atmospheric transmittance; in the microwave, the atmospheric transmissivity
$t_D(\lambda)$	—	Diffuse transmittance
U	m s^{-1}	The scalar wind speed at a 10-m height
U_0	m s^{-1}	Spacecraft velocity
U_{LOS}	m s^{-1}	Line-of-sight wind speed, the wind speed in the azimuthal look direction of a passive microwave radiometer
u, v	m s^{-1}	x- and y-components of an ocean current
V	mm	Equivalent height in liquid water of the columnar water vapor
v	m s^{-1}	Local phase speed of light
w	m	Width of an imaging radar
$\boldsymbol{x}$	length	Vector position (x, y, z)
X, Y	—	Coefficients used in discussion of particulate scattering properties
X_S	length	Imaging radar cross-track swathwidth
Y_S	length	Imaging radar along-track swathwidth
z_H	km	Reference height for the top of the atmosphere
α	deg	Scattering angle relative to the forward direction
α	—	Ångström exponent used to describe aerosols
α_S	sr	Solid angle resolution of an ideal optical instrument
$\beta(\alpha, \lambda)$	km^{-1} sr^{-1}, m^{-1} sr^{-1}	Atmospheric and oceanic volume scattering function
$\tilde{\beta}(\alpha, \lambda)$	sr^{-1}	Oceanic scattering phase function
β_0	km^{-1} sr^{-1}, m^{-1} sr^{-1}	Isotropic scattering phase function
$\beta_T, \beta_w, \beta_p, \beta_\phi$	m^{-1} sr^{-1}	Total, pure seawater, particulate, and phytoplankton volume scattering function
$\Delta\hat{E}$	J	Energy difference associated with a change in the internal state of a molecule or atom
Δf	Hz, MHz	Instrument bandwidth, also used to describe Doppler shift

Symbol	Unit	Definition
Δh_{ion}	m	Range delay caused by atmospheric free electrons
ΔT_{45}	K	Temperature difference between AVHRR channels 4 and 5, $\Delta T_{45} = T_4 - T_5$
ΔT_{53}	K	Temperature difference between AVHRR channels 5 and 3, $\Delta T_{53} = T_5 - T_3$
$\Delta x, \Delta y$	m	Radar resolution in the cross-track and along-track direction
$\Delta \theta_{1/2}$	deg	Half-power beamwidth; for imaging radars, the half-power beamwidth in the cross-track direction
$\Delta \phi_{1/2}$	deg	Half-power beamwidth in the along-track direction
ε	farad m^{-1}	Electrical permittivity
$\varepsilon(\lambda, \lambda_0)$	—	Single scattering color ratio for aerosols
ε_0	farad m^{-1}	Permittivity in vacuum
ε_r	—	Complex dielectric constant, $\varepsilon_r = \varepsilon' + i\varepsilon''$
ζ	m	Sea surface height relative to the geoid
ζ_D	m	Dynamic height, or the oceanographic height calculated from the vertical density structure
η	—	Complex index of refraction, $\eta = n + i\chi$
η	m	Vertical displacement of ocean surface waves
η_M	—	Main beam efficiency of a microwave antenna
θ	deg	Angle
θ_S	deg	Solar zenith angle
θ_Z	deg	Zenith angle
$\kappa_A, \kappa_E, \kappa_S$	km^{-1}	Atmospheric absorption, extinction and scattering coefficients
κ_R	km^{-1}	Rayleigh scattering attenuation coefficient
κ_{oxy}	km^{-1}	Oxygen absorption coefficient
κ_{vap}	km^{-1}	Water vapor absorption coefficient
λ	nm, μm	Radiation wavelength
λ_w	mm, m	Wavelength of ocean surface waves
μ	henry m^{-1}	Magnetic permeability
μ_0	henry m^{-1}	Vacuum permeability
Π	W m^{-4} sr^{-1}	The atmospheric radiative source term
ρ	kg m^{-3}	Density of seawater
ρ_a	kg m^{-3}	Density of air
ρ_H, ρ_V	—	Horizontal, vertical reflection coefficients
ρ_{ion}	TECU	Free-electron columnar density

Symbol	Unit	Definition
$\rho_w(\lambda)$	—	Extraterrestrial reflectance generated by the water-leaving radiance
$[\rho_w(\lambda)]_N$	—	Normalized extraterrestrial reflectance
σ	siemens m^{-1}	Electrical conductivity
σ	m^2	Radar scattering cross section
σ^2	—	Mean-square sea surface slope
σ_0	—	Normalized radar scattering cross section (pronounced sigma-zero)
σ_N	—	Standard deviation of noise
$\sigma_{VV}, \sigma_{HH}, \sigma_{HV}, \sigma_{VH}$	—	Normalized radar scattering cross section for VV, HH, HV, and VH transmitting and receiving
σ_η	m	Root-mean-square sea surface height
τ	s	Pulse duration or length
$\tau(\lambda)$	—	Optical depth
τ_A	—	Optical depth associated with aerosol scattering
τ_{OZ}	—	Optical thickness of the ozone layer
$\tau_R(\lambda)$	km	Rayleigh optical thickness
Φ	W	Radiant flux or power
Φ_N	W	Noise generated internally to an instrument
Φ_T	W	Total radiant flux or power transmitted by an antenna
$\Phi_{(V, H)}$	W	V-pol or H-pol radiant flux received by an antenna
Φ_λ	W μm^{-1}	Spectral form of the radiant flux
Φ_σ	W	Received power corrected for atmospheric attenuation
ϕ	deg	Azimuth angle
ϕ_R	deg	Azimuthal angle relative to the wind direction
ϕ_W	deg	Azimuthal wind direction
χ	—	Imaginary part of the index of refraction
χ, ψ	deg	Latitude, longitude
Ω	sr	Solid angle
Ω_E	s^{-1}	Angular rotation of the Earth
Ω_M	sr	Main beam solid angle of a microwave antenna
Ω_p	sr	Pattern solid angle of a microwave antenna
ω	s^{-1}	Radian frequency of an electromagnetic wave
$\omega_0(\lambda)$	—	Single scattering atmospheric albedo
$\omega_A(\lambda)$	—	Aerosol single scattering albedo
$\omega_R(\lambda)$	—	Rayleigh single scattering albedo

Abbreviations and acronyms

AATSR	Advanced ATSR (ESA)
ADEOS-1, -2	Advanced Earth Observing Satellite (Japan)
AGC	Automatic Gain Control (Altimeter function)
AHRPT	Advanced High Resolution Picture Transmission (METOP)
ALOS	Advanced Land Observing Satellite (Japan)
ALT	Altimeter on TOPEX/POSEIDON
AMSR	Advanced Microwave Scanning Radiometer (Japan) on ADEOS-2
AMSR-E	AMSR-EOS (Japan) on AQUA
AOML	Atlantic Oceanographic and Meteorological Laboratory (NOAA)
AOP	Apparent Optical Properties
APC	Antenna Pattern Correction
APT	Automatic Picture Transmission (data transfer mode for AVHRR)
AQUA	Second major EOS satellite (Not an abbreviation)
ASAR	Advanced SAR (ENVISAT)
ASCAT	Advanced Scatterometer (METOP)
ATSR	Along-Track Scanning Radiometer (ESA)
AVHRR	Advanced Very High Resolution Radiometer (US)
AVISO	Archiving, Validation and Interpretation of Satellite Oceanographic data (France)
CalTech	California Institute of Technology
C-band	Frequencies of about 5 Ghz
CDOM	Colored Dissolved Organic Material
CHAMP	CHAllenging Minisatellite Payload (German gravity mission)
Chl-a	Chlorophyll-a
CMIS	Conical-scanning Microwave Imager/Sounder (US passive microwave imager on NPP and NPOESS)
CNES	Centre National d'Études Spatiales (National Center for Space Studies, France)
CryoSat	ESA satellite for ice sheet investigation
CSA	Canadian Space Agency
CZCS	Coastal Zone Color Scanner

dB	Decibels
DMSP	Defense Meteorological Satellite Program (US), also name of a satellite
DOD	Department of Defense (US)
DORIS	Doppler Orbitography and Radiopositioning Integrated by Satellite (France)
ECMWF	European Centre for Medium-range Weather Forecasts
EFOV	Effective Field-Of-View; shape of the FOV after time-averaging
EM	ElectroMagnetic
EMR	ElectroMagnetic Radiation
ENVISAT	Environmental Satellite (ESA)
EOS	Earth Observing System (US, with international components)
ERS-1, -2	European Remote-sensing Satellite
ESA	European Space Agency
ESMR	Electrically Scanned Microwave Radiometer (US)
EUMETSAT	European Organization for the Exploitation of Meteorological Satellites
FLH	Fluorescence Line Height
FM	Frequency Modulation
FOV	Field-Of-View, see also EFOV, IFOV
FY	Feng Yun (Wind and Cloud) as in FY-1C and FY-1D; name of satellite (China)
FY	First Year, as in first year sea ice
GAC	Global Area Coverage (AVHRR data mode)
Gbit	Gigabit or 10^9 bits
GCOM	Global Change Observation Missions (Japan)
GLAS	Geoscience Laser Altimeter System (US)
GLI	Global Imager, ocean color instrument on ADEOS-2 (Japan)
GOCE	Gravity Field and Steady-State Ocean Circulation Explorer (ESA)
GOES	Geostationary Operational Environmental Satellite (US)
GHz	Gigahertz
GRACE	Gravity Recovery and Climate Experiment
HH	Antenna that transmits and receives with an H-polarization
H-pol	Horizontally polarized
HRD	High Rate Data (NPOESS data transfer mode)
HRPT	High Resolution Picture Transmission (AVHRR data transfer mode)
HV	Antenna that transmits with an H-polarization and receives with a V-polarization
HY	Haiyang (Ocean) satellite as in HY-1 (China)
IAPSO	International Association for Physical Sciences of the Ocean
ICESat	Ice, Cloud and land Elevation Satellite (US)

IEEE	Institute of Electrical and Electronics Engineers
IFOV	Instantaneous Field-Of-View, or Instrument Field-Of-View
IJPS	Initial Joint Polar-orbiting operational satellite System (US, EUMETSAT)
IOP	Inherent Optical Properties
IPO	Integrated Project Office (NPOESS)
IR	Infrared
ITCZ	Inter-Tropical Convergence Zone
JASON-1	TOPEX follow-on (Not abbreviation)
JAXA	Japan Aerospace Exploration Agency (replaced NASDA)
JERS-1	Japanese Earth Resources Satellite
JMA	Japan Meteorological Agency
JMR	Jason Microwave Radiometer
JPL	Jet Propulsion Laboratory (NASA), operated by CalTech
K-band	Frequencies between 11 and 36 GHz
K_u-band	Frequencies of about 14 GHz
KOSMOS	USSR satellite series
LAC	Local Area Coverage (data mode for AVHRR)
L-band	Frequencies of about 1 GHz
LRA	Laser Retroreflector Array
LRD	Low Rate Data (NPOESS data transfer mode)
M-AERI	Marine-Atmosphere Emitted Radiance Interferometer (US)
Mbps	Megabits-per-second
MCSST	Multi-Channel Sea Surface Temperature (algorithm)
MEDS	Maritime Environmental Data Service (Canada)
MERIS	Medium Resolution Imaging Spectrometer (ENVISAT)
METEOSAT	Geosynchronous Meteorology Satellite (EUMETSAT)
METOP	MÉTéorologie OPérationnelle (Operational Meteorology) (EUMETSAT satellite)
MHz	Megahertz
MOBY	Marine Optical Buoy (Ocean color calibration buoy near Hawaii)
MODIS	Moderate Resolution Imaging Spectroradiometer on TERRA, AQUA
MODTRAN	Program for calculation of atmospheric transmissivity
MOS	Modular Optical Scanner (Germany)
MSL	Mean Sea Level
MVIRSR	Multispectral Visible-Infrared Scanning Radiometer (China)
MY	Multiyear, as in multiyear sea ice
NASA	National Aeronautics and Space Administration (US)
NASDA	National Space Development Agency (Japan)
NDBC	National Data Buoy Center (US)
NDT	Nitrate-Depletion Temperature

NESDIS	National Environmental Satellite Data and Information Service (US)
NIR	Near-infrared
NLSST	NonLinear SST (algorithm)
NOAA	National Oceanic and Atmospheric Administration (US)
NOAA-10, -11, ...	Names of NOAA operational polar orbiting satellites
NPOESS	National Polar-orbiting Operational Environmental Satellite System (US)
NPP	NPOESS Preparatory Project (US)
NRCS	Normalized Radar Cross Section
NSCAT	NASA Scatterometer (ADEOS-1)
NWP	Numerical Weather Prediction
OC3M	Ocean Chlorophyll Version 3 MODIS bio-optical algorithm
OC4	Ocean Chlorophyll Version 4 SeaWiFS bio-optical algorithm
OCTS	Ocean Color and Temperature Sensor (Japan)
OKEAN	Russia/Ukraine series of satellites
OLS	Optical Line Scanner (Visible/infrared instrument on DMSP)
OSTM	Ocean Surface Topography Mission (US/France, JASON-1 follow-on)
OVWM	Ocean Vector Wind Mission
OW	Open Water (sea ice algorithms)
PALSAR	Phased Array L-Band SAR (Japan)
Pixel	Abbreviation for picture element
PMEL	Pacific Marine Environmental Laboratory (NOAA)
POD	Precision Orbit Determination
POES	Polar Operational Environmental Satellite (US)
POLDER	Polarization and Directionality of the Earth's Reflectances (France), ocean color instrument on ENVISAT
POSEIDON	Premier Observatoire Spatial Étude Intensive Dynamique Océan et Nivosphère, French contribution, TOPEX/POSEIDON satellite.
PRF	Pulse repetition frequency
psu	Precision salinity units (Units of oceanic salinity)
RA-2	Radar Altimeter-2 (ENVISAT altimeter)
RADARSAT-1, -2	SAR satellites (Canada)
RGB	Red-Green-Blue color mixing
RGPS	RADARSAT Geophysical Processing System (US)
rms	Root-mean-square
rss	Root-sum-of-the-squares
RTE	Radiative Transfer Equation
SAR	Synthetic Aperture Radar
SASS	SEASAT-A Satellite Scatterometer (US)

ScanSAR	Wide-swath SAR imaging mode (partial abbreviation)
SeaBAM	SeaWiFS Bio-optical Algorithm Mini-Workshop
SEASAT	First ocean observing satellite (1979, US)
SeaWiFS	Sea-viewing Wide Field-of-view Sensor (US)
SeaWinds	Radar vector wind instrument (Not an abbreviation)
SGLI	Second-generation Global Imager (Japan)
SIRAL	SAR Interferometric Radar Altimeter (ESA)
SLAR	Side-Looking Airborne Radar
SLR	Side-Looking Radar
SLR	Satellite Laser Ranging
SMMR	Scanning Multichannel Microwave Radiometer (US)
SMOS	Soil Moisture and Ocean Salinity instrument (ESA)
SSALT	Solid State Altimeter on TOPEX (France)
SSH	Sea Surface Height
SSM/I	Special Sensor Microwave/Imager (US)
SSMI/S	Special Sensor Microwave Imager/Sounder (SSM/I upgrade)
SSS	Sea Surface Salinity
SST	Sea Surface Temperature
SST4	MODIS SST algorithm at 4 μm
SWH	Significant Wave Height ($H_{1/3}$)
TECU	Total Electron Content Unit (1 TECU = 10^{16} electrons m^{-2}), columnar concentration of free electrons
TERRA	First major EOS satellite (Not an abbreviation)
TIR	Thermal-Infrared
TIROS-N	Television Infrared Observation Satellite-N (Early version of POES satellite)
TIW	Tropical Instability Waves
TMI	TRMM Microwave Imager (Japan)
TMR	TOPEX Microwave Radiometer
TOA	Top of the Atmosphere
TOGA-TAO	Tropical Ocean Global Atmosphere–Tropical Atmosphere Ocean
TOMS	Total Ozone Mapping Spectrometer
TOPEX	TOPography EXperiment (US/France)
TRMM	Tropical Rainfall Measuring Mission (US/Japan)
TRSR	Turbo Rogue Space Receiver BlackJack GPS receivers (Satellite GPS receivers used on JASON-1)
UKMO	United Kingdom Meteorological Office
UTC	Universal Time Coordinated
UV	Ultraviolet
VH	Antenna that transmits with a V-polarization and receives with an H-polarization

VIIRS	Visible/Infrared Imager/Radiometer Suite (NPP and NPOESS instrument)
VIR	Visible/Infrared
VNIR	Visible/Near-Infrared
V-pol	Vertically polarized
VV	Antenna that transmits and receives with a V-polarization
WindSat	Polarimetric radiometer for vector wind measurements (Not an abbreviation)
WSOA	Wide Swath Ocean Altimeter (Part of OSTM)
WVSST	Water Vapor Sea Surface Temperature (algorithm)
X-band	Frequencies of about 10 GHz

1
Background

1.1 Introduction

During the past thirty years, rapid technological growth has advanced the ability of satellites to observe and monitor the global ocean and its overlying atmosphere. Because similar advances occurred in computer systems and software, it is now possible rapidly to acquire and analyze large satellite data sets such as the global distribution of ocean waves, the variations in sea surface height associated with large scale current systems and planetary waves, surface vector winds, and regional and global variations in ocean biology. The immediate availability of these data allows for their assimilation into numerical models, where they contribute to the prediction of future oceanic weather and climate.

The ocean covers approximately 70% of the Earth's surface, is dynamic on a variety of scales, and contains most of the Earth's water as well as important marine ecosystems. For its role in biology, the ocean contains about 25% of the total planetary vegetation, with much of this restricted to a few limited coastal regions (Jeffrey and Mantoura, 1997). Regions of high biological productivity include the Grand Banks off Newfoundland, the Bering Sea and Gulf of Alaska, the North Sea and the Peruvian coast. Between 80 and 90% of the world fish catch occurs in these and similar regions. For its role in climate, determination of the changes in ocean heat storage and measurement of the vertical fluxes of heat, moisture and CO_2 between the atmosphere and ocean are critical to understanding global warming and climate change. Large scale ocean currents carry about half of the heat transported between the equator and the poles; the atmosphere transports the remainder. Away from the polar regions, the combination of these transports with the large oceanic heat capacity relative to the atmosphere means that the ocean moderates the global climate and improves the habitability of the continents (Wunsch *et al.*, 1981). Because some numerical models predict that global change will preferentially impact the polar regions, the ability to monitor the extent and thickness of the Arctic and Antarctic ice covers is important both for short-term navigation and for longer-term climate studies. All these examples illustrate the need to monitor and observe the ocean on a range of local to global scales.

The growth in satellite systems has been driven in part by technology and in part by societal concerns. Societal concerns include the importance of the ocean to national security and naval operations, global commerce, the generation of extreme weather and climate variability, fisheries management, the extraction of offshore gas, oil and minerals, and public health and recreation. Additionally, about 50% of the global population lives in cities within 50 km of the coast, where these cities are experiencing rapid growth. These coastal populations are vulnerable to natural hazards such as sea level rise, tsunamis, hurricanes and typhoons. There are also public heath considerations associated with the oceanic disposal of urban runoff, sewage and garbage, and in the monitoring and prediction of the growth of pathogenic organisms such as red tides. Satellite observing systems play a central role in addressing these concerns.

In the 1970s, the United States launched the first ocean remote sensing satellites. Since that time, the oceanic variables available from satellites include sea surface temperature (SST), the height and directional distribution of ocean swell, wind speed and direction, atmospheric water content and rain rate, the changes in sea surface height associated with ocean tides, currents and planetary waves, concentrations of phytoplankton, sediments, and suspended and dissolved material, and the areal extent and types of polar sea ice. Prior to the 1980s, such properties were determined from dedicated and expensive ship expeditions, or in the polar regions from surveys made from aircraft, drifting ships and ice islands. Such techniques meant that the ocean could only be surveyed slowly and incrementally. At present, satellite imagers can make simultaneous observations of motions with scales of 1–1000 km that are difficult to observe even from multiple ships. For variables such as the near surface air temperature that are not retrievable by remote sensing, satellites relay data from moored and drifting buoys that make direct measurements of such quantities. Even for those ocean depths that are inaccessible to satellite observations, there exist instruments that profile the ocean interior and periodically come to the surface, where they report their observations by satellite.

Because satellites survey a variety of oceanic properties with near global coverage and at intervals of 1–10 days, then rapidly transmit these observations to national and international forecast centers, these data are of great operational importance. In addition, the observations contribute to long-term studies of global climate change, sea level rise, and the decadal-scale atmospheric and oceanographic oscillations, including the Pacific Decadal Oscillation (PDO), the North Atlantic Oscillation (NAO), the El Niño/Southern Ocean Oscillation (ENSO), and the Arctic Oscillation (AO).

In the following, Section 1.2 defines remote sensing and describes its oceanographic applications. Section 1.3 describes the satellite orbits used in remote sensing, and Sections 1.4 and 1.5 describe the geosynchronous and sun-synchronous satellites. Section 1.6 discusses the imaging techniques used by satellites in sun-synchronous and other low earth orbits. Section 1.7 describes the different processing levels of satellite data. Section 1.8 describes past, present and pending satellite missions through 2007.

1.2 Definition of remote sensing

Earth remote sensing is primarily defined as the use of electromagnetic radiation to acquire information about the ocean, land and atmosphere without being in physical contact with the object, surface or phenomenon under investigation. Unlike shipboard measurements of quantities such as SST or wind speed, which are direct measurements made at a point by a thermometer or anemometer, remote sensing measurements of such quantities cover broad areas and are indirect, in that the geophysical quantity of interest is inferred from the properties of the reflected or emitted radiation. The sensors can range from a simple camera mounted on a pigeon to a multispectral satellite scanner.

Because the satellite instrument is not in physical contact with the phenomena under investigation, its properties must be inferred from the intensity and frequency distribution of the received radiation. This distribution depends on how the received radiation is generated and altered by its propagation through the atmosphere. This radiation has three principal sources: blackbody radiation emitted from the surface, reflected solar radiation, and the energy backscattered from directed energy pulses transmitted by satellite radars. The properties of the received radiation also depend on the sensor, which must be designed so that its observing wavelengths are appropriate for the phenomenon in question. Finally, the received data must be organized into images or data sets so that, if desired, the spatial distributions of the quantity under investigation can be viewed. This is the generally accepted definition of remote sensing; recently, it has been expanded to include the use of satellite measurements of gravity to infer land and ocean properties.

Because of the atmospheric interference that Chapters 4 and 9 describe, the ocean can only be viewed in three electromagnetic wavelength bands or windows, called the *visible*, *infrared* and *microwave*. In the visible and extending into the near infrared observations depend on reflected sunlight, and are restricted to daytime cloud-free periods. Because the visible spectrum contains the only wavelengths at which light penetrates to oceanic depths greater than 10 m, visible observations can yield information on the depth-averaged color changes associated with phytoplankton concentrations. In the infrared, the observations measure the blackbody radiation emitted from the top few micrometers of the sea surface, so that although these observations are independent of daylight, they are still restricted to cloud-free periods.

In the microwave and especially at the longer microwave wavelengths, the surface can be viewed through clouds and is only obscured by heavy rain. Microwave observations divide into *passive* and *active*. *Passive* instruments observe either reflected solar radiation or the naturally emitted blackbody radiation. Passive observations at multiple frequencies can retrieve such atmosphere and ocean surface properties as the amount of ice cover, the atmospheric water vapor and liquid water content, SST and wind speed. In contrast, active measurements are made by radars that transmit pulses of energy toward the ocean, then receive the backscatter, so that the radar provides its own illumination. The active microwave instruments include imaging radars (the Synthetic Aperture Radar or SAR), directed, pulsed vertical beams (altimeter), several pulsed fan beams at oblique angles to the satellite orbit (scatterometer), and an oblique rotating pulsed beam (also scatterometer). These scatterometers are highly directional non-imaging radars that receive the backscatter from relatively small surface areas, then correct it for atmospheric interference and instrument noise. Together, these instruments provide information on the roughness and topography of the sea surface, wind speed and direction, wave heights, directional spectra of ocean surface waves and the distribution and types of sea ice.

1.3 Satellite orbits

The orbit of an Earth observing satellite divides into two parts, the satellite motion in its orbit plane relative to the Earth's center of mass, and the satellite orbit relative to the rotating Earth. The time-dependent position of the satellite in its orbit is called the satellite *ephemeris*. For the rotating Earth, the orbit is frequently described in terms of its *ground track*, which is the time-dependent location of the surface intersection of the line between the satellite and the Earth's center of mass. The point directly beneath the satellite is called the satellite *nadir*. The following first considers the satellite motion in its orbit plane, then describes how the addition of the Earth's rotation determines the satellite ground track.

Rees (2001, Chapter 10), Elachi (1987, Appendix B) and Duck and King (1983) survey the commonly used, near circular remote sensing orbits. These are described relative to a rectangular coordinate system with its origin at the Earth's center of mass. The z-axis is in the northerly direction and co-located with the Earth's rotation axis, the x-axis is in the equatorial plane and in the direction γ of a star in the constellation Aries, and the y-axis is in the direction appropriate for a right-handed coordinate system. Relative to these axes, the six Keplerian Orbital Elements describe the satellite location. Because two of these are specific to elliptical orbits, for circular orbits, the six elements are reduced to four. As Figure 1.1 shows, these are the *right ascension of the ascending node* or simply the *ascending node* Ω, which is the angle between the x-axis and the point at which the orbit crosses the equator; the *radial distance H*, which is the satellite height above the Earth's center of mass; the *orbit true anomaly* θ, which is the angular position of the satellite in its orbit relative to Ω; the *inclination I*, which is the angle between the Earth's axis and the normal to the orbit plane with the convention that I is always positive. Of these variables, I and Ω specify the orientation and position of the orbit plane relative to the fixed stars; H and θ specify the satellite position within the orbit plane. The advantage of this description is that I, Ω and H are either fixed or slowly varying, so that over short periods, θ describes the instantaneous satellite position. Based on the magnitude of I, there are three kinds of orbits. If $I = 90°$, the orbit is polar; if $I < 90°$, the orbit is prograde and precesses in the same direction as the Earth's rotation as in Figure 1.1; if $I > 90°$, the orbit is retrograde and precesses in the opposite direction.

In remote sensing, interest is generally not in the satellite position in its orbit, but rather in its location on its surface ground track. For a non-rotating spherical Earth, the orbit track is a great circle, or on the Mercator map shown in Figure 1.2a, a simple sine wave (Elachi, 1987, Section B-1-4). Because of the Earth's rotation, the orbit track

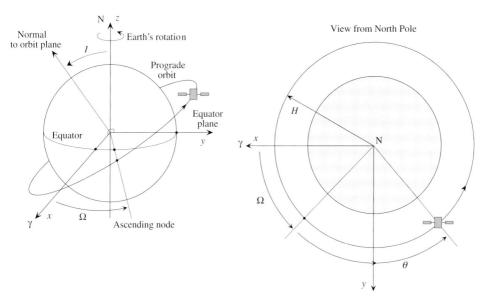

Figure 1.1. For a circular orbit, the Keplerian parameters used to describe the orientation of the orbit plane and the satellite position along the orbit.

is steadily displaced to the west, yielding the succession of tracks shown in Figure 1.2b. On the tracks, the numbers *i, ii, iii* mark the beginning and end of each orbit, where, for example, the points marked *ii* are at the same time and geographic location. Another orbit property concerns the equatorial separation L_E between successive orbits. If division of a multiple of the equatorial circumference by L_E is an integer, the orbit is an exact repeat orbit, so that after a given period of time, the satellite repeats the same track lines. This property is particularly valuable for instruments such as the altimeter, since it allows successive measurements of sea surface height along the same ground track.

The three principally employed types of Earth observation orbits are called *geosynchronous*, *sun-synchronous*, and *near equatorial low inclination* (Figure 1.3). There is also a fourth altimeter orbit that is at a slightly higher altitude than the sun-synchronous orbits. The following summary of the properties of these orbits shows that each particular orbit has advantages and disadvantages. Because no single orbit allows for coverage of all space and time scales, there is no such thing as a "perfect" satellite orbit or system. Instead, the choice of orbit depends on the phenomenon under investigation.

The geosynchronous orbits are located at an altitude of about 36 000 km above the Earth's surface. The geostationary orbit is a special case; it lies in the Earth's equatorial plane ($I = 0°$). In this orbit, the satellite remains over a fixed equatorial location so that it continuously observes the same surface area. The plane of the more general geosynchronous orbit is tilted relative to the equator ($I \neq 0°$), so that, although the mean surface position of this

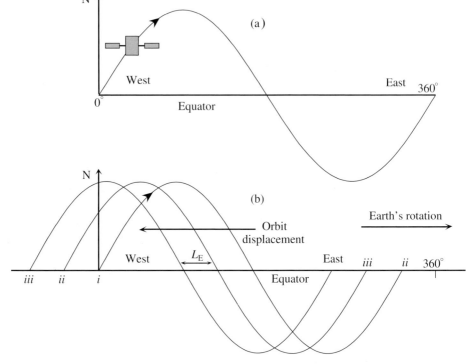

Figure 1.2. Mercator map of the satellite ground track for the orbit in Figure 1.1 and for (a) non-rotating Earth; (b) rotating Earth. See text for further description (Adapted from Elachi, 1987, Figure B-6).

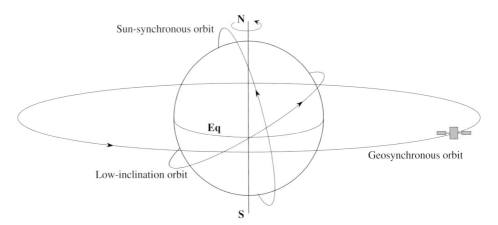

Figure 1.3. Examples of the sun-synchronous, geosynchronous and low-inclination orbits, where 'Eq' is the equator (Adapted from Asrar and Dozier, 1994, Figure 3).

satellite is stationary, its ground path is described by a figure eight centered on the equator (Elachi, 1987). In both cases, the satellite passes in and out of the Earth's shadow. The geosynchronous satellite period equals 23.93 hours, which is the time in which the Earth rotates around its axis relative to the fixed stars. In contrast, the 24-hour day is the time between successive noons, and thus is a combination of the Earth's rotation about its axis and the Earth's rotation in its orbit. Geosynchronous satellites are used for weather observation, for equatorial observations of SST and for data relay and communication.

The sun-synchronous orbit is retrograde with $I > 90°$, at an altitude of about 800 km, and is thus much lower than the geosynchronous orbits. For comparison, the Space Shuttle operates at an altitude of about 500–600 km. The sun-synchronous period is about 90 minutes, corresponding to 16 orbits per day. The reason this orbit is called sun-synchronous is that throughout the year each orbit crosses the equator at the same local time of day. Consequently, Ω is not constant, but changes slowly with time. The drift in Ω occurs because of the Earth's equatorial bulge, which causes the plane of a near polar orbit to rotate slowly around the pole (Rees, 2001). For a retrograde orbit, the inclination and orbit height can be set so that the orbit rotates about 1° per day in the ecliptic or Earth–sun plane, and in an equal but opposite direction to the orbital motion of the Earth around the sun. Relative to the fixed stars, the sun-synchronous orbit plane rotates once per year, so that its orbit plane remains at a constant angle to the line between the sun and Earth. Figure 1.4 shows the change in the angular position of the orbit in the Earth–sun plane as the Earth moves an angular distance of 90° in its orbit, during a period of approximately 90 days.

Because the sun-synchronous equator crossings always occur at the same local time of day, satellites in this orbit can make daily observations of SST or ocean chlorophyll at the same time in their diurnal cycle. Consequently, the observations remove the effect of the

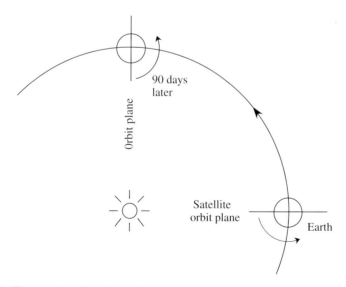

Figure 1.4. The rotation of the plane of a sun-synchronous orbit in the Earth–sun orbit plane.

diurnal cycle from the variable of interest. Also, since cloudiness over the ocean generally increases throughout the day, the crossing time can be chosen to minimize cloudiness under the satellite. Sun-synchronous satellites are the most common of the ocean observing satellites and are often referred to as polar orbiters. Sun-synchronous orbits are described in terms of their daytime equatorial crossing times, as in a 0730 descending or a 1330 ascending orbit, where descending refers to a southward satellite velocity, ascending refers to a northward velocity, and the crossing time is local. The orbits are also described as "early morning", "mid-morning" and "early afternoon". One difficulty with this orbit is that, because of the tilted orbit plane, the satellite does not pass directly over the poles. This means that the poles may be excluded from instrument coverage, where this lack of coverage is called the "hole at the pole". Plates 2 and 10 give examples of the swath coverage for this orbit, and show that, depending on the instrument, a single sun-synchronous satellite can provide near global coverage at 1–2 day intervals.

The near equatorial low inclination orbit used for the Tropical Rainfall Measuring Mission (TRMM) is circular with an altitude of 350 km and an inclination angle of 35°. This orbit covers approximately half the globe, and in a one-month period, observes any specific area at every hour of the day with a tropical sampling rate that is roughly twice that of a polar orbiter. The advantage of this orbit is that it allows TRMM to determine the dependence of tropical rainfall throughout its diurnal cycle. Finally, the altimeter measures sea surface height. Because the tidal bulge associated with the 12- and 24-hour tides always lies directly beneath a satellite in a sun-synchronous orbit, the altimeter orbit is generally at a higher non-synchronous altitude of 1200 to 1400 km. Consequently, the orbit is not in phase with the tides and the satellite experiences a smaller atmospheric drag. Altimeter satellites in this orbit include the US-French TOPEX/POSEIDON and JASON-1 missions, which are discussed in Chapter 12.

1.4 Geosynchronous satellites

The geosynchronous satellites important to oceanography include the weather and data relay satellites. Figure 1.5 shows one of the Meteosat series of European geosynchronous satellites; the Japanese Geostationary Meteorological Satellite (GMS) has a similar appearance. The satellite is oriented so that its long axis is parallel to the Earth's rotation axis, with its large lower cylindrical part spinning about this axis at about 100 revolutions per minute. On each spin, a single visible/infrared sensor sweeps across the Earth's disk and the data collected are stored or broadcast. On the next revolution, the north–south sensor view angle changes slightly, and the scan is repeated. From such multiple scans, it takes about 20 minutes to create an image of the Earth's disk. This technique is called *spin-scan*. The upper part of the satellite does not spin; antennas in this non-spinning section broadcast the data to the surface. The spinning helps keep the satellite in thermal equilibrium and stabilizes the satellite in its orbit. In contrast, the US GOES satellites, which were previously spin-scan, now use a different technique where the images are acquired by a scanner that employs two mirrors, one sweeping across the Earth's disk, the other stepping north-to-south.

Figure 1.5. The European Organization for the Exploitation of Meteorological Satellites (EUMET-SAT) geostationary spin-stabilized Meteosat satellite. Solar panels cover the satellite; the sensor points out of the page (Courtesy Coordination Group for Meteorological Satellites, EUMETSAT, used with permission).

A network of geosynchronous satellites provides global coverage between ±60° latitude. NOAA maintains two geosynchronous weather satellites, called Geostationary Operational Environmental Satellites (GOES). These satellites are located over the equator at approximately 75° W and 135° W, or at the longitudes of the east and west coast of the United States (Figure 1.6). Europe maintains two geosynchronous weather satellites, one over the Atlantic at approximately 0°; the other over the Indian Ocean at 75° E. Russia and India also maintain satellites at 75° E, although India generally reserves its data for domestic use. Japan maintains a weather satellite at 140° E; China at 105° E. Consequently, the globe is covered by five overlapping fields-of-view, placed approximately equally around the globe, with a sixth from China at 105° E. Even though none of the geosynchronous imagers view the polar regions, they provide sequential visible and infrared imagery of clouds and SST patterns for the equatorial and temperate latitudes at 20–30 minute intervals. The second class of geosynchronous satellites are the data relay satellites, which transfer data from the polar orbiters to the ground. The US maintains the Tracking and Data Relay Satellite System (TDRSS) that consists of about four active satellites and three on standby. TDRSS is the primary communication link between the TERRA and AQUA

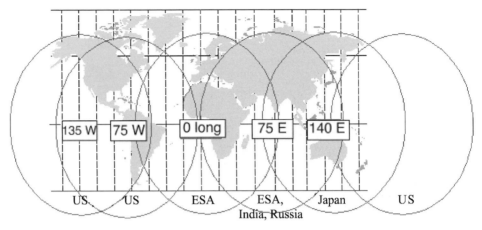

Figure 1.6. The fields-of-view of five geostationary satellites, which provide near global coverage. The longitudes in the boxes give the coverage center; the names at the bottom identify the responsible countries; the US 135° W satellite is shown at both sides of the figure. The Chinese satellite coverage is not shown. See text for more detail (Courtesy Coordination Group for Meteorological Satellites, EUMETSAT, used with permission).

spacecraft and the surface. The European Space Agency (ESA) and Japan maintain similar systems.

1.5 US sun-synchronous satellites

A number of countries maintain operational sun-synchronous satellites with oceanographic instrumentation, where the term *operational* means that the data from these satellites are regularly used in oceanographic or atmospheric forecasting. In the United States, three government agencies operate such satellites. The National Aeronautics and Space Administration (NASA) maintains a series of research satellites, the National Oceanic and Atmospheric Administration (NOAA) maintains the operational meteorological and oceanographic satellites, and the Department of Defense (DOD) maintains military satellites with oceanographic applications. Other operational sun-synchronous satellite programs include the Russian Meteor series and the Chinese Feng Yun (Wind and Cloud) FY-1C and FY-1D series.

In the US, the NOAA satellites are launched by NASA, administered by NOAA and carry instruments from France and the United Kingdom. The cooperative Polar Operational Environmental Satellite (POES) program administers these satellites, which are called POES or NOAA satellites (Figure 1.7). During construction and before launch, these satellites are described by letters, as in NOAA-K; after launch they are described by numbers, so that for example, NOAA-K became NOAA-15. In addition to a variety of instruments used to gather atmospheric data as input to numerical weather forecasts, the principal oceanographic instrument on the NOAA satellites is the visible/infrared Advanced Very High

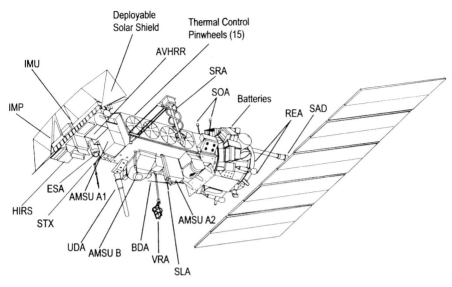

Figure 1.7. A schematic diagram of the NOAA-15 satellite. AMSU, Advanced Microwave Sounding Unit; BDA, Beacon Transmitting Antenna; ESA, Earth Sensor Assembly; HIRS, High Resolution Infrared Sounder; IMP, Instrument Mounting Platform; REA, Reaction Engine Assembly; SAD, Solar Array Drive; SRA, Search-and-Rescue Receiving Antenna; STX, S-Band Transmitting Antenna; SLA, Search-and-Rescue L-Band Antenna; SOA, S-Band Omni Antenna; UDA, Ultra-High Frequency Data Collection Antenna; VRA, Very-High Frequency Real-time Antenna (Courtesy NOAA).

Resolution Radiometer (AVHRR) used for SST retrieval and for studies of land and clouds. AVHRR observations began in 1978 with the launch of the Television Infrared Observation Satellite-N (TIROS-N); the first AVHRR specifically designed for SST retrieval was the AVHRR/2 launched in 1981 on NOAA-7. The AVHRR data are continuously broadcast in an open format, so that with the use of a relatively simple ground station these data can be downloaded over most of the globe. As Chapter 7 discusses in more detail, AVHRR observations have provided a two-decade time series of global SST. The AVHRR observations will continue through about 2010. After this time and as Chapter 14 describes, the Visible/Infrared Imager/Radiometer Suite (VIIRS) will replace the AVHRR.

The NOAA satellites operate at altitudes between 830 km and 870 km, where historically the morning satellite occupied a 0730 descending orbit; the afternoon satellite, a 1330 ascending orbit. The orbit parameters are such that the satellites have a 12-hour repeat cycle. Because the crossing times of the two satellites are approximately 6 hours apart, with nighttime equator crossings of approximately 1930 ascending and 0130 descending, they acquire imagery from almost every point on the Earth's surface at 6-hour intervals.

In a separate program, the US Department of Defense maintains two Air Force Defense Meteorological Satellite Program (DMSP) sun-synchronous satellites that operate at a nominal altitude of 830 km with dawn–dusk crossing times. These carry the Optical Line

Scanner (OLS), which operates similarly to the AVHRR, and the passive microwave Special
Sensor Microwave/Imager (SSM/I). As Chapters 8 and 9 discuss, SSM/I and the 2003 launch
of the Special Sensor Microwave Imager/Sounder (SSMI/S) modification of the SSM/I provide time series of sea ice extent. Unlike AVHRR data, the DMSP data are encrypted and
require a specially designed ground station to read them. These data are not released to the
public until 60 days after acquisition.

As part of the Earth Observing System (EOS), NASA maintains the third pair of US Earth
observing sun-synchronous satellites. These are called TERRA and AQUA; TERRA is in
a mid-morning orbit, AQUA in an early afternoon orbit. In addition to other instruments,
both satellites carry the visible/infrared Moderate Resolution Imaging Spectroradiometer
(MODIS) for investigation of ocean color, sea surface temperature and clouds; AQUA also
carries the Japanese passive microwave Advanced Microwave Scanning Radiometer-EOS
(AMSR-E).

In recognition that the four US NOAA and DMSP satellites comprise two redundant
systems, in 1994, a presidential decision transferred the management of these satellites to
the new National Polar-orbiting Operational Environmental Satellite System (NPOESS).
The purpose of NPOESS is to reduce the number of operational satellites from four to
three, of which the US will provide two satellites, the Europeans will provide one. As part
of this transition, the European MÉTéorologie OPérationnelle (Operational Meteorology
or METOP) satellite will replace the current mid-morning POES satellite in 2005. Chapter
14 describes the POES to NPOESS transition in detail and shows that NPOESS will also
continue some of the observations made by TERRA and AQUA.

1.6 Imaging techniques

Satellites use several scanning methods to generate images. As Section 1.4 describes, certain
of the geosynchronous satellites use spin-scan to generate images. For the sun-synchronous
and other low Earth orbits, in the visible/infrared several different but related scanning
techniques are used to generate images. As Chapters 8, 10 and 14 show, different scanning
methods are used by passive and active microwave instruments. Section 1.6.1 describes the
geometry used for a sensor viewing the Earth's surface, then shows for a simple telescope
how the surface field-of-view changes with view angle. Sections 1.6.2–1.6.4 discuss three
scanning techniques used with low Earth orbits called *cross-track* or *whiskbroom*, *along-
track* or *pushbroom*, and what this book calls *hybrid whiskbroom*, where each of these
depends on the satellite motion along its trajectory. Section 1.6.5 concludes with a discussion
of resolution.

1.6.1 Viewing the Earth's surface

Figure 1.8 shows the terminology and geometry for a satellite sensor viewing the Earth's
surface. On this figure, the point on the surface beneath the satellite is its nadir point; the
point observed by the instrument is its scan point. Zenith means directly overhead. The

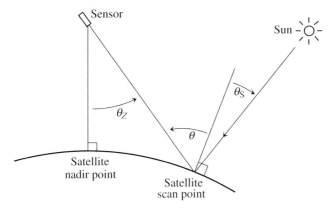

Figure 1.8. The angles used to describe the sensor view direction and the solar angle relative to a spherical Earth. The angle θ_Z is the zenith or view angle associated with the satellite sensor and is defined relative to satellite nadir, θ is the incidence or look angle and θ_S is the solar zenith angle, both defined relative to the local vertical at the satellite scan point.

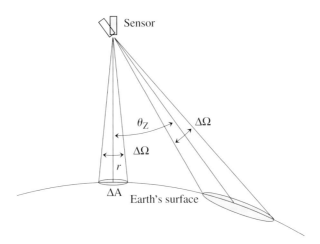

Figure 1.9. The surface area observed by an optical instrument with a constant solid angle field-of-view, for nadir and off-nadir view angles.

angle between the nadir line and the instrument look direction is the zenith angle θ_Z and the angle between the view direction and the local vertical at the scan point is the incidence or look angle θ. At off-nadir view angles, θ and θ_Z differ because of the Earth's curvature. The figure shows that the solar zenith angle θ_S is also measured relative to the local vertical. Because oceanic surface properties are functions of θ, the following chapters will primarily use θ to describe the satellite view angle.

Many optical instruments employ telescopes with circular lenses and apertures to view the Earth at a variety of view angles (Figure 1.9). For this case, the instrument solid angle $\Delta\Omega = \Delta A/r^2$ is a constant, where ΔA is the surface area observed by the telescope at nadir and r is the distance from the instrument to the surface. The surface area is also called the *instrument field-of-view* or equivalently the *instantaneous field-of-view* (IFOV), or often simply the field-of-view (FOV). For a nadir view, the FOV is a circle; at off-nadir view angles, because of this angle and the Earth's curvature, the FOV is an ellipse.

1.6.2 *Cross-track or whiskbroom scanners*

Whiskbroom scanners construct images from the combination of the satellite motion along its trajectory and the rotation of a telescope/mirror combination relative to the spacecraft. For these instruments, the scan is described in terms of three directions: *along-track* is in the direction of the satellite trajectory, *cross-track* is at right angles to the trajectory and *along-scan* is in the scan direction of the sensor on the surface (Figure 1.10). Examples of such instruments include the AVHRR and the Sea-viewing Wide Field-of-View Sensor (SeaWiFS).

Figure 1.10 shows a schematic drawing of the surface scanning pattern and operation of an idealized single and multichannel AVHRR. The single channel scanner in Figure 1.10a collects radiation from the FOV at a single wavelength band; the multichannel scanner in Figure 1.10b collects radiation from the same FOV at several wavelength bands. The instrument operates as follows: for each wavelength band, the detectors are focused on a mirror mounted at a 45° angle to its axis of rotation that rotates uniformly around 360°. At the same time that the rotating mirror sweeps the FOV across the surface, the satellite motion moves it along the satellite trajectory, so that an image is constructed from the successive parallel scans. Because the mirror rotates as the satellite advances, the scan lines lie at an

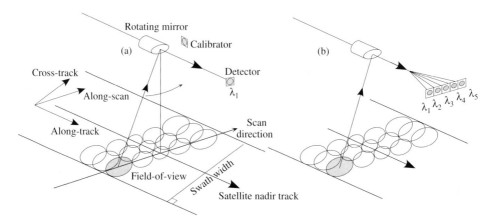

Figure 1.10. Schematic drawing of a cross-track or whiskbroom scanner. The circles show the fields-of-view; the gray ellipse shows the instrument FOV; the radiation from the FOV is focused on the detector, also shown in gray. (a) Single wavelength scanner; (b) multiwavelength scanner. The λ_i are the center wavelengths of the detectors.

oblique angle to the satellite trajectory. The figure also shows a calibration source that is held at a constant radiance. The source is located such that after completion of a surface scan, each channel views and stores a calibration value. A great advantage of the cross-track scanners is that the sensors are calibrated once per rotation.

A property of the whiskbroom scanners shown in Figure 1.10 is that as the off-nadir angle increases, the FOV size increases and its shape changes from a circle to an ellipse. The growth in FOV size can be large. For a sun-synchronous satellite at an altitude of 800 km, the FOV area at $\theta_Z = 45°$ exceeds its nadir value by a factor of 1.5 in the along-track direction and 3.5 in the along-scan direction; at 55°, the area exceeds its nadir value by respectively 2 and 6. For these scanners, the mirror rotation rate is set so that on successive scans the nadir FOVs are adjacent to one another. Consequently, as the off-nadir FOVs increase in area they overlap. Because of this growth in FOV area, the overall shape of a scan resembles a bowtie, where the problems associated with this growth in FOV area with increasing off-nadir scan angle are called the *bowtie effect*.

The received data are also averaged over short periods of time into a series of successive time blocks. This further increases the FOV area, where the time-averaged FOV is called the *effective field-of-view* or *EFOV*. As Section 1.7 describes in more detail, on the ground the data are resampled to a uniform grid, where each cell in the grid has the area of the nadir FOV. Given the increase in both atmospheric interference and EFOV area with increasing zenith angle, data taken at θ_Z greater than 45–55° are noisier than data taken near nadir. Finally, some sensors such as the OLS (Optical Line Scanner) on the DMSP satellite, use a variety of techniques such as a variable focus telescope to adjust the solid angle so that the FOV area is independent of look angle.

1.6.3 Along-track or pushbroom scanners

In contrast to the whiskbroom scanner, the pushbroom scanner uses long linear arrays of sensors to observe the surface in the cross-track direction, where each sensor, or for multiple bands, each set of sensors is focused on a specific track line beneath the satellite (Figure 1.11). For this instrument, the nadir FOV is a circle; the off-nadir FOVs are ellipses. The advantage of this technique is that the *dwell time*, or time interval for which the sensor is focused on a specific surface area, is greater than for the whiskbroom. Because it allows for a greater signal-to-noise ratio and a higher spatial resolution than is possible for whiskbroom sensors, this increased dwell time is one of the most useful properties of the pushbroom instruments. Examples include the 30-m resolution Enhanced Thematic Mapper Plus (ETM+) on the LANDSAT-7 satellite, the German Modular Optical Scanner (MOS) on the Indian IRS-P3, and the ESA Medium Resolution Imaging Spectrometer (MERIS) on ENVISAT. The advantages of the pushbroom scanner are longer dwell time and better spatial resolution; the disadvantages are that multiple sensors can lose their relative calibrations, making the instrument less accurate. Additionally, for a wide-swath

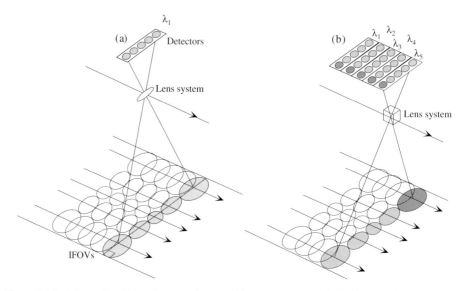

Figure 1.11. Schematic of the along-track or pushbroom scanner. (a) Single wavelength scanner; (b) multi-wavelength scanner. The ellipses show the FOVs; the gray ellipses are simultaneously viewed by the strip of detectors. (b) shows how the dark gray ellipse is viewed at multiple bands by the strip of dark gray detectors. See text for further description.

high-resolution instrument, an unwieldy number of sensors may be required in the cross-track direction.

1.6.4 Hybrid cross-track scanner

The need for wide-swath, high spatial resolution scanners led to the development of hybrid cross-track scanners that combine the properties of the whisk and pushbroom scanners. The hybrid scanner uses linear arrays of sensors with their long axis oriented to make observations in the along-track dimension. These arrays receive radiation from within a large aspect ratio elliptical FOV with its along-track length much longer than its cross-scan length (Figure 1.12). The advantage of this scanner is that it provides a way to increase dwell time and obtain high resolution from a wide-swath instrument while still permitting calibration of the sensors at each rotation. Examples include the MODIS on TERRA and AQUA with its 2300-km swathwidth, and the forthcoming VIIRS with its 3000-km swathwidth. Specifically, the MODIS nadir FOV dimensions are 10 km in the along-track direction and 1 km in the cross-scan direction (Barnes, Pagano and Salomonson, 1998). In the along-track direction, and depending on the observational wavelength, the number of detectors is 10, 20 or 40, corresponding to an along-track FOV dimension at nadir of respectively 1.0, 0.5 and 0.25 km. The advantage of this scanning technique is that, if it were replaced with a simple whiskbroom sensor, the mirror would have to spin ten times as fast for the

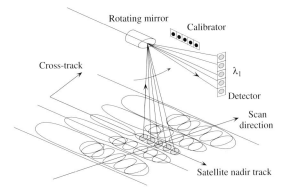

Figure 1.12. A schematic diagram of a single wavelength hybrid cross-track scanner. The large ellipses on the surface show the instrument FOVs; within the large ellipses, the smaller ellipses show the FOVs viewed by each detector; the gray ellipses show the FOVs viewed at a particular instant. For simplicity, the figure only shows the single wavelength instrument.

same spatial resolution, reducing the dwell time and increasing the noise both by a factor of ten.

1.6.5 Resolution

As the next section describes in detail, the data from these instruments are resampled into a uniform grid, where the grid spacing approximately corresponds to the nadir FOV diameter. Each element in the grid is called a *pixel*, which is the abbreviation for picture element. Typically for the AVHRR and SeaWiFS, the pixel measures 1 km by 1 km, referred to as a 1-km pixel, where the pixel area equals that of the nadir FOV. For this case, the instrument is also described as having a 1-km *resolution*, meaning that objects smaller than 1 km cannot be distinguished by the imager. This definition of resolution as equal to the nadir FOV is used in the visible, infrared and passive microwave. For radars and as Section 13.2.2 describes, the relation between resolution and pixel size is different, in that the smallest pixel size used in the image equals half the resolution.

1.7 Processing of the imager data

Processing of the swath data from these scanners is usually described as a series of steps or processing levels. Using AVHRR or SeaWiFS as an example, the levels consist of the following (Parkinson and Greenstone, 2000, p. V):

Level 0: The unprocessed engineering outputs from the sensor are presented at full resolution. All communication artifacts such as headers have been removed.

Level 1A: The unprocessed sensor data are presented at full resolution in units of digital counts in a swath format that contains in appended files time references, calibration coefficients and geo-referencing information.

Level 1B: The swath data are presented in digital counts plus appended files that contain geolocation data, sensor calibration data for conversion of digital counts to radiances or brightness temperatures, and information on data quality. Unlike Level 1A, Level 1B files are in a computer-friendly format. The data are organized into one scan line per record, where the geographic distance between data points corresponds to the cross-track dimension of the nadir FOV. Many of the AVHRR and SeaWiFS data are distributed in this format. Not all instruments have a Level 1B equivalent.

Level 2: The Level 1 data are processed to produce geophysical data products such as brightness temperatures, radiances or additional products in a swath format. For AVHRR, these additional products might include SST. As Chapter 7 shows, this SST calculation involves the use of data from multiple channels, application of a cloud mask, correction of atmospheric emission and attenuation, and interpretation of the radiances or brightness temperatures in terms of the physical properties of the ocean surface or atmosphere.

Level 3: The geophysical data product is mapped to a uniform grid in space and time. For AVHRR, these data might consist of a two-day global average of gridded sea surface temperature.

Level 4: The output from a numerical model is combined with data from multiple satellite measurements. An example might be the evolution of El Niño in the Pacific.

1.8 Past, present and pending satellite missions

Ocean remote sensing began in the United States in the 1970s. The success of the early US satellite missions led the Soviet Union, Japan and the multinational European Space Agency (ESA) to join the US by the 1980s and early 1990s. In 2002, ESA included members from fifteen countries: Austria, Belgium, Denmark, Finland, France, Germany, Ireland, Italy, Netherlands, Norway, Portugal, Spain, Sweden, Switzerland, and the United Kingdom. The second European space agency is the European Organization for the Exploitation of Meteorological Satellites (EUMETSAT). EUMETSAT was created within ESA in 1986 to manage the geosynchronous Meteosat satellites, and is currently involved with the development of the METOP polar orbiters. Within Europe, France, Germany and Italy also have individual space programs.

In the early 2000s, there are at least eleven countries or groups of countries with ocean observing instruments or satellites: Brazil, Canada, ESA, India, Japan, People's Republic of China (China), Republic of China (Taiwan), Republic of Korea, Russia, Ukraine and the United States. Of these, Canada maintains its RADARSAT Synthetic Aperture Radar (SAR) satellites; India, its OCEANSAT satellites; and China, its Feng Yun series. At least one private company also maintains ocean satellites. In the following, Section 1.8.1 gives a brief history of the US oceanographic satellite research programs and Section 1.8.2 reviews the past, present and future through 2007 of all oceanographic satellite missions.

1.8.1 The growth of US oceanographic research missions

Since 1970, there have been three generations of NASA oceanographic satellite missions (Wilson, 2001; Wilson, Apel and Lindstrom, 2001), with the transition to a fourth generation taking place in 2001–2010. Following Wilson (2001), the first generation consisted of the scatterometer experiments on the 1973 Skylab, the launch and operation of the US Navy 1975 GEOS altimeter satellite, and the operation between 1973 and 1976 of the single channel Electrically Scanned Microwave Radiometer (ESMR) on the NIMBUS-5 satellite. These missions demonstrated the potential for scatterometer wind retrieval, satellite altimetry and passive microwave retrieval of sea ice properties. Based on these and on a large variety of aircraft experiments, the second generation consisted of the 1978 launch of the TIROS-N, SEASAT and NIMBUS-7 satellites. As described in Section 1.5, TIROS-N carried an early version of AVHRR. SEASAT carried four pioneering instruments; a multichannel passive microwave radiometer, a wind scatterometer, a SAR and a radar altimeter. SEASAT failed after 99 days of operation, but was immediately followed by the launch of NIMBUS-7. Although NIMBUS-7 lacked the altimeter, SAR and scatterometer, it carried the Coastal Zone Color Scanner (CZCS) and a microwave imager similar to that on SEASAT. Even though some of its instruments failed earlier, NIMBUS-7 operated for about a decade, or until 1988.

The first- and second-generation satellite observations provided successful demonstrations of the retrieval of ocean surface height, ocean color, surface winds and sea ice properties to an accuracy appropriate to oceanography. For the third generation missions, Wilson shows that NASA took a different approach. Briefly, NASA set two requirements: that the oceanographic community contribute to the mission justification, planning and support, and that each future mission have a non-NASA partner, where the second requirement was in part prompted by cost overruns on the Space Shuttle. The first requirement led to a series of joint studies with the oceanographic community, the results of which led NASA to focus on four areas: sea surface height or altimetry, biological ocean color or productivity, the properties of the sea ice cover, and the land, ice and open ocean applications of SAR.

These studies and the partnership requirement led to the following missions: (1) the 1987 continuation of the passive microwave observations on NIMBUS-7 by the US Department of Defense on the DMSP satellites; (2) completion of the NASA Alaska SAR Facility (ASF), which became operational in September 1991 and the concomitant NASA agreements with Canada, ESA and Japan for ASF reception of their satellite SAR data; (3) the 1991 French launch of the US/France TOPEX/POSEIDON altimeter mission; (4) the 1995 US launch of the Canadian RADARSAT SAR satellite; (5) the 1995 launch of the NASA wind scatterometer (NSCAT) on the Japanese Advanced Earth Observing Satellite (ADEOS); (6) the contract with Orbital Science Corporation for the purchase of ocean color data from the SeaWiFS instrument launched by the company in 1997.

The NASA demand for scientific justification and for partners in future missions is one reason for the long delay between the launch of NIMBUS-7 in 1978 and the TOPEX altimeter in 1991. Another reason is that it took years to analyze all of the ocean color, SAR and scatterometer data collected by the second generation missions, so that parallel to the third generation planning and instrument development, there was an additional effort to prepare computationally for the new missions.

Beginning in the mid 1980s, there were also increasing concerns about the effects of climate change on global food supplies, and the increased vulnerability of coastal populations to violent weather and sea level rise. These created a demand for (1) better forecast capability for ocean-generated atmospheric storms such as typhoons and hurricanes, (2) long-term time series of oceanic and sea ice variables. These served as the basis for the planning and launch of recent large multinational projects such as the NASA Earth Observing System (EOS) program, with the 1999 TERRA launch and the 2002 AQUA launch. The 2002 ESA launch of ENVISAT and Japanese launch of ADEOS-2 complemented these EOS missions. EOS was originally conceived as a 15-year program, where 15 years was assumed sufficient to observe the onset of global change, with the lifetime of each satellite and its replacements being approximately five years. Instead, the EOS missions represented by AQUA and TERRA have been shortened to a five-year program, where some of their observations will be incorporated into NPOESS, some will be taken over by dedicated satellites, and some will be abandoned. Finally, as Chapter 14 describes in detail, the fourth generation of ocean observing satellites consists of the transition of certain aspects of these research missions to operational programs.

1.8.2 Satellite missions through 2007

Table 1.1 lists the major past, present and pending ocean satellite missions through 2007, giving the mission name, lifetime, country of origin and purpose. Chapter 14 discusses the transition to NPOESS and those missions scheduled after 2007. To simplify the table, most operational satellites are excluded. For 2004 and beyond, satellite missions that are planned but not in orbit are identified by the lack of a dash (–) following their launch year. The table lists the launch date and, if known, the lifetime, the country providing the satellite, its name and purpose. Under the satellite name, the table lists the instruments and if the instrument is provided by another country, adds that country name in parentheses.

   The table clearly shows the growth in the number and diversity of satellite missions since 1995. During the 1990s, ocean satellites were built or launched by ESA, US, Japan, Russia, China, and India. Some of these missions are joint, including the US/France TOPEX/POSEIDON and JASON-1 altimeter project, and the US/Japan Tropical Rainfall Measuring Mission (TRMM). Another form of international mission occurs when different countries provide a single instrument to a specific national project. These include the NASA NSCAT wind instrument on the Japanese ADEOS-1, the Japanese Advanced Microwave Scanning Radiometer-EOS (AMSR-E) on AQUA, the German Modular Optical Scanner (MOS) ocean color sensor on the Indian IRS-P3 and the US SeaWinds instrument on ADEOS-2.

   The table also shows that there are two kinds of satellite missions; those designed around multiple instruments, and those designed around a single instrument such as the altimeter. The multi-instrument philosophy places many instruments on one satellite; this allows the many instruments to take advantage of a common power supply and communication links, and permits simultaneous measurements. Examples include SEASAT, NIMBUS-7, ADEOS-1, ERS-1 and -2, TERRA, AQUA and ENVISAT. A further advantage of the

Table 1.1. *Past, present and near future ocean satellite missions through 2007*

Launch year and lifetime	Country or agency	Mission name and instruments	Oceanographic measurements
1978 (3 months)	US	SEASAT Altimeter SAR, Synthetic Aperture Radar SASS, SEASAT-A Satellite Scatterometer SMMR, Scanning Multichannel Microwave Radiometer	Sea surface height, SAR, vector winds, passive microwave
1978–1987	US	NIMBUS-7 CZCS, Coastal Zone Color Scanner SMMR	Passive microwave, ocean color
1985–1990	US/Navy	GEOSAT Altimeter	Sea surface height
1991–2001	ESA	ERS-1, -2, Earth Resources Satellite AMI, Advanced Microwave Instrument (combination of SAR, scatterometer) ATSR, Along-Track Scanning Radiometer	Winds, SAR, SST
1992–1998	Japan	JERS-1, Japanese Earth Resources Satellite SAR	SAR
1992–	US/France	TOPEX/POSEIDON NASA altimeter Poseidon altimeter (France)	Sea surface height
1995–	Canada	RADARSAT-1 SAR	SAR

(*cont.*)

Table 1.1. *(Cont.)*

Launch year and lifetime	Country or agency	Mission name and instruments	Oceanographic measurements
1996–	India	IRS-P3, Indian Resource Satellite MOS, Modular Optical Scanner (Germany) WiFS, Wide Field-of-view Sensor	Ocean color
1996–1997	Japan	ADEOS-1, Advanced Earth Observing Satellite NSCAT, NASA Scatterometer (US) OCTS, Ocean Color and Temperature Sensor	Ocean color, winds
1997–	US/Japan	TRMM, Tropical Rainfall Measuring Mission TMI, TRMM microwave imager	Rainfall, SST
1997–	US/Orbital Science Corp.	SeaStar SeaWiFS, Sea-viewing Wide Field-of-view Sensor	Ocean color
1999	India	OCEANSAT-1 (IRS-P4) OCM, Ocean Color Monitor (Germany)	Ocean color
1999	China	FY-1C, Feng Yun-1C (Wind and Clouds-1C) MVIRSR, Multispectral Visible/IR Scanning Radiometer	SST
1999–	US	QuikSCAT SeaWinds	Vector winds
1999–	Republic of China	ROCSAT, Republic of China Satellite OCI, Ocean Color Instrument	Ocean color

22

1999–	US	TERRA	
		ASTER, Advanced Spaceborne Thermal Emission and Reflection Radiometer (Japan)	Ocean color, SST
		CERES, Clouds and the Earth's Radiant Energy System	
		MISR, Multi-angle Imaging Spectroradiometer	
		MODIS, Moderate Resolution Imaging Spectroradiometer	
		MOPITT, Measurements of Pollution in the Troposphere (Canada)	
1999–	Korea	KOMPSAT, Republic of Korea Satellite	Ocean color
		OSMI, Ocean Scanning Multispectral Imager	
2001–	US/France	JASON-1	Sea surface topography
		Poseidon altimeter	
2002–	ESA	ENVISAT, Environmental Satellite	SST, SAR, ocean color
		AATSR, Advanced Along-Track Scanner Radiometer	
		ASAR, Advanced SAR	
		GOMOS, Global Ozone Monitoring by Occultation of Stars	
		MERIS, Medium Resolution Imaging Spectrometer	
		MIPAS, Michelson Interferometric Passive Atmospheric Sounder	
		RA-2, Radar Altimeter-2	
		SCIAMACHY, Scanning Image Absorption Spectrometer for Atmospheric Cartography	

(cont.)

Table 1.1. (*Cont.*)

Launch year and lifetime	Country or agency	Mission name and instruments	Oceanographic measurements
2002–	US	AQUA AIRS, Atmospheric Infrared Sounder AMSU-A, Advanced Microwave Sounding Unit-A CERES, Clouds and the Earth's Radiant Energy System MODIS, Moderate Resolution Imaging Spectroradiometer HSB, Humidity Sounder for Brazil (Brazil) AMSR-E, Advanced Microwave Scanning Radiometer-EOS (Japan)	Ocean color, passive microwave
2002–	US/Germany	GRACE, Gravity Recovery and Climate Experiment	Gravity mission
2002–	China	FY-1D MVIRSR, Multispectral Visible/IR Scanning Radiometer	SST
2002–	China	HY-1, Haiyang-1 (Ocean-1)	Ocean color
2002–2003	Japan	ADEOS-2 AMSR, Advanced Microwave Scanning Radiometer GLI, Global Imager ILAS-2, Improved Limb Atmospheric Spectrometer-2 POLDER, Polarization and Directionality of Earth's Reflectance (France) SeaWinds, (NASA)	Passive microwave, vector winds, ocean color
2003–	US	Coriolis/WindSat (US Navy/NPOESS)	Passive microwave vector winds

Year	Country	Satellite / Instruments	Application
2003–	US	ICESat GLAS, Geoscience Laser Altimeter System	Ice sheet properties
2004	Canada	RADARSAT-2 SAR	SAR
2004	ESA	CryoSat SIRAL, SAR Interferometric Radar Altimeter	Ice sheet properties
2004	India	OCEANSAT-2	Ocean color
2005	US	NPP, NPOESS Preparatory Project VIIRS, Visible/Infrared Imaging/Radiometer Suite CMIS, Conical Microwave Imager/Sounder CrIS, Cross-track Infrared Sounder	Ocean color, SST, passive microwave, atmospheric sounders
2005	Japan	ALOS, Advanced Land Observing Satellite PALSAR, Phased Array L-band SAR	SAR
2006	ESA	SMOS, Soil Moisture and Ocean Salinity	Sea surface salinity
2006	France/US	OSTM, Ocean Surface Topography Mission Poseidon altimeter WSOA, Wide Swath Ocean Altimeter	Sea surface topography
2007	Japan	GCOM-B1, Global Change Observation Mission-B1 AMSR F/O, Advanced Microwave Scanning Radiometer Follow On SGLI, Second-generation Global Imager OVWM, Ocean Vector Wind Mission, also known as AlphaScat (US)	ADEOS-2 successor

To shorten the table, operational systems such as POES and DMSP are not included.

25

multi-instrument missions is that they can consist of groups of complementary instruments focusing on a single aspect of the Earth's behavior. For example, TERRA, AQUA and ENVISAT focus on those properties of the atmosphere, oceans, and land surface related to global environmental change.

A risk associated with large flight projects is illustrated by the premature failure of multi-instrument missions such as SEASAT, ADEOS-1 and ADEOS-2. Because many instruments were lost in these failures, large gaps occurred in oceanic coverage and in the efforts of the research community. Given these risks, there are advantages to placing some individual instruments on their own dedicated satellites, so that if the satellite fails, only one instrument is lost. Also, if a backup instrument is available, it can be rapidly launched into orbit. Such a replacement occurred following the failure of ADEOS-1 and the resultant loss of NSCAT, after only nine months of operation. Because of the importance of the wind observations, about two years later, the US launched the dedicated SeaWinds scatterometer on the QuikSCAT satellite. Other examples of dedicated missions include the SeaWiFS ocean color sensor on the SeaStar satellite, the TOPEX/POSEIDON and JASON-1 altimeters, and the RADARSAT SAR. Another advantage of single missions is that the instrument performance is not restricted by the presence of other instruments; for example, to accommodate space limitations on ADEOS-2, the SeaWinds dish antenna is smaller than desired. Availability of power and data storage impose similar restrictions. Physical reasons also exist for the choice of separate missions. For example, because the altimeter missions require an accurate orbit determination, the use of a physically small satellite yields a reduced atmospheric drag. Also, the TOPEX and JASON-1 altimeters avoid tidal aliasing by not being in a sun-synchronous orbit. The disadvantage of instruments flying alone is that the scheduling of simultaneous measurements of the same phenomena by different instruments is more difficult.

The table shows that there are many new flight projects launched, planned and proposed for 2001–2007. The second EOS AQUA satellite was launched in May 2002, where the first, TERRA, was launched in December 1999. The European ENVISAT was launched in early 2002 and the Japanese ADEOS-2 was launched in late 2002. Other smaller and equally important missions include the 2001 US–French JASON-1 altimeter mission and the 2002 US–German GRACE gravity mission. As Chapter 14 discusses in more detail, the later part of this decade includes the US NPOESS and ESA METOP satellites, the Japanese GCOM-B missions, and a number of innovative altimetry, sea surface salinity and vector wind missions.

Given this diversity of satellite missions, the purpose of this book is to describe the satellites and their instruments, the physical principles underlying the instrument operations, and to give examples of their data and applications. The book divides into five parts. Chapters 2, 3 and parts of 4 and 5 introduce the properties of radiation and the ocean surface that are common to all chapters; Chapters 4, 5, 6 and 7 discuss the visible/infrared atmospheric properties, instruments and applications. Chapters 8 and 9 describe the passive microwave instruments, atmospheric properties and applications, and Chapters 10, 11, 12 and 13 discuss the active microwave. Chapter 14 concludes with a discussion of future missions.

1.9 Further reading and acknowledgements

For more information on POES, see http://poes.gsfc.nasa.gov/ or http://www.noaa.gov under the US National Environmental Satellite Data and Information Service (NES-DIS) at http://www.nesdis.noaa.gov/. Information about satellite orbits is available at http://celestrak.com. The NPOESS system is discussed in Chapter 14 and at the Integrated Project Office (IPO) website http://www.ipo.noaa.gov. Much of the information on the flight projects was taken from websites found from searches based on the satellite name. Kramer (1994) also provides a comprehensive survey of all satellites and their instruments through 1993. Vaughan (2001) surveys current and future European satellite programs, and Vaughan and Wilson (2001) discuss ENVISAT. The worldwide schedule of satellite launches is available at http://spaceflightnow.com/tracking/. Sabins (1987, Chapter 1) further discusses sensors and scanning techniques. I thank Michael Alfultis, Stan Wilson and Michael Van Woert for their help with the material in this chapter.

2

Ocean surface phenomena

2.1 Introduction

This chapter summarizes those open ocean and floating ice properties that modify the surface and affect the emitted and reflected radiation. For the open ocean, these include wind-generated capillary and gravity waves, breaking waves, the generation and decay of foam, and the modulation of short waves by long waves and currents. Short waves are also suppressed by natural and human-generated slicks. At longer time periods and over larger spatial scales, ocean currents, eddies and Rossby and Kelvin planetary waves generate large scale changes in ocean surface elevation. Polar ice properties that affect the radiation include the areal extent and type of pack ice, and the presence and size of icebergs.

In the following, Section 2.2 discusses the oceanic winds and the ocean surface wave properties important to remote sensing: in particular the difference between the short-period capillary waves and the longer-period gravity waves, the changes that occur in the gravity wave profile with increasing wave amplitude, the growth of capillary waves on the surface of the longer-period waves, the effect of wave breaking and the generation of foam. The section also discusses the distribution of wave surface slopes as a function of azimuth angle relative to the wind direction. Although this topic seems obscure, it is essential for determination of sun glint, which can overwhelm the satellite observations, and for the measurement of vector wind speeds at microwave frequencies. The section concludes with a discussion of surface slicks. Section 2.3 discusses the changes in sea surface height induced by ocean currents and long-period planetary waves, and Section 2.4 discusses sea ice.

2.2 Ocean surface winds and waves

Surface winds play a dominant role in the modification of the temperate ocean surface. The most prevalent process is the wind generation of ocean waves. Another is that the surface wind stress and the atmospheric heat exchange drive the ocean circulation. Figure 2.1 shows the distribution of the global wind speeds over ice-free waters derived by two methods: first, from satellite passive microwave Special Sensor Microwave/Imager (SSM/I) observations of wind magnitudes using the techniques described in Chapter 9; second, from winds co-located with satellite observations and derived from National Center for Environmental

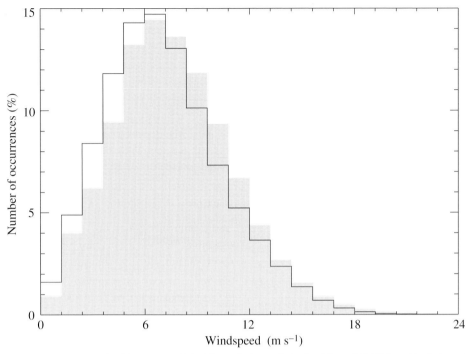

Figure 2.1. Comparative histograms of the 10-m wind speed obtained from SSM/I data (shaded area) and from co-located NCEP pressure data (line). Both data sets consist of 6.8×10^{10} measurements taken between January 1992 and December 1997. The bin sizes for the winds are 1.2 m s^{-1} (Courtesy Remote Sensing Systems, Santa Rosa, CA, used with permission).

Prediction (NCEP) gridded surface pressures. For this figure and throughout the book, the surface wind velocity corresponds to the wind measured at a 10-m height, called the 10-m wind speed U. An important difference between the description of ocean currents and atmospheric winds is that currents are described by their direction of flow; winds in terms of their upwind direction. Figure 2.1 shows that for both distributions, the peak in the wind speed distribution lies between 5 and 8 m s^{-1}, where about 40% of the wind speeds lie in this range, with a mean wind speed of about 7 m s^{-1}. Although wind speeds greater than 12 m s^{-1} strongly contribute to the generation of waves, foam and to the transfer of momentum to currents, they occupy only 10% of the histogram.

As Phillips (1977) describes, the wind-driven wave amplitudes and range of excited wavelengths depend on the turbulent energy flux from the atmosphere to the sea surface. This flux in turn strongly depends on the temperature stratification above the sea surface; if the atmosphere is warmer than the surface, then the atmosphere is stably stratified so that for the same wind speed, the turbulent flux is less than for an unstable stratification. Consequently, for the same wind speed, a stronger flux yields more waves and roughness, while a weaker flux yields less. Slicks also affect the surface response. In summary, the frequency and amplitude distribution of the wind-induced surface waves depends not only on U, but also on the ocean/atmosphere temperature difference and on the presence or absence of slicks.

Winds excite waves ranging in length from less than a centimeter to hundreds of meters where, depending on the observational window, all lengths are important to remote sensing. Long ocean waves are dominated by gravity, but for centimeter-scale waves, the effects of surface tension or *capillarity* become important. For a surface tension appropriate to seawater, Figure 2.2 compares the phase speed of pure gravity waves with that of capillary-gravity waves (Phillips, 1977). The figure shows that the gravity wave phase speed increases with wavelength, while the capillary-gravity phase speed has a minimum at a wavelength of 1.8 cm. The figure also shows that for the same wavelengths, capillary-gravity waves propagate faster than gravity waves and that surface tension is important up to wavelengths of about 7 cm. Although these capillary-gravity waves are very short relative to long gravity waves, their presence and distribution relative to the wind direction strongly contribute to microwave remote sensing.

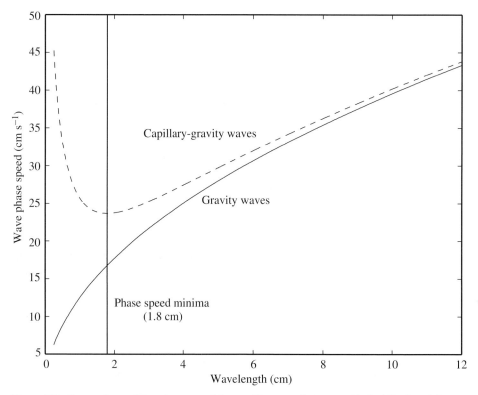

Figure 2.2. Comparison of the phase speed for capillary-gravity waves (dashed line) and for pure gravity waves (solid line) plotted versus wavelength for seawater. The vertical line marks the phase speed minimum for capillary waves. See text for additional information.

As observation of any pond or puddle shows and as Kawai (1979) demonstrates in his laboratory experiments, capillary-gravity waves with wavelengths close to the phase speed minimum immediately form and grow following the onset of a gust. These waves achieve an equilibrium distribution around the minimum wavelength within a few seconds of the wind onset, are independent of position, and rapidly decay when the wind ceases. If the wind continues blowing, the frequency of the largest amplitude or dominant wave shifts to lower frequencies and longer wavelengths. This is also the case for capillary wave growth on an existing swell field (Donelan and Pierson, 1987, p. 4975).

The generation of ocean swell differs from capillary waves, in that ocean swell can be generated at great distances from the observation site where the swell properties are only slowly modified by changes in the wind speed. The evolution of long-period wind-generated waves can be described as a function of either time or fetch, where fetch is defined as the downwind distance from a coast. The time description applies to the case of a uniform wind turned on at a specific time over an initially flat water surface far from any coast. For this case, as time proceeds, capillary waves appear first, then gravity waves form at lower frequencies, longer wavelengths and with greater amplitudes, so that the width and size of the wave frequency distribution increases with time. This wave growth continues until the energy input from the wind equals the energy dissipation by breaking and viscosity, at which time an equilibrium is reached that is independent of position. In contrast, the fetch description applies to a steady wind blowing off a coast, where the wave spectra are independent of time and depend primarily on wind speed and fetch (Huang, Tung and Long, 1990, especially their Figure 1). Consequently, as the fetch increases, the waves increase in amplitude and length. At distances far from the coast, the wave spectrum again reaches a wind speed-dependent equilibrium.

Seasonally, the strongest winds and largest waves in the Northern Hemisphere occur in the winter North Atlantic and North Pacific. In the Southern Hemisphere, the strongest winds and largest waves occur during the austral winter in the Southern Ocean, which is a circumpolar sea unobstructed by land masses. Kinsman (1984) states that a fetch of 1500 km is sufficient for the development of the largest observed storm waves. Of these, the largest peak-to-trough wave amplitude observed to date was about 34 m, as recorded by the *USS Ramapo* in 1934 in the central North Pacific (Kinsman, 1984, p. 10). Characteristic wavelengths within storms range from 150 m in the North Atlantic, to 240 m in the Southern Ocean. Long-period swell has been observed with lengths as long as 600 m (Kinsman, 1984).

2.2.1 Change in the wave profile with increasing amplitude

Ocean surface waves are described in terms of their amplitude a_w, which is defined as half the peak-to-trough wave height, their wavenumber $k_w = 2\pi/\lambda_w$ and their radian frequency $\omega_w = 2\pi/T_w$, where T_w is the wave period, λ_w is the wavelength, and the subscript w distinguishes these terms from those used to describe electromagnetic waves. If η is the wave height measured from the mean free surface and x is parallel to the wave propagation

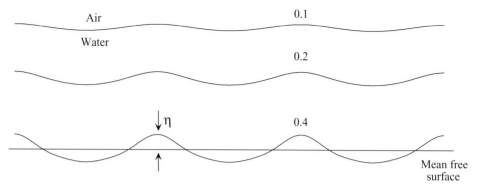

Figure 2.3. Comparison of the profiles of a single frequency gravity wave for the different values of the wave slope $a_w k_w$ given above each wave. The vertical axis is exaggerated by 60% to emphasize the change in wave shape with amplitude.

direction, the small amplitude waves are described as follows:

$$\eta = a_w \sin(k_w x - \omega_w t) \tag{2.1}$$

The non-dimensional form of the wave amplitude is the wave slope $a_w k_w$. At small amplitudes or for $a_w k_w \ll 1$, gravity waves are pure sinusoids; as $a_w k_w$ increases, the wave shape is described by the addition of higher order harmonics. For three different values of $a_w k_w$, Figure 2.3 shows the change in the shape of wave profile from (2.1), where the profiles are derived from a third-order expansion of the classic Stokes wave solution (Lamb, 1945, Section 250, Equation 3). The top curve in the figure corresponds to $\lambda_w = 100$ m and $a_w = 1.6$ m, and is a nearly perfect sinusoid. For the middle curve and the same wavelength, a_w increases to 3.2 m, and for the bottom curve, to 6.4 m. Comparison of these curves shows that the addition of the nonlinear terms creates a wave with a broad trough and a narrow sharp crest, so that the wave tends toward a trochoidal shape (Kinsman, 1984, p. 255). This change in shape with increasing amplitude from a pure sinusoid to a trochoidal shape has important implications for passive microwave and altimeter observations of the surface and, as the following shows, for wave damping and breaking.

From Lighthill (1980, pp. 453–454), the largest possible amplitude a_{max} that a gravity wave can attain is given by

$$a_{max} k_w = 0.444 \quad \text{or} \quad a_{max} = 0.0706 \lambda_w \tag{2.2}$$

so that a 100-m long wave has a maximum amplitude of 7 m. As the waves approach this maximum height, theoretical investigations show that the crest remains symmetric and tends toward a $120°$ interior angle. Measured over a quarter wavelength, the maximum wave slope is about $15°$.

For large amplitude gravity and capillary-gravity waves, one effect of this curvature increase at the crest is that, as Kinsman (1984, p. 538) describes, capillary waves form just ahead of the crest on the forward, downwind face of the wave (Figure 2.4). Because capillary

Wind and wave propagation direction

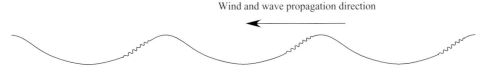

Figure 2.4. The growth of the parasitic capillary waves just beyond the crest and on the forward face of an ocean wave. The vertical scale is exaggerated.

waves are rapidly damped by viscosity, the energy transferred to them from the long waves is dissipated, yielding a decrease in the long wave amplitude. For this reason, these waves are called *parasitic capillaries*. Because these capillaries form asymmetrically about the wave crest, instruments sensitive to wave roughness have a greater response looking upwind than downwind, which, as Chapters 9 and 11 show, allows microwave instruments to determine the wind speed and direction.

2.2.2 Wave breaking, energy absorption, and the properties of foam

If energy continues to be added to a wave and as its amplitude increases toward its maximum, long waves break. In contrast, instead of breaking, capillary-gravity waves lose energy to shorter parasitic capillaries and to non-breaking turbulence. For example, Figure 2.5 is a photograph and drawing of a large amplitude, 10-cm long wave in a wavetank, and shows a turbulent, but non-breaking region at the crest and parasitic capillaries on the downwind face.

Long wave, deep water breaking occurs as follows: if the winds are strong enough, as the crest approaches a 120° wedge shape, the crest spills forward down the front face of the wave and breaks (Donelan and Pierson, 1987). This is called *whitecapping*, and occurs for wind speeds greater than about 3 m s^{-1} (Melville, 1996). Wave breaking restores equilibrium to the surface by reducing the wave amplitude, expels small seawater droplets into the air, and entrains air bubbles into the water column, generating a transient layer of foam. Figure 2.6 shows three photographs of the wave breaking and foam generation associated with a North Atlantic storm. The winds are gusting to 25–30 m s^{-1}, the reported wave heights are 12–15 m. The photographs illustrate the surface roughness and foam that accompany these strong winds. In another example, Plate 1 is an oblique low-altitude aircraft photograph of the Japan Sea showing, for a wind speed of 17 m s^{-1}, that foam covers an appreciable fraction of the surface. Each of these photographs shows that waves and breaking must be considered in remote sensing of the sea surface. Perkowitz (2000) states that at any time, 2–3% of the ocean surface is covered by foam, an area equivalent to that of the United States.

Two factors associated with breaking, droplet expulsion and foam generation, must also be considered. First, droplet expulsion transfers sea salt into the marine boundary layer at a global rate of 10^9 metric tons per year (Perkowitz, 2000). The aerosol generated by the sea salt reduces the transmittance of the marine boundary layer, so that for ocean color retrieval the presence of this aerosol must be determined. Second, foam changes the reflective and

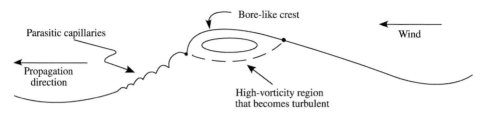

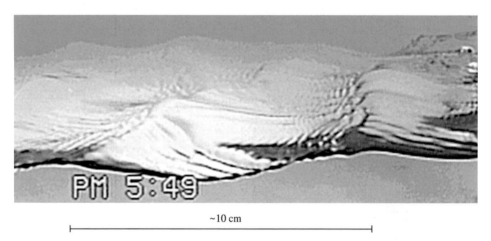

~10 cm

Figure 2.5. An interpretative drawing and a video frame of a small wind-generated gravity wave losing energy through parasitic capillaries (Figure 1 from Jessup and Zappa, 1997, © 1997 American Geophysical Union, reproduced/modified by permission of AGU, courtesy Andrew Jessup).

emissive properties of the sea surface. In the visible, because foam is much more reflective than seawater, it can falsify the ocean color retrieval. In the microwave, because foam has different emissive properties than seawater, increases in areal extent with wind speed, and preferentially forms on the downwind face of breaking waves, foam contributes to the retrieval of the scalar and vector wind speed.

Frouin, Schwindling and Deschamps (1996) and Moore, Voss and Gordon (2000) summarize the physical properties of whitecap foam. It consists of two parts, surface foam that is made up of small volumes of air surrounded by thin layers of seawater, and subsurface bubbles that result from the injection of air into the water column by breaking waves. From field observations at wind speeds of 8 m s^{-1}, Lamarre and Melville (1996) show that the bubbles in the water column occur to depths of at least 3 m, where the void fraction immediately below the surface is about 20%, and where the air bubble concentrations fall off exponentially with an e-folding depth of 0.18 m (for examples, see Melville, 1996, Figure 3; and Baldy, 1993). This combination of bubbles on the surface and bubbles rising slowly at depth means that the surface bubble presence is sustained for about half a wave period or for as long as 10–20 s (Lamarre and Melville, 1996; Koepke, 1984). Although the area and duration of the foam patches depend on fetch, wind speed and air and water temperature, the

Figure 2.6. Examples of the ocean waves and wave breaking associated with a storm in the North Atlantic during December 1991. The winds are gusting 25–30 m s^{-1}, the reported wave heights are 12–15 m. Large breaking waves (top); shorter waves breaking while riding on longer waves (middle); short, strongly wind-forced breaking waves (bottom) (Photographs by E. Terrill and W. K. Melville; Figure 1 from Melville 1996, with permission, from the *Annual Review of Fluid Mechanics*, Volume 28, © 1996 by Annual Reviews, courtesy W. K. Melville).

foam areal extent has a nonlinear dependence on wind speed. For example, a foam coverage of 0–1% occurs for winds of 1.4–7 m s^{-1}, a coverage of at least 4% for winds greater than 14 m s^{-1}, and a coverage as large as 20% for wind speeds of 20 m s^{-1} or greater (Melville, 1996; Monahan and O'Muircheartaigh, 1986). Monahan and O'Muircheartaigh show that above onset wind speeds of about 4 m s^{-1}, the areal foam extent increases approximately as $U^{3.5}$, although field studies of this dependence produce very noisy results (Moore *et al.*, 2000).

2.2.3 Root-mean-square amplitude and significant wave height

In many instances, the wave field can be described as the sum of a collection of waves with random amplitudes, wavelengths and propagation directions. Similar to the single frequency wave, the resultant wave amplitude is described in terms of the wave height $\eta(x, y, t)$, where x and y lie in the plane of the mean free surface, and where for use in the next section the x-axis points downwind, the y-axis crosswind. For these definitions, $\bar{\eta} \equiv 0$, where the overbar indicates an average over a long period of time, and the root-mean-square (rms) displacement σ_η is given from

$$\sigma_\eta^2 = \bar{\eta}^2 \tag{2.3}$$

This parameter is frequently used to describe the amplitude of a field of random waves. For the simple sine wave in (2.1), $\sigma_\eta^2 = a_w^2/2$.

The wave amplitude can also be described in terms of the *significant wave height* (*SWH*) or $H_{1/3}$. Significant wave height has an unusual definition; it is defined as the average crest-to-trough height of the one-third largest waves. For remote sensing, $H_{1/3}$ is used to describe the ocean swell properties observed by the satellite altimeter. The definition of $H_{1/3}$ is apparently based on how a mariner might estimate the wave height from a ship; it was used in the early wave forecast models (Kinsman, 1984). Wunsch and Stammer (1998, p. 233) describe $H_{1/3}$ as "an archaic, but historically important" term for wave height. Paraphrasing Kinsman (1984, p. 302), there is nothing particularly significant about $H_{1/3}$, it is just another average. Chelton *et al.* (2001b) review $H_{1/3}$ and state that it can be written in terms of σ_η, where

$$H_{1/3} = 4\sigma_\eta \tag{2.4}$$

Figure 2.7 illustrates σ_η and $H_{1/3}$ for a numerically generated wave field. For a narrow wavelength bandwidth, Equation (2.4) is exact. For broader bandwidths, the coefficient in (2.4) decreases from 4 to 3. Because Chelton *et al.* (2001b) show that this change has a negligible effect on the altimeter retrieval, for the present purposes and over a wide range of bandwidths, (2.4) is a reasonable approximation.

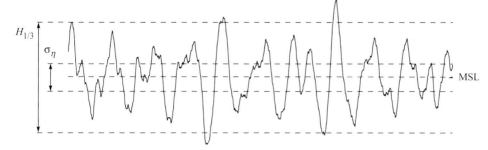

Figure 2.7. A field of random waves with a nearly Gaussian amplitude distribution, and greatly exaggerated amplitudes. The figure shows the mean sea level (MSL) and the rms and significant wave heights.

2.2.4 Azimuthal distribution of sea surface slopes

The azimuthal distribution of wave slopes relative to the wind direction affects remote sensing in three ways. First, at all frequencies, the wave slopes can reflect sunlight directly into the instrument and overwhelm the desired observations. Viewing a water surface from a hillside or building shows that, in the visible, the solar reflection from the wind-roughened surface forms a bright diffuse spot composed of many transient reflecting facets. This phenomenon is called *sun glint* or *sun glitter*. Second, the wave slopes diffuse the sunlight transmitted across the ocean interface into the interior and alter the magnitude of the water-leaving radiances contributing to the ocean color retrieval. Third, because the wave slopes have an azimuthal distribution relative to the wind direction, active and passive microwave measurements can retrieve both wind speed and direction.

In the discussion of this azimuthal dependence, the total mean-square slope σ^2 and the alongwind and crosswind rms components of the wave slopes σ_L, σ_C are defined as

$$\sigma_L^2 = \overline{\eta_x^2}, \qquad \sigma_C^2 = \overline{\eta_y^2}, \qquad \sigma^2 = \sigma_L^2 + \sigma_C^2 \tag{2.5}$$

For the sine wave in (2.1), the mean-square slope is $\sigma_L^2 = a_w^2 k_w^2 / 2$. In what may be the most cited paper in the remote sensing literature, Cox and Munk (1954) use aerial photographs of sun glint taken under different wind conditions near Hawaii to describe the angular distribution of the reflecting slopes as a function of wind speed. Their results show that the largest slopes occur in the upwind and downwind direction, the smallest in the crosswind direction and that the slope magnitude varies smoothly with azimuth angle. They also find that large slopes are slightly more probable in the upwind than the downwind direction. Cox and Munk (1954, p. 206) suggest that the source of this asymmetry is due to parasitic capillary formation on the forward wave faces. Consequently, the reflection of the sun on the sea surface forms an ellipse, with its long axis parallel to the wind, its short axis at right angles and where the ellipse is slightly broader in the upwind direction. For $1 < U < 12 \text{ m s}^{-1}$, Cox and Munk (1954) and Wu (1990), who reanalyzes their data with the addition of many modern studies, show that the ratio of the crosswind to alongwind mean-square slopes σ_C^2/σ_L^2 varies from 0.6 to 1.0 with a mean of 0.8.

In terms of Mobley's notation (1994, Section 4.3), Cox and Munk (1954) find a linear dependence of the mean slopes on wind speed,

$$
\begin{aligned}
\sigma_L^2 &= AU, & A &= 3.5 \times 10^{-3} \text{ s m}^{-1} \\
\sigma_C^2 &= BU, & B &= 2.8 \times 10^{-3} \text{ s m}^{-1}
\end{aligned}
\tag{2.6}
$$

In his reanalysis, Wu (1990) finds that the slopes vary as the logarithm of U; for $U < 7 \text{ m s}^{-1}$, the various components of σ^2 increase slowly, for $U > 7 \text{ m s}^{-1}$, they increase more rapidly.

2.2.5 Surface slicks

From field experiments, Cox and Munk (1954) also show that the addition of an oil slick with a thickness of order 1 μm causes a reduction of the surface slopes from their clean water values by a factor of 2 or 3, and the disappearance of waves with lengths less than about 0.3 m. As Chapter 13 shows, this damping makes it possible for radars to observe oil slicks. The oceanic sources of these slicks divide into natural and man-made, and into biogenic and petroleum slicks (Clemente-Colón and Yan, 2000). Man-made oil slicks result from accidental spills, the illegal discharge of petroleum products from ships and from harbor runoff. There are also natural petroleum seeps in the Gulf of Mexico and in the Santa Barbara Channel off southern California. Plankton and fish produce biogenic slicks, which also result from waste discharged from factory fishing vessels. Because these slicks greatly reduce the short wave amplitudes, they are visible in radar imagery.

2.3 Ocean currents, geostrophy and sea surface height

The upper layers of the ocean are dominated by wind-driven features such as the Gulf Stream, Kuroshio, Antarctic Circumpolar Current and coastal upwelling. Combined with evaporation in the tropics, cooling in the north and south, and seasonal heating and cooling, the wind stress determines the vertical mass fluxes in the upper ocean and maintains its density structure (Wunsch, 2002).

For the rotating Earth, the *geostrophic flow* approximation describes the relation between geostrophic currents, density structure and sea surface height. In the vertical, the approximation assumes that the ocean is in near hydrostatic balance, so that

$$
\frac{dp}{dz} = -g\rho(p, S, T)
\tag{2.7}
$$

In (2.7), p is pressure, g is the acceleration of gravity, ρ is density, S is salinity and T is temperature. The variables p, S and T are measured by oceanographic instruments and are often given in terms of the local rectangular coordinates x, y, z and t that rotate with the Earth (Figure 2.8). From Cushman-Roisin (1994), these coordinates follow the convention

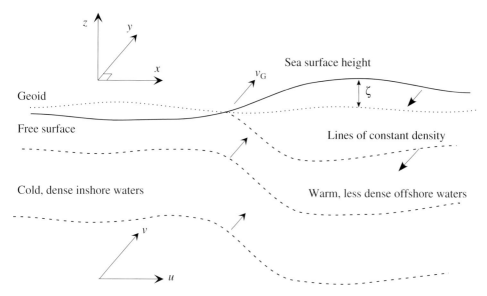

Figure 2.8. Geostrophic flow in the Northern Hemisphere along a line of constant latitude where the figure is based on the Gulf Stream. The figure shows the free surface (solid line), the geoid or equipotential surface (dotted line), the lines of constant density (dashed lines) and the velocities resulting from the geostrophic balance. The variable ζ is the sea surface height as defined for the altimeter (Adapted from Figure 1, Stommel, 1966).

that z is parallel to gravity and increases upward, while x and y lie in the horizontal plane with x parallel to longitude and y parallel to latitude.

In the horizontal, the approximation neglects the time-dependent and nonlinear terms in the equations of motion (Cushman-Roisin, 1994). Geostrophic flow is then derived from a balance between the horizontal pressure gradients and the Coriolis forces generated by the ocean velocities measured in the Earth's rotating frame. In the following, $f = 2\Omega_E \sin \chi$ is the Coriolis parameter, where $\Omega_E = 7.727 \times 10^{-5}\ \mathrm{s}^{-1}$ is the Earth's angular rotation and χ is latitude. For the x and y components of the geostrophic velocity u_G and v_G, the approximation yields

$$\rho f v_G = \frac{\mathrm{d}p}{\mathrm{d}x}, \qquad \rho f u_G = -\frac{\mathrm{d}p}{\mathrm{d}y} \tag{2.8}$$

Combination of Equations (2.7) and (2.8) gives for v_G at a height z, relative to a reference level z_0,

$$v_G(x, y, z) = \frac{g}{f} \int_{z_0}^{z} \frac{\mathrm{d}\rho}{\mathrm{d}x}\, \mathrm{d}z + v_0(x, y, z_0) \tag{2.9}$$

with a similar equation for u_G. In (2.9), v_0 is an unknown reference velocity depending on z_0. Measurement of v_0 at any depth, including the free surface, combined with knowledge of the interior density distributions permits calculation of the absolute velocity profile.

For these flows, Figure 2.8 shows a schematic drawing of the distribution of density and sea surface height on a constant latitude line across a flow similar to the Gulf Stream. The arrows show the v-components of geostrophic velocity, the dashed lines are constant density surfaces, the solid line is the sea surface height and the dotted line is the geoid. As Chapter 12 describes in more detail, the geoid is the equipotential surface along which there are no parallel components of acceleration, corresponding to mean sea level in the absence of external forcing. In the ocean, the uneven density distributions lead to the displacement of the sea surface height above and below the geoid. For example, from the schematic in Figure 2.8, the onshore waters are cold and dense while the offshore waters are warm and less dense. Within these water masses, if two columns of seawater are defined such that they extend between the sea surface and the same deep surface of constant pressure, the columns have the same mass. But because the onshore column is more dense than the offshore, its height is less than the offshore. This height difference ranges from 1 m across the Gulf Stream to 10 cm or less for ocean eddies.

There is an important difference between the relative measurement of sea surface height in the classical oceanographic analysis, and the altimeter measurement. In the classic oceano-graphic analysis, sea surface height is a relative measurement called *dynamic height* $\zeta_D(x, y)$. Dynamic height is calculated relative to a reference depth or pressure from integration of the vertical density anomalies derived from individual oceanographic stations and sections (Pond and Pickard, 1986, Chapter 8). The stations consist of deep CTD (conductivity–temperature–depth) casts, from which the density anomalies are determined. The reference depth is called the *level of no motion*; if there is motion at this depth, the surface displace-ment and the geostrophic velocities are measured relative to arbitrary constants. At the sea surface, the geostrophic velocities can be written in terms of the slope of the dynamic height (Knauss, 1997):

$$v_G(x, y, 0) - v_0(x, y, z) = \frac{g}{f}\frac{d\zeta_D}{dx} \qquad u_G(x, y, 0) - u_0(x, y, z) = -\frac{g}{f}\frac{d\zeta_D}{dy} \qquad (2.10)$$

On the left-hand side of (2.10), the height at which the surface velocities are evaluated is approximated as $z = 0$, and v_0 and u_0 are the arbitrary constant velocities. Because of the uncertainty concerning the depth or even in some cases the existence of the level of no motion, it is very difficult to determine absolute geostrophic velocities from oceanographic observations.

In the altimeter analysis and as Chapter 12 shows, the altimeter measures the absolute height of the sea surface relative to the Earth's center of mass. The sea surface height is then defined as the difference $\zeta(x, y, t)$ between the sea surface and the geoid. Because this is an absolute measurement, changes in the height measured by the altimeter are generated not only by geostrophic flows, but also by other geophysical processes including tides, seasonal heating and cooling, and changes in atmospheric pressure. To calculate the ζ-contribution due only to geostrophy, these other sources must be removed. Following their removal and substituting for the Coriolis parameter, Equation (2.10) can be written

$$v_G(x, y, 0) = g(2\Omega_E \sin \chi)^{-1}\frac{d\zeta}{dx} \qquad u_G(x, y, 0) = -g(2\Omega_E \sin \chi)^{-1}\frac{d\zeta}{dy} \qquad (2.11)$$

Because Equation (2.11) is written in terms of the absolute surface displacement ζ, the unknown u_0 and v_0 in Equation (2.10) no longer appear. This means that the direct measurement of sea surface slope yields the surface geostrophic velocity. Given this result and provided that coincident surveys of the ocean interior are available, then from (2.9) the interior geostrophic velocity profile can be calculated.

There are at least three qualifications concerning the geostrophic balance. First, in the vicinity of the equator where χ goes to zero, the denominators of Equation (2.11) also approach zero and the geostrophic approximation breaks down. Consequently, the dynamics of equatorial flows differ from those at higher latitudes. Second, real oceanographic flows are not steady, but vary with time. For this case, the geostrophic balance is supplemented by a small acceleration term. Because this imbalance in the geostrophic equations is very small and generally unobservable by direct oceanographic measurements, the geostrophic velocities are still derived from Equations (2.9) and (2.10). Third, even given an altimeter measurement of ζ in Equation (2.11), the derived geostrophic surface velocities are not necessarily the true surface velocities. The reason for this is that, in the surface boundary layer, the velocity tends not to be in geostrophic balance, rather the flow responds to the turbulent stresses generated by the wind and waves. Even though these wind effects dominate the upper 100 m of the ocean, because geostrophic velocities are dominant below this depth, their associated pressure gradients determine the sea surface slope so that the relation between the interior flows and the surface slopes still holds (Wunsch and Stammer, 1998).

Table 2.1 gives typical length and time scales for a variety of oceanographic flows. The shortest spatial scales over which these flows occur correspond to the Rossby radius of deformation, which ranges from about 10 km in the Arctic to 60 km at mid-latitudes to 200 km in the tropics; the shortest time scale is 1–2 days (Cushman-Roisin, 1994). The largest spatial scales are basin-wide, which in the Pacific correspond to 10 000 km. The different scales are related, for example western boundary currents such as the Gulf Stream and Kuroshio consist of a vigorous northern flow with a cross-stream length scale of about 100 km, and a diffuse return flow that occupies the rest of the basin with scales of order 10^4 km. Superimposed on the time-averaged flows are a large variety of time-dependent flows, which occur at different time and space scales, are often referred to as mesoscale eddies, and can have energy levels that are one or more orders of magnitude greater than the mean flows (Wunsch and Stammer, 1998).

Finally, the alternation between La Niña and El Niño is an important example of planetary flow phenomena. The transition between them creates dramatic changes in the tropical atmosphere, generates equatorial Kelvin and Rossby waves, alters the physical and biological properties of the equatorial Atlantic and Pacific, and affects

Table 2.1. *Space and time scales of oceanic phenomena*

Phenomenon	Surface length scales	Period	Comments
Western boundary currents (Gulf Stream, Kuroshio)	130 cm/100 km	Days to years	Position is variable, with a 25% variability in transport
Basin-scale gyres (North Atlantic, North Pacific)	50 cm/3–10 × 10³ km	One to many years	25% variability
Mesoscale eddies	10–25 cm/100 km	100 days	100% variability
Small eddies	10 cm/10–100 km	1–2 days	100% variability
Eastern boundary currents	30 cm/100 km	Days to years	100% variability, with possible reversals in direction
Equatorial currents	30 cm/5000 km	Months to years	100% variability
Internal tides	10 cm/1–100 km	1 day	100% variability
Coastal upwelling	10 cm/10–100 km	1 day–1 week	100% variability
Internal Rossby and Kelvin waves	10 cm/1000 km	Months	100% variability

For the surface length scale column, the first number is a characteristic height, the second is the horizontal scale of the motion.
Adapted from Wunsch *et al.* 1981, Table 1, and Chelton 2001.

global climate. During the past century, La Niña conditions have been interrupted at three to seven year intervals by an El Niño, of which the 1997–98 El Niño was the strongest on record (McPhaden, 1999). Because many satellites observed the 1997–98 El Niño/La Niña transition, several of the examples used in this book illustrate their differences.

2.4 Sea ice

The sea ice covers of the Arctic and Antarctic Oceans experience strong seasonal cycles and play a major role in the modification of the heat and salt flux to the underlying ocean. Wadhams (2000) provides an excellent introduction to sea ice with many photographs. In their Arctic and Antarctic atlases, Zwally *et al.* (1983), Parkinson *et al.* (1987) and Gloersen *et al.* (1992) also describe the ice properties and provide many photographs.

Because the Arctic Ocean is a nearly enclosed basin surrounded by land, while the Antarctic Ocean surrounds the Antarctic continent and is itself surrounded by open ocean, the kinds of sea ice that form in each region differ from one another. In the north, the

Arctic Ocean has a small oceanic vertical heat flux, and an annual snowfall of about 200 mm (Wadhams, 2000). Its major ice types include young ice, first-year ice, which is less than one year old, has not survived a summer, and has thicknesses of 1–2 m, and multi-year ice, which is older than one year with thicknesses of 2–4 m. As Cavalieri (1994) describes, young ice is characterized by a high salinity surface layer that gives it a distinctive microwave signature. First-year ice has a less saline surface layer, with salt and air inclusions in its near surface layers, and multiyear ice has a hard upper surface consisting of nearly fresh water. In the summer, all categories of Arctic sea ice approach the freezing point, so that as the ice desalinates, the upper surface melts and fresh water melt ponds form on the surface. The ice that survives the summer refreezes, forming multiyear ice.

In contrast, around the Antarctic continent and at the ice margins in the North Atlantic and Pacific where the adjacent ocean contains large ocean waves, sea ice forms differently. When sea ice forms in the presence of waves, the combination of the surface heat loss with the wave-induced mixing cools the upper ocean to the freezing point and sometimes even induces a slight supercooling. This means that once freezing begins, ice formation occurs throughout the upper layer as small millimeter-scale crystals, called frazil crystals, that float to the surface. As these crystals collect on the surface, the resultant slurry, called grease ice, damps out the short-period waves in a manner similar to an oil slick, and gives the surface a smooth appearance (Figure 2.9a). When this ice reaches a thickness of about 100 mm, its surface begins to freeze, which, combined with the long-period ocean swell propagating through the ice, breaks the surface into floes with diameters of 0.3 to 0.5 m, called pancake ice (Figure 2.9b). Because of wave-induced collisions, the pancakes grow raised rims. The presence of these rims causes an increase in both their atmospheric drag coefficients and their radar reflectivity. Around Antarctica, this is the dominant mechanism for ice formation, where as time goes on these pans aggregate into collections of large floes. In the North Atlantic, this mode of ice formation also occurs in an ice edge feature called the Odden (Wadhams, 2000). This formation of frazil and pancake ice also occurs in the wind-generated regions of open water called polynyas. As Chapter 13 will show, both ice types are visible in SAR imagery.

Because the Antarctic oceanic heat flux is about five times that in the Arctic, the first-year sea ice thicknesses are only about 0.7 m (Wadhams, 2000, Section 2.3.2). Also, because of the snow blowing off the continent and the snow generated by moisture flux from the adjacent open ocean, snow accumulations on Antarctic ice are much greater than on Arctic, with characteristic thicknesses of 0.5–0.7 m. The combination of thinner ice and much greater snow accumulation means that for much of this ice, its interface is depressed below sea level. Consequently, seawater intrudes into the snow above the ice surface, leading to seawater freezing at the upper interface. As Chapter 9 discusses, this freezing process may explain why Antarctic sea ice has different microwave signatures compared to Arctic ice. Finally, because most of this ice melts in summer, there is much less Antarctic multiyear ice.

Figure 2.9. Shipboard photographs of sea ice in the Greenland Sea. (a) A slurry of frazil ice crystals called grease ice; (b) pancake ice (Courtesy Richard Hall and Peter Wadhams, used with permission).

2.5 **Further reading and acknowledgements**

Lighthill (1980), Lamb (1945), Phillips (1977) and Kinsman (1984) present the basic theory of water waves, and Kinsman (1984) provides an excellent collection of wave photographs. There are many textbooks that describe the ocean circulation and the planetary Kelvin and Rossby waves; Cushman-Roisin (1994) is a good introductory text, Pedlosky (1987) is more advanced. I thank Andrew Jessup, William Melville, Ellen Lettvin and William Plant for help with the material in this chapter.

3

Electromagnetic radiation

3.1 Introduction

This chapter describes those properties of electromagnetic radiation (EMR) relevant to remote sensing. Specifically, Section 3.2 gives a brief description of the nature of electromagnetic radiation, its propagation in different media, and its polarization. Section 3.3 describes several different ways of describing radiation fluxes. Section 3.4 discusses blackbody radiation, Planck's equation and the concepts of emission and absorption. Section 3.5 discusses the basic optics applicable to an instrument operating in the visible and infrared, then describes the operation and spatial resolution of an ideal instrument. The section concludes with a discussion of terms such as bandwidth and signal-to-noise ratio that are applicable to real instruments.

3.2 Descriptions of electromagnetic radiation

As many textbooks describe, EMR has a dual nature, in that it behaves both as discrete quanta of radiation and as electromagnetic waves (Jackson, 1975; Born and Wolf, 1999). In the quantum description, radiation propagates as photons, which are massless, discrete bundles of energy released by atomic or molecular changes of state.

The energy $\hat{E}$ carried by each packet is

$$\hat{E} = \hbar f \tag{3.1}$$

where f is the frequency in cycles-per-second or Hz, and $\hbar = 6.625 \times 10^{-34}$ J s is the Planck constant. When the radiation is generated by only a small number of molecular sources, the quanta are discrete; when the number of sources is increased, the classical wave solution describes the radiation.

In the wave description, Maxwell's equations govern the radiation, where the parameters that describe the medium through which the radiation propagates are the magnetic permeability μ, the electrical permittivity ε and the electrical conductivity σ. The three different media of oceanographic concern are vacuum, atmosphere and ocean. These are each assumed to be homogenous isotropic media that are non-magnetic and contain no free

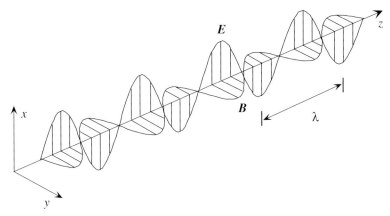

Figure 3.1. The electric and magnetic field components of an electromagnetic wave for a plane-polarized wave.

charges. For each of these, $\mu = \mu_0$, where μ_0 is the vacuum permeability and ε and σ are constant. In the atmosphere and vacuum, $\sigma = 0$, while in the ocean, σ is nonzero.

The plane wave solution to Maxwell's equations is given in terms of an electric field vector $\boldsymbol{E}$ and a magnetic field vector $\boldsymbol{B}$, where the bold type indicates a vector and where $\boldsymbol{B}$ and $\boldsymbol{E}$ are perpendicular both to each other and to the propagation direction (Figure 3.1). With this notation, the electric field component of the plane wave solution has the form

$$\boldsymbol{E} = \boldsymbol{E_0}\exp[i(\boldsymbol{k}z - \omega t)] \qquad (3.2)$$

with an analogous form for $\boldsymbol{B}$. In (3.2), $\boldsymbol{E_0}$ is the complex wave amplitude, $\boldsymbol{k} = k + ik_{\mathrm{im}}$ is the complex wavenumber with real and imaginary parts k and k_{im}, and ω is the wave frequency in radians-per-second. The wavenumber k is related to the wavelength λ by $k = 2\pi/\lambda$; ω is related to the wave period T by $\omega = 2\pi/T$ and to the wave frequency by $\omega = 2\pi f$.

The wavelength λ has units of length, which depending on the observing window is expressed in m, μm (micrometers or 10^{-6} m) or nm (nanometers or 10^{-9} m); the wave frequency f is in units of Hz, MHz (megahertz or 10^6 Hz) or GHz (gigahertz or 10^9 Hz). Because the early optics experimenters worked with light at different wavelengths, wave propagation in the visible/infrared is generally described in terms of wavelength; because the early microwave experimenters worked with frequency, propagation in the microwave is described in terms of frequency. Also, because of the secrecy surrounding the development of radar during World War II, the microwave frequencies are often described by letters, the most common used in this book being C-band (5 GHz) and K_u-band (13.6 GHz). Table A.1 in the Appendix gives this terminology in more detail.

3.2.1 Uses of the electromagnetic spectrum

Satellite remote sensing takes place in a crowded electromagnetic spectrum that, especially in the microwave, restricts the location and width of the observational frequencies. Figure 3.2 shows some of the allocations of the US electromagnetic spectrum as a function of frequency and wavelength. At 10^5 Hz, the Amplitude Modulated (AM) radio band is characterized by km-long wavelengths that are not used in satellite remote sensing. The higher frequencies of 10^7–10^8 Hz contain the Frequency Modulated (FM), TV and cellular phone bands. The frequencies between 10^9 and 10^{11} Hz (1–100 GHz) contain passive and active microwave

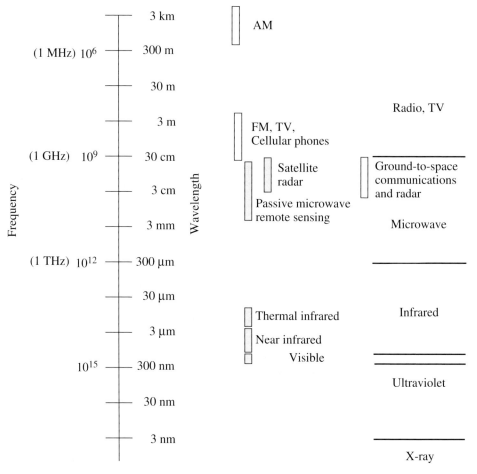

Figure 3.2. The electromagnetic spectrum and its uses as a function of frequency and wavelength. The gray bars show the bands used in satellite remote sensing.

Table 3.1. *Subregions of the spectrum for the ultraviolet through infrared wavelengths*

Name	Abbreviation	Wavelength band
Ultraviolet	UV	10–400 nm
Ultraviolet-B	UV-B	280–320 nm
Visible	V	400–700 nm
Near Infrared	NIR	0.7–3.5 μm
Visible/Near Infrared	VNIR	0.4–3.5 μm
Thermal Infrared	TIR	3.5–20 μm
Visible/Infrared	VIR	0.4–20 μm

Adapted from Kramer, 1994; Thomas and Stamnes, 1999

remote sensing plus a large variety of commercial and military communications and ground radar operations. The infrared bands occur at $10^{13}–10^{14}$ GHz; the narrow visible band occurs at 10^{15} GHz; the ultraviolet (UV) region is at higher frequencies.

Although the frequency allocation presented in Figure 3.2 appears relatively uncrowded, this is because of the small figure scale. A more detailed frequency allocation may be viewed at the website (http://www.ntia.doc.gov/osmhome/osmhome.html) maintained by the Office of Spectrum Management, National Telecommunications and Information Administration, US Department of Commerce. The information presented by this office shows the large variety of users and the resultant pressures on the remote sensing bands, especially in the frequency range 1–10 GHz. Because at these frequencies and as Chapter 9 shows, radiation propagates through the atmosphere and clouds with little or no attenuation, there is a great demand for frequency assignments from military and civilian radars, aircraft navigation, satellite direct broadcast and communications, and cellular telephone systems. These users place enormous pressures on the frequencies used by remote sensing and have forced changes in their assigned frequencies and restricted their bandwidths.

The visible and infrared (VIR) wavelengths occur between approximately 0.4 and 20 μm. These wavelengths are heavily used in remote sensing, but because of interference by clouds and the atmosphere, they do not experience the same pressures from other users. Table 3.1 shows the terminology and abbreviations used to describe the wavelength bands within and adjacent to the VIR. The visible spectrum lies between 0.4 and 0.7 μm, and approximately divides into the following colors: 400–440 nm, violet; 440–500 nm, blue; 500–550 nm, green; 550–590 nm, yellow; 590–630 nm, orange; 630–700 nm, red. The UV band occurs at shorter wavelengths than the visible; for completeness, the table lists the wavelength range of the biologically important UV-B radiation, which destroys DNA and causes skin cancer. The Near Infrared (NIR) band occurs at longer wavelengths than the visible, and like the visible, is dominated by reflected solar radiation. The Thermal Infrared (TIR) band includes those wavelengths dominated by thermal emission from the Earth's surface, which as Chapter 7 shows, are used in the retrieval of sea surface temperature.

3.2.2 Dispersion relation and index of refraction

The dispersion relation governs the propagation of EMR in different media and its attenuation in the ocean. From Jackson (1975, Chapter 10), the general form of the dispersion
relation is

$$k^2 = \omega^2 \mu \varepsilon - i\omega\mu\sigma \tag{3.3}$$

For vacuum, the conductivity $\sigma = 0$, so that (3.3) becomes

$$k^2 = \omega^2 \mu_0 \varepsilon_0 \tag{3.4}$$

where the subscript 0 indicates vacuum quantities. The propagation velocity or phase speed
v of the radiation is

$$v = \omega/k = \lambda f \tag{3.5}$$

so that in vacuum, the speed of light c can be written as

$$c = 1/\sqrt{\mu_0 \varepsilon_0} = 3 \times 10^8 \, \text{m s}^{-1} \tag{3.6}$$

In other materials, v is the local lightspeed that can differ from c; for example, in water
the lightspeed is about $0.75c$. Unlike ocean waves in deep water, where the phase speed
increases with wavelength, for a homogenous medium, v is constant.

Radiation in the atmosphere propagates at approximately the same speed as in vacuum, and in all three media, $\mu = \mu_0$. But because in seawater $\sigma > 0$, EMR propagation
becomes more complicated, and the dispersion relation is written as follows (Born and Wolf,
1999):

$$k^2 = \omega^2 \mu_0 \varepsilon_0 [(\varepsilon/\varepsilon_0) + i\sigma/\omega\varepsilon_0] \tag{3.7}$$

To rewrite Equation (3.7) into a more useful form, the complex dielectric constant ε_r is
defined as

$$\varepsilon_r = \varepsilon' + i\varepsilon'' \tag{3.8}$$

where $\varepsilon' = \varepsilon/\varepsilon_0$, so that, $0 \le \varepsilon' \le 1$, and $\varepsilon'' = \sigma/\varepsilon_0\omega$. Substitution of c from (3.6) and ε_r
from (3.8) into (3.7) gives

$$k = (\omega/c)\sqrt{\varepsilon_r} \tag{3.9}$$

To simplify this expression further, the *index of refraction* η is defined as $\eta = \sqrt{\varepsilon_r}$ and
written as $\eta = n + i\chi$, where n is the real part, χ the imaginary. From this definition,

$$k = (\omega/c)(n + i\chi) \tag{3.10}$$

From Born and Wolf (1999), substitution of Equation (3.10) into the plane wave propagation equation gives

$$E = E_0 \exp[i(kz - \omega t)] \exp(-\omega \chi z/c)$$

$$\text{(a)} \qquad\qquad\qquad \text{(b)} \qquad\qquad\qquad \text{(3.11)}$$

In Equation (3.11), (a) is an oscillatory wave solution where $k = \omega n/c$ so that the wave propagates with a phase speed c/n; (b) is a damped exponential.

Because the wave energy is proportional to E^2 and from term (b) above, as the radiation propagates through water, the energy decays as $\exp(-2\omega \chi z/c)$, or equivalently as $\exp(-4\pi \chi z/\lambda)$. The *absorption coefficient* $a(\lambda)$ is therefore defined as

$$a(\lambda) = 4\pi \chi /\lambda \qquad\qquad (3.12)$$

For the energy decay, the absorption depth d_a is defined as the $1/e$ decay distance,

$$d_a = [a(\lambda)]^{-1} = \lambda/4\pi \chi \qquad\qquad (3.13)$$

For distilled water, which has similar properties to seawater, and for λ ranging from the ultraviolet to the microwave, Figure 3.3 shows the real and imaginary parts of η. The most striking feature of the figure occurs for χ in the vicinity of the visible wavelengths, where it experiences a 10^{10} change in magnitude. This region, the physics of which is discussed in Jackson (1975) and Mobley (1994), means that light only propagates to appreciable oceanic depths within a narrow band centered on the visible. Specifically, for blue light of $\lambda = 440$ nm propagating in water, $\chi = 9 \times 10^{-10}$, yielding a d_a of about 40 m, whereas for $\lambda = 10$ μm in the infrared, $\chi = 0.05$, so that $d_a = 16$ μm, which is much smaller than in the visible. Chapter 5 uses η to describe the reflection and refraction of radiation incident on the interface, and further discusses the dependence of d_a on λ.

3.2.3 Review of solid geometry

Much of this book uses the spherical coordinate system shown in Figure 3.4, where r is the radial distance, θ is the zenith angle and ϕ is the azimuth angle. The figure also shows the differential area dA generated by small changes in θ and ϕ. The definition of the differential solid angle $d\Omega$ is

$$d\Omega = dA/r^2 = \sin \theta \, d\theta \, d\phi \qquad\qquad (3.14)$$

The solid angle has units of steradians, or sr, where there are 4π sr in a sphere. NASA often uses orthogonal radians to describe the solid angle measured by an optical instrument so that the AVHRR for example, has a solid angle resolution of 1.3 mr $\times$ 1.3 mr, where the milliradian, mr, equals 10^{-3} radians. For comparison, Section 3.5.1 shows that the resolution of the human eye is about 0.25 mr, or five times smaller than the AVHRR resolution.

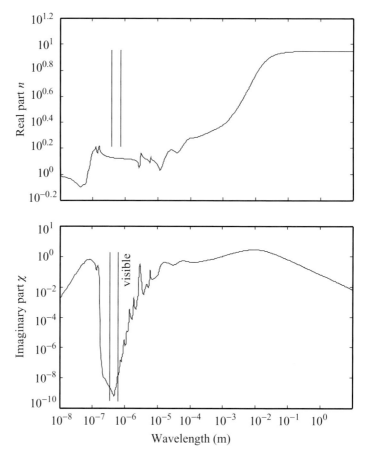

Figure 3.3. Plots of the real and imaginary parts of the index of refraction η over the wavelength domain of interest. The two figures have different vertical scales; the pair of vertical lines marks the visible spectrum (Adapted from Mobley, 1995; data from Segelstein, 1981).

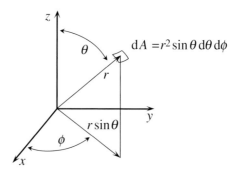

Figure 3.4. The spherical coordinate system.

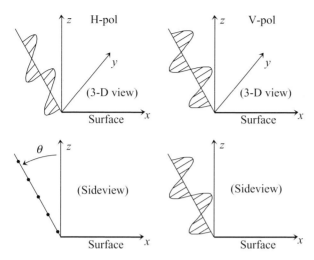

Figure 3.5. The difference between the plane of the electric field vector for vertically polarized (V-pol) and horizontally polarized (H-pol) radiation, determined relative to the Earth's surface.

3.2.4 Polarization and the Stokes parameters

For remote sensing, the intensity and frequency distribution of the radiation that is emitted or reflected from a surface permits inference of the surface properties. As next discussed, the *polarization* of the radiation is equally important. Following Jackson (1975, Chapter 7), the most general electromagnetic plane wave can be represented as the vector sum of two waves with frequency f propagating in the z-direction, the first with an electric field vector of magnitude E_x parallel to the x-direction, the second of magnitude E_y parallel to the y-direction. For the resultant wave, description of the time dependence of the vector field $\boldsymbol{E}$ is an important part of remote sensing. First, for the special case when the resultant $\boldsymbol{E}$ points in a single direction, the wave is *linearly polarized*. Second, if the two components have equal magnitudes but different phases, then $\boldsymbol{E}$ rotates around the z-axis at the frequency f and the wave is *circularly polarized*. Looking into the wave, if the rotation is counter-clockwise, the wave is *left-circular* polarized, with the reverse definition for *right-circular* polarized. Third, if the two components have unequal magnitudes, the wave is *elliptically* polarized. Fourth, radiation from sources such as the sun can be *randomly* polarized, meaning that $\boldsymbol{E}$ takes different directions at random.

As Jackson (1975) shows, the properties of a plane, arbitrarily polarized electromagnetic wave are completely described by the four Stokes parameters. The first two Stokes parameters are the vertically (V-pol) and horizontally (H-pol) polarized components of the radiation; the other two are discussed below. For Earth remote sensing, and as Figure 3.5 shows in a three-dimensional perspective view and in sideview, V-pol and H-pol are defined relative to the Earth's surface. The H-pol component of $\boldsymbol{E}$ lies in the plane that is parallel to the surface; the V-pol component is at right angles to the H-pol and lies in the vertical plane. Given these definitions, the energy in the electric field $\boldsymbol{E}^2$ is proportional to $E_V^2 + E_H^2$, where the subscripts indicate polarization. This division into V- and H-pol is only possible for $0 < \theta < \pi/2$; at $\theta = 0$ or vertical incidence, E lies completely in the horizontal plane.

As Chapters 9–13 show, because the polarization of the emitted or reflected radiation depends on the state of the ocean surface, instruments sensitive to V- and H-pol are often used in microwave remote sensing. For example, measurement of the frequency, polarization and intensity of the radiation emitted or reflected from the ocean surface allows determination of the surface temperature and salinity, the wind speed and direction, and whether the surface is ice covered or ice free. Although the third and fourth Stokes components are less commonly used in remote sensing than V- and H-pol, certain of the newer and proposed passive microwave radiometers measure all four Stokes parameters. As Jackson (1975) and Yueh (1997) describe, the third Stokes parameter is proportional to the real part of the correlation between the V- and H-pol components of E; the fourth Stokes parameter is proportional to the imaginary part of this correlation. An alternative way to describe these components is that the third parameter is the difference between the electromagnetic components offset by $\pm 45°$ from the V- and H-pol components in the x-y plane; the fourth parameter is proportional to the difference in relative intensity between the left-circular and right-circular components of the electromagnetic wave. As Chapter 9 shows, under certain conditions, simultaneous passive microwave measurements of all four Stokes components allow retrieval of the vector wind speed.

3.3 Ways to describe EMR

There are a variety of ways to describe the propagation and intensity of EMR. These descriptions are specifically concerned with the *flux* of energy, or power, in units of joules-per-second or watts, and in the radiation incident on or emitted from a surface. The discussion begins with the radiant flux Φ.

1. The *radiant flux* Φ is the rate at which energy is transported toward or away from a surface, with units of watts (W). For example, the total radiant flux or power emitted by the sun is $\Phi_S = 3.9 \times 10^{26}$ W.
2. The *radiant intensity* $I = d\Phi/d\Omega$ is the radiant flux per-unit-solid-angle, with units of W sr^{-1} and is used in the description of radiation propagating from a point source. From the definition of Φ, and because there are 4π steradians in a sphere, the sun has a radiant intensity of $I = 3.1 \times 10^{25}$ W sr^{-1}.
3. The *flux density* $d\Phi/dA$ has units of W m^{-2}, and is the radiant flux per-unit-area that is either incident on or emitted from a unit surface area. The incident flux density is called the *irradiance* E; the emitted or outgoing, the *exitance* M. The book follows the oceanographic convention and uses the irradiance E for both incident and outgoing radiation. As an example, consider a 1-m^2 square panel at right angles to the sun at the Earth's orbit. For a mean Earth–sun distance of 1.5×10^8 km, the solid angle subtended by this panel is 4.4×10^{-23} sr, so that the incident irradiance on the panel is $E = 1400$ W m^{-2}.

   The use of the symbol E for irradiance is unfortunate, since it can be confused with the electric field vector E, even though E is proportional to E^2. There are three forms of the irradiance, *scalar*

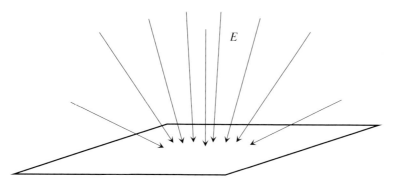

Figure 3.6. The irradiance incident on the half plane.

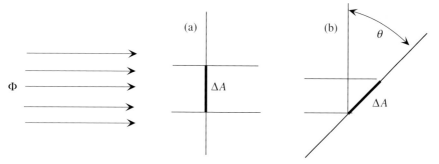

Figure 3.7. A beam of plane parallel radiation incident on (a) a plane at right angles to the propagation direction, and (b) a plane tilted at an angle θ. The tilt reduces the incident irradiance on ΔA.

irradiance, vector irradiance, and *plane irradiance* (Mobley, 1994). The scalar irradiance is the irradiance incident on a spherical sensor from all directions; the vector irradiance is irradiance divided into its orthogonal spatial components; the plane irradiance, which is primarily used in this book, consists of the fluxes that are incident or emitted in all directions above or below the half plane, weighted by the cosine of their angle to the vertical (Figure 3.6). The flat plate collector shown in Figure 3.6 measures the plane irradiance and is often referred to as a *cosine collector.* The reason for the cosine weighting is as follows: If a beam of plane parallel radiation is normally incident on a plane, then an element of area ΔA receives an irradiance $\Delta \Phi / \Delta A$. But if the plane is tilted at an angle θ relative to the normal, then the component of ΔA normal to the beam is reduced by an amount $\cos \theta$, so that the irradiance incident on ΔA is also reduced by $\cos \theta$ (Figure 3.7). For this reason, although a flat plate detector collects radiation from all angles in the upper half plane, the off-normal radiation is weighted by $\cos \theta$. Finally, in the treatment of visible radiation in and above the ocean, the plane irradiance is divided into its upward E_u and downward E_d components.

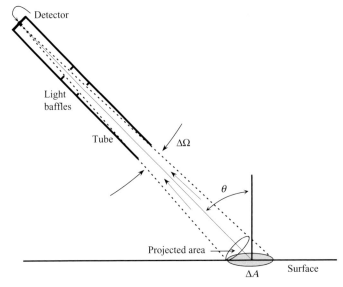

Figure 3.8. Schematic of a radiance meter viewing the surface (Radiance meter adapted from Figure 5.6 of Kirk, 1996).

4. The *radiance L* has units of W m^{-2} sr^{-1} and is defined as the radiant flux propagating toward or away from a surface in a specified direction within a solid angle dΩ. Radiance is a difficult but important concept. The flux is emitted from or incident on a differential unit area dA inclined at an angle θ to the direction of energy propagation, and is written

$$L \equiv \frac{d^2 \Phi}{d\Omega \, dA \cos \theta} \tag{3.15}$$

Figure 3.8 shows a schematic of a radiance meter pointed at a surface. The instrument captures the radiance that propagates within the solid angle $\Delta\Omega$ and is emitted or reflected from the area ΔA, which is inclined at an angle θ relative to the incident radiation, and where $\Delta A \cos \theta$ is the component of the surface in the look direction. From Equation (3.15), this means that the radiance incident on the detector is

$$L = \frac{\Delta \Phi}{\Delta\Omega \Delta A \cos \theta} \tag{3.16}$$

Because several surfaces of oceanographic interest have radiances that are independent of look angle, there are great advantages to this terminology, so that the concept of radiance will be used frequently in the following chapters.

Traditionally, emitted radiances are called *radiance* with the symbol L; incident radiances are called *brightness* with the symbol B. These terms are often used independent of direction, with the brightness B used in passive microwave radiometry for both incoming and outgoing radiation, and the radiance L similarly used in ocean color observations. As Figure 3.8 shows, radiance is the appropriate description of radiation to use with telescope-like instruments that observe the ocean surface at oblique angles and gather light or radiation within a specified solid angle, and as shown below, it is also appropriate for radiation propagating from an extended surface toward an observing instrument. For propagation in free space, or when radiance is used with $\theta = 0$, it is called a *field radiance*.

The concept of radiance can be difficult to understand. To provide a better understanding of radiance, the following sections first discuss some useful approximations, definitions and properties of radiating surfaces, use the radiance to describe the operation of an ideal optical instrument, then conclude with examples.

3.3.1 Lambert surfaces

Lambert surfaces have the useful property that their emitted radiance is independent of direction; as this and subsequent chapters show, these surfaces are of particular importance for ocean remote sensing in the visible/infrared. The term can also be used with reflectors, so that foam and clouds, for example, can be considered as Lambert or Lambertian reflectors of sunlight, such that for a large range of look angles in the VIR, the reflected radiance is independent of direction. For a plane Lambert surface, a simple relation exists between the irradiance E and the radiance L. Calculation of E by integration of L over the upper half plane yields

$$E = \frac{d\Phi}{dA} = L \int_0^{2\pi} d\theta \int_0^{\pi/2} \cos\theta \, \sin\theta \, d\theta = \pi L \qquad (3.17)$$

Chapter 5 uses this relation in discussion of the radiances backscattered from the ocean interior.

3.3.2 Spectral properties

Because satellite instruments observe the ocean at specific frequencies or wavelengths and within specific bandwidths of Δf or $\Delta\lambda$, the behavior of the electromagnetic radiation within these narrow windows must be determined. To deal with these windows, the *spectral* forms of the radiant flux and the radiance are next defined, where the adjective spectral means "per-unit-wavelength or per-unit-frequency" (Mobley, 1994, Chapter 1).

The spectral form of the radiant flux with regard to wavelength is

$$\left. \frac{d\Phi}{d\lambda} \right|_\lambda = \Phi_\lambda \qquad (3.18)$$

which has units of W μm^{-1}, so that for a narrow wavelength band with center wavelength λ_c, the energy received in a spectral window is approximately $\Phi_\lambda(\lambda_c)\Delta\lambda$. The equivalent form in frequency is

$$\left.\frac{d\Phi}{df}\right|_f = \Phi_f \tag{3.19}$$

with units of W s^{-1}. The spectral radiance in terms of wavelength and frequency is written as

$$\left.\frac{dL}{d\lambda}\right|_\lambda = L_\lambda \quad \text{and} \quad \left.\frac{dL}{df}\right|_f = L_f \tag{3.20}$$

where L_λ has units of W m^{-3} sr^{-1} and L_f has units of J m^{-2} sr^{-1}. Similar forms exist for the spectral irradiance and intensity. In the literature, the adjective "spectral" and the subscripts λ and f are frequently omitted for brevity, even though their absence implies that the non-spectral quantities are integrated over a range of wavelengths or frequencies (Mobley, 1994, Chapter 1). The following initially uses the spectral notation, while subsequent chapters do not.

3.4 Radiation from a perfect emitter

In 1900, Max Planck showed that for a perfect emitter or radiator held at a constant uniform temperature, the spectral radiance is a function of only temperature and wavelength, or equivalently of temperature and frequency. Such a radiator is called a *blackbody* radiator, or blackbody. This concept can be applied to opaque surfaces such as the ocean in the thermal infrared and microwave, and to small uniform regions of volume emitters such as gases. Planck's equation gives the spectral radiance emitted by a blackbody,

$$L_\lambda(\lambda, T) = \frac{2\hbar c^2}{\lambda^5 \exp[(\hbar c / k_B \lambda T) - 1]} \tag{3.21}$$

Equation (3.21) has units of W m^{-3} sr^{-1}, where these units can be interpreted as watts per-unit-area per-unit-solid-angle per-unit-wavelength. In (3.21), $\hbar$ is the previously defined Planck constant, c is the speed of light, and $k_B = 1.38 \times 10^{-23}$ J K^{-1} is the Boltzmann constant. For later use, the right-hand side of Equation (3.21) will be defined as the Planck function $f_P(\lambda, T)$.

Figure 3.9 compares the spectral irradiance derived from Planck's equation for an ideal-ized sun with a 5900 K blackbody temperature, the measured solar irradiance at the top of the atmosphere (TOA), and the solar irradiance measured at the surface for a solar zenith angle $\theta_S = 60°$. The 5900 K solar irradiance is calculated as follows. The solar radius is 7.0×10^5 km, and the Earth–sun separation is 1.5×10^8 km, so that at the TOA, the solar disk subtends a solid angle of 6.8×10^{-5} sr. If the solar disk is assumed to be a blackbody Lambert radiator, the irradiance follows from the definition of E and Equation (3.16). The TOA solar irradiance data is from the best current estimate of the irradiance spectrum, based on line-by-line computations by Robert Kurucz of the Harvard-Smithsonian Observatory, from data posted by Robert F. Cahalan (http://climate.gsfc.nasa.gov). At the TOA, the fine

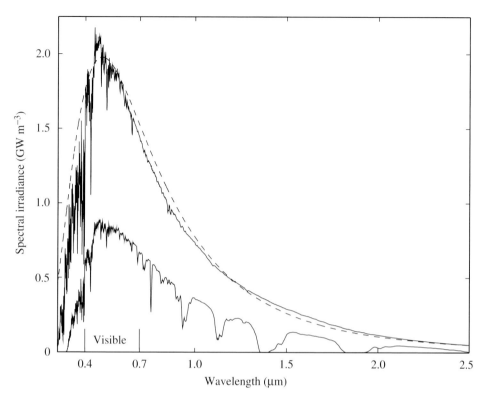

Figure 3.9. Comparison of the solar irradiance at the top of the atmosphere (upper solid curve) with Planck's equation at $T = 5900$ K (dashed line), and with the solar irradiance at the surface for a 60° solar zenith angle (lower solid curve). The visible spectrum lies within the pair of vertical lines; see text for further description and for data sources.

structure in the solar irradiance is caused by Fraunhofer absorption lines associated with the solar photosphere. As the figure shows, the 5900 K curve nearly matches the TOA irradiance, and the solar irradiance peak lies within the visible spectrum. The bottom solid curve shows the solar irradiance at the Earth's surface, where to separate the curves, θ_S is set equal to 60° so that solar irradiance is reduced by a factor of two. The source of the additional gaps and fine structure in the surface irradiance is attenuation by atmospheric gases, which will be discussed further in Chapter 4. Finally, each of the curves shows the asymmetry in Planck's equation relative to its maximum; at the shorter UV wavelengths, the radiance decreases very rapidly with decreasing λ; at the longer thermal wavelengths it falls off more slowly as λ increases.

3.4.1 Properties of Planck's equation

For several different blackbody temperatures, Figure 3.10 shows the dependence of Planck's equation on wavelength and displays several interesting properties.

1. The *Wien displacement law* gives the dependence on temperature of the wavelength of maximum radiance in Planck's equation. This wavelength is proportional to $1/T$, so that warm bodies emit their maximum radiation at shorter wavelengths than cold bodies.
2. The Stefan–Boltzmann law describes the temperature dependence of the total radiance or energy under the curve for Planck's equation, and shows that the total increases as T^4.
3. At any wavelength, there is always some emitted radiation, so that, if a specific radiance is observed at a particular wavelength, there is only one possible associated brightness temperature. This means that, ignoring the atmosphere and if the ocean radiance is measured at any wavelength with a sufficient degree of accuracy, the surface brightness temperature is uniquely determined.

If there were no atmosphere, then from Planck's equation, the sea surface temperature could be inferred from a single measurement of the surface radiance at almost any wavelength. In the real world, because the atmosphere absorbs, radiates and scatters radiation, in most cases the radiances received at the satellite differ from those emitted at the surface, which greatly complicates the retrieval.

3.4.2 Frequency form of Planck's equation

Substitution of the invariant $\lambda f = c$, and its differential $d\lambda = -(c/f^2)\,df$ into Equation (3.21) yields the frequency form of Planck's equation:

$$I_f(f) = \frac{2\hbar f^3}{c^2 \exp[(\hbar f/k_B T) - 1]} \tag{3.22}$$

with units of J m^{-2} sr^{-1}. Also, quantities such as the solar irradiance and the atmospheric attenuation are often given as functions of inverse wavelength in units of cm^{-1}; when written in terms of inverse wavelength, the resultant form of Planck's equation is similar to (3.22).

3.4.3 Limiting forms of Planck's equation

There are two limiting forms of Planck's equation, the long wavelength or Rayleigh–Jeans approximation that is applicable to the microwave and the short wavelength approximation. Because of its application to the microwave, the Rayleigh–Jeans approximation is the most important; it is valid at low frequencies or long wavelengths, where λ or f must satisfy the following inequality:

$$\hbar f/k_B T = \hbar c/\lambda k_B T \ll 1 \tag{3.23}$$

Substitution of (3.23) into (3.22) yields

$$L_f = 2k_B T f^2/c^2 = 2k_B T/\lambda^2 \tag{3.24}$$

At long wavelengths, Equation (3.24) shows that the spectral radiance L_f is a linear function of temperature. In the microwave, because of this linear dependence, the brightness temperature and brightness are often used interchangeably.

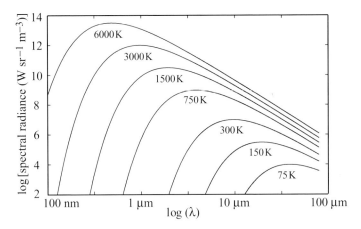

Figure 3.10. Comparison of the spectral blackbody radiances at the specified temperatures.

For high frequencies or short wavelengths, $hf/k_BT \gg 1$. Substitution of this limit into Planck's equation gives

$$B_f \sim f^3 \exp(-\hbar f/k_B T) \tag{3.25}$$

Equation (3.25) shows that the brightness decreases exponentially as frequency increases. In summary, at high frequencies or short wavelengths, the brightness falls off exponentially with increasing frequency or decreasing temperature; at low frequencies or long wavelengths, it decreases as f^{-2} and at a fixed frequency, increases linearly with temperature. This difference in behavior at the short and long wavelengths is consistent with the asymmetry shown in Figures 3.9 and 3.10, and makes remote sensing possible at the long microwave wavelengths.

3.4.4 Absorption and emission

As Thomas and Stamnes (1999, Chapter 5) show, a surface interacts with radiation in four ways. It can emit radiation into the surrounding environment, and can also absorb, reflect, or transmit the radiation incident upon it. Regarding emission and absorption, most objects radiate less efficiently than a blackbody and have a directional dependence to their radiation. Since the blackbody is by definition the most efficient possible radiator, other objects must radiate either less or equally efficiently. Because of this difference, non-blackbody radiators are called *gray bodies*, where their radiation properties are defined in terms of an *emissivity* $e(\lambda; \theta, \phi)$. The emissivity is defined as the ratio of gray body radiance to the blackbody, and is written

$$e(\lambda; \theta, \phi) = L_\lambda(\lambda, T; \theta, \phi)/f_P(\lambda, T) \tag{3.26}$$

By definition, the blackbody radiances have no directional dependence, so that blackbodies are Lambert emitters and absorbers. Since the blackbody is the most efficient emitter, $0 \leq e \leq 1$.

The emissivity has several important properties. First, emissivity generally depends on λ, so that emissivity magnitude and directional properties vary across the electromagnetic spectrum. For example, in the infrared and for θ less than about 45°, the emissivity of open water and sea ice are both approximately given by $e = 0.98$. In contrast, in the microwave and at the commonly used 50° look angle, the ocean emissivity is 0.4, and the sea ice emissivity is approximately 0.8. Thus at microwave frequencies, sea ice can have a greater brightness temperature than seawater. Second, the emissivity can be regarded as a physical surface property that is nearly independent of temperature and depends on the nature of the surface or substance. Even if two surfaces have the same physical temperature, when observed at the same λ, they can be distinguished by the differences in their emitted or reflected radiances.

For a blackbody emitter, Figures 3.11a and 3.11b compare the angular dependence of the intensity and radiance. The intensity has a spherical envelope and the radiance has a hemispherical envelope so that its distribution is Lambertian. The difference between the two occurs because of the $\cos\theta$ term in Equation (3.15). Figure 3.11c also shows the radiance distribution for an arbitrary gray body at the same temperature. Gray body radiances can depend on θ and are always less than or equal to the blackbody radiance at the same temperature.

For the radiation absorbed by a gray body, the *absorptance* $\hat{a}(\lambda; \theta, \phi)$ is defined in a manner identical to Equation (3.26), except that $\hat{a}$ is the ratio of the spectral radiance absorbed by a gray body to that absorbed by a blackbody. In the microwave, the absorptance is called the *absorptivity*. For a blackbody, all of the energy incident on its surface is absorbed, whereas for a gray body, only part is absorbed. For the gray body, the remaining energy can be either reflected, transmitted, or partially reflected and partially transmitted. Examples of nonemitting bodies include a perfect reflector, such as an ideal mirror, or a perfect transmitter, such as an ideal sheet of glass through which radiation passes without losses.

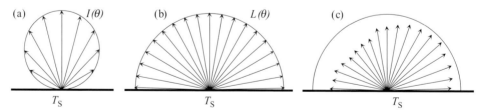

Figure 3.11. Comparison of (a) the angular dependence and magnitude of the intensity and (b) the radiance emitted from a blackbody with (c) the radiance emitted from an arbitrary gray body, all at the same surface temperature T_S. For the gray body, the semicircular line shows the blackbody radiance at the same temperature.

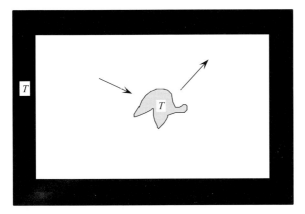

Figure 3.12. A hypothetical gray body inside a black box.

3.4.5 Kirchoff's law

The relationship between $\hat{a}$ and e is given by Kirchoff's law, which states that a surface in equilibrium with its surroundings must absorb and emit energy at the same rate, so that

$$\hat{a}(\lambda;\theta,\phi) = e(\lambda;\theta,\phi) \tag{3.27}$$

Why is this so? Consider Figure 3.12, which shows a hypothetical gray body located inside a black box, where both the gray body and the surrounding blackbody are at temperature T. Suppose that the gray body has an emissivity $e = 0.5$, but an absorptance $\hat{a} = 1$. Consequently, the black box temperature would decrease with time, and that of the gray body would increase. For a closed system with no work done on it, this temperature behavior violates the second law of thermodynamics. To avoid this violation, the emissivity must equal the absorptance so that matter must absorb and emit radiation in the same way. Therefore, for a specific substance, measurement of its absorption properties also determines its emission properties. For the situation shown in Figure 3.12 and for radiances incident on the body, a fraction $\hat{a}$ will be absorbed, and a fraction $(1 - \hat{a})$ will be reflected. Simultaneously, the body emits radiances with an emissivity $e = \hat{a}$, so that the system remains in thermal balance. Chapter 4 applies this concept to gases, which permits derivation of an important source term in the radiative transfer equation.

3.5 The ideal instrument

The previous material is valid for all remote sensing wavelengths. In contrast, this section examines the properties of a simple telescope that operates in the VIR, where Chapter 9 provides a related discussion of microwave antennas. In the following, Section 3.5.1 describes the *Rayleigh criterion* and its role in determination of the instrument resolution. Section 3.5.2 calculates the resolution of an ideal, vertically oriented instrument and the energy flux it receives from the surface. Section 3.5.3 repeats the calculation for an instrument at an arbitrary orientation and shows that for radiation from a Lambert surface the received energy flux is independent of look angle. Finally, Section 3.5.4 discusses bandwidth and the treatment of noise.

3.5.1 The Rayleigh criterion

This book makes use of two different Rayleigh criteria, both called the Rayleigh criterion. The first concerns the resolving power of lenses; the second concerns the scattering of radiation from surfaces. This section discusses the first case, Section 5.2 discusses the second. For the first, all optical instruments have apertures, where the aperture is the area of the light-gathering lens or opening that separates the sensor from the environment. As the following discusses, the diffraction of light at the aperture edge determines the minimum angular resolution of the instrument. The assumption that the aperture is a two-dimensional slit simplifies the discussion; except for a multiplicative constant, the analysis is also correct for a circular opening.

Figure 3.13 shows the relevant two-dimensional geometry. Consider two line sources of light with an angular separation $\Delta\theta$, where the sources are separated from an image plane by an aperture of width D. Because of the wave nature of light, each line source generates a diffraction pattern on the image plane, where the vertical extent of this pattern varies inversely with D. For a very small $\Delta\theta$, the patterns overlap, making it impossible to discriminate between the images of the two light sources. Light diffraction at the aperture edge thus sets an inescapable lower limit to the angular resolution of the instrument. Quantitatively, the Rayleigh criterion states that the two sources can be distinguished only if the following relation holds:

$$\Delta\theta \geq \lambda/D \tag{3.28}$$

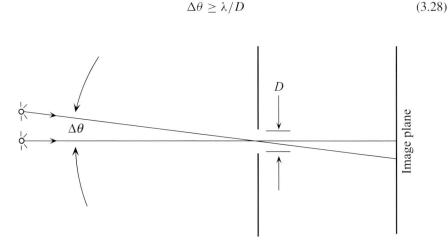

Figure 3.13. The diffraction of two line sources of light by a slit.

For two point sources and a circular aperture, Charman (1995) shows that Equation (3.28) becomes

$$\Delta\theta \geq 1.22\lambda/D \tag{3.29}$$

These relations give the limiting minimum angular resolution of an optical instrument. For example, Charman states that the Rayleigh criterion approximately gives the resolution of a healthy human eye. For a pupil diameter of $D = 3$ mm, and for blue light or $\lambda = 0.45$ μm, Equation (3.29) gives $\Delta\theta = 0.2$ mr, which for a 1 m separation between the eye and object corresponds to a surface resolution of about 0.2 mm.

3.5.2 The simple telescope

This and the next section derive the radiant flux received by an idealized nadir-looking and slant-looking instrument viewing a Lambert surface, and show that the received flux is independent of look angle. Figure 3.14 shows the idealized optical instrument. In Figure 3.14a, the angles are greatly exaggerated for clarity; for the human eye or the AVHRR, Figure 3.14b shows the instrument solid angles and fields-of-view drawn approximately to scale.

The figure shows that the instrument consists of a lens with a focal length f_L, an aperture area A and a sensor or detector area A_S. For photographic film, the detector area is determined

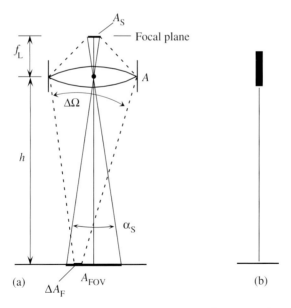

Figure 3.14. A schematic drawing of an ideal telescope. (a) The instrument drawn to an exaggerated scale; (b) the instrument resolution and solid angles drawn approximately to scale for the AVHRR or human eye.

by the diameter of the individual grains of silver nitrate on the film; for an eye, by the size of the nerve endings in the retina; for a satellite sensor, by the area of a charge-coupled-device on the focal plane. The instrument is at a height h above the surface, and the FOV from which the sensor collects radiation is A_{FOV}. From geometric considerations, the solid angle resolution of the instrument is

$$\alpha_S = A_S/f_L^2 = A_{FOV}/h^2 \tag{3.30}$$

Equation (3.30) only applies if the Rayleigh criterion in (3.29) is satisfied. Given the desire in some cases to make the FOV as small as possible, Equation (3.30) shows that this can be done by reducing the size of the sensor element and by increasing the focal length. However, if the Rayleigh criterion is not satisfied, then no matter how long the focal length or how small the sensor, the resolution cannot be improved.

If the Rayleigh criterion is satisfied, then the radiance that is emitted or reflected from each element of surface area ΔA_F and focused on the detector propagates within the solid angle $\Delta\Omega = A/h^2$, as outlined by the dashed lines on Figure 3.14a. The instrument is then defined by two solid angles, the instrument resolution α_S and $\Delta\Omega$. The solid angle α_S determines the FOV; $\Delta\Omega$ determines the magnitude of the energy flux incident on the detector. Given the nadir-looking instrument in Figure 3.14, the incident radiant flux Φ_{IN} is calculated on the assumption of no atmospheric interference, a narrow beam instrument with $\alpha_S \ll 1$, and a Lambert reflecting or emitting surface. For these assumptions, Φ_{IN} follows from the definition of L in Equation (3.15) with $\cos\theta \simeq 1$ for the narrow beam instrument, so that in differential form,

$$d^2\Phi_{IN} = L\,d\Omega\,dA_F \tag{3.31}$$

From (3.31), calculation of Φ_{IN} involves integrating the radiation from each element of surface area over the entire FOV, and over the solid angle subtended by the aperture. On the further assumption that the aspect ratio of the beam is so small that each differential surface area subtends the same $\Delta\Omega$, then Φ_{IN} can be written

$$\Phi_{IN} = L\int_{A_{FOV}} dA_F \int_{\Delta\Omega} d\Omega = LA_{FOV}(A/h^2) = LA\alpha_S \tag{3.32}$$

For the nadir-viewing instrument, Equation (3.32) shows that the energy flux received at the detector is a product of the surface radiance, the aperture area, and the instrument solid angle, where the last two parameters are determined by the instrument design. As the next section shows, this is also true for an instrument viewing the surface at an off-nadir angle.

3.5.3 Slant-looking instrument

Figure 3.15 shows the same instrument pointed at the surface at an off-nadir angle θ, again in both an exaggerated view and with the resolution approximately to scale. The distance from the telescope to the surface is $h_1 = h/\cos\theta$, where at h_1, the field-of-view at right

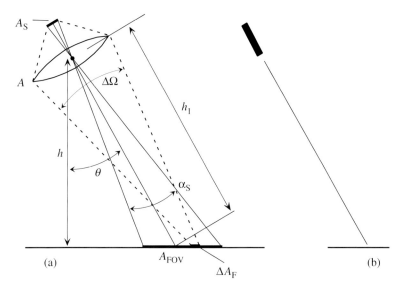

Figure 3.15. Schematic drawing of the slant-looking instrument. (a) Exaggerated scale; (b) approximately to scale.

angles to the look direction is $A_I = \alpha h_I^2$, and the projection of this area onto the surface plane is

$$A_{\text{FOV}} = \alpha h_I^2 / \cos \theta \tag{3.33}$$

As with the nadir-looking instrument, the radiation from each differential surface area ΔA_F received by the instrument lies within the solid angle $\Delta \Omega = A / h_I^2$. For this case, because the aperture A is now further from the surface, $\Delta \Omega$ is smaller than its nadir-looking value.

From the definition of radiance,

$$d^2 \Phi_{\text{IN}} \equiv L \, d\Omega \, dA_F \cos \theta \tag{3.34}$$

Integration of (3.34) over the surface field-of-view and the solid angle defined by the aperture gives

$$\Phi_{\text{IN}} = L \left(A / h_I^2 \right) \alpha_S h_I^2 = \alpha_S A L \tag{3.35}$$

For the off-nadir view angles, Equation (3.35) shows that an increase in FOV compensates for the reduction in $\Delta \Omega$, so that the radiant flux received at the sensor is identical to the nadir-viewing result in Equation (3.32). Consequently, for a Lambert surface, a narrow beam instrument and a transparent atmosphere, the radiance received by the instrument is independent of θ.

Equation (3.35) shows the advantages of working in terms of radiance, especially for Lambert surfaces. Here are two additional examples. First, consider a sheet of bond paper that is illuminated by fluorescent lights. Because of its microscale-roughened surface, when the paper is viewed at a variety of distances and look angles, the distribution of light scattered from the paper is approximately Lambertian. From Equation (3.35) and for an eye with a constant pupil dilation, as long as the solid angle defined by each nerve ending within the eye is smaller than the solid angle defined by the paper, then consistent with daily experience, the radiant flux received from the paper is approximately independent of look angle and distance.

Second, consider the light from the sun and stars. Because the solar disk subtends an angle from Earth of about 0.5° or 10 mr, which is much greater than the 0.2 mr resolution of the human eye, it is very dangerous to look directly at the sun. This occurs because in Equation (3.35), the eye resolution α_S is less than the angle subtended by the sun, so that the entire solar radiance is focused on a single nerve ending, yielding the potential for severe eye damage. But for the same sun at a distance several light-years away so that it subtends a solid angle less than 0.2 mr, most of what the eye views is empty space. The result of this is that, even though the sun and the distant star have the same radiance, the power received from the star is much less than from the sun.

3.5.4 Finite bandwidth instrument and treatment of noise

For a real instrument, the center wavelengths and bandwidths are tailored to the phenomena under investigation and to the atmospheric windows. If the detector is characterized by a center wavelength λ_c and a bandwidth $\Delta\lambda$, where $\Delta\lambda$ is sufficiently small that the surface radiance is approximately constant at $L(\lambda_c)\Delta\lambda$, then the radiant flux Φ_{IN} incident on the detector becomes

$$\Phi_{IN} = A\alpha_S L(\lambda_c)\Delta\lambda \tag{3.36}$$

The detector and its accompanying circuitry convert Φ_{IN} into the electrical energy Φ_{OUT}. Within the instrument and as Figure 3.16 shows, a noise power Φ_N is defined so that at any instant,

$$\Phi_N = \Phi_{OUT} - \Phi_{IN} \tag{3.37}$$

The standard deviation or rms uncertainty of the noise σ_N is easily calculated from the time series of Φ_N so that

$$\Phi_{IN} = \Phi_{OUT} \pm \sigma_N \tag{3.38}$$

Given σ_N and following Stewart (1985, Section 8.1), the *signal-to-noise* ratio S_N of the instrument is defined as

$$S_N = \Phi_{IN}/\sigma_N = L(\lambda_c)\Delta\lambda A\alpha_S/\sigma_N \tag{3.39}$$

To make Φ_{OUT} as noise-free as possible, S_N must be large. Examination of Equation (3.39) suggests several ways to do this. The first is to make the aperture A large, which is easy

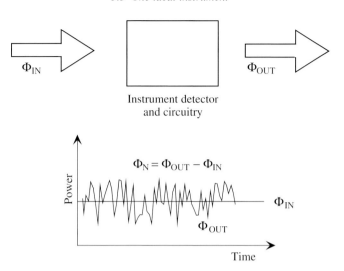

Figure 3.16. The generation of noise within the detector and its accompanying circuitry.

in principle, but difficult in reality. For example, the lens diameter of the Hubble Space Telescope was limited by the size of the cargo bay on the Space Shuttle. Any increase in A means that the instrument becomes bigger and heavier, all of which increases the difficulty and expense of placing the instrument in orbit. The second is to make the solid angle α_S as large as possible. In most cases however, the desirable solution is to make α_S small, so that the surface FOV is small.

The third is to choose λ_c and $\Delta\lambda$ in such a way to maximize the received power. Because $\Delta\lambda$ and λ_c are generally chosen to provide specific environmental information, this may not be possible. As the next chapter shows, the location of λ_c is also in part determined by the location of the atmospheric windows and their properties. Similarly, $\Delta\lambda$ is set either by the phenomenon under investigation, as occurs in biological studies in the visible, or by the width of the atmospheric windows, as occurs in the thermal-infrared. Given these restrictions, the easiest way to reduce the S_N is to reduce σ_N. This is done in two ways: first, by insuring that the instrument has a low thermal noise, which involves cooling the instrument and shielding it from the sun; second, by averaging many measurements of the same area over a short enough time that the radiance L does not change. As the following chapters show, both techniques are used to reduce noise.

Finally, because σ_N is an instrument-specific, nonlinear function of Φ_{IN}, σ_N is generally defined relative to the magnitude of the input radiant flux, radiance or blackbody temperature. Further, to aid in the physical interpretation of the uncertainty, when Φ_{OUT} is converted to a radiance or temperature, σ_N is converted to an uncertainty in the same units. For these conditions, σ_N is written as a *noise-equivalent-delta-radiance* (*NEΔL*), or equivalently, a *noise-equivalent-delta-temperature* (*NEΔT*), where the noise is expressed as an rms uncertainty in the received radiance or temperature. For example, an AVHRR observation of an ocean surface temperature of 300 K might have an $NE\Delta T$ of ± 0.5 K. For this case, the instrument would be described as having a $NE\Delta T$ of ± 0.5 K at 300 K.

3.6 Further reading and acknowledgements

Rees (2001) provides a valuable introductory text, and Jackson (1975) and Born and Wolf (1999) give a comprehensive survey of the properties of electromagnetic plane waves. Mobley (1994) discusses the different terms used to describe radiation in the ocean; Thomas and Stamnes (1999) discuss these terms for the atmosphere. Some of the material in the early part of this chapter is adapted from class notes of Dudley Chelton.

4

Atmospheric properties and radiative transfer

4.1 Introduction

The atmosphere lies between the ocean surface and the satellite sensor, and greatly affects the transmission of radiation. The presence of fixed concentrations of atmospheric gases such as oxygen, carbon dioxide, ozone and nitrogen dioxide, plus the variable concentrations of water vapor means that only a few windows exist in the visible, infrared and microwave for Earth observations. Even within these windows, the atmospheric absorption varies with the concentrations of water vapor and the liquid water droplets and ice particles that make up clouds. The absorption is also affected by atmospheric aerosols, which include the water droplets and salt nuclei in the marine boundary layer, and the particulate matter generated over land by urban pollution, biomass burning and volcanic eruptions that is advected over the oceans.

In the following, Section 4.2 describes the vertical structure of the atmosphere and the molecular and aerosol constituents that affect the transmission of radiation. Sections 4.3 and 4.4 describe the propagation, absorption and scattering of a narrow beam of radiation. For the different atmospheric constituents, Section 4.5 discusses the dependence of transmissivity on wavelength and the role of these constituents in defining the atmospheric windows. To prevent the chapter from becoming overly long, this discussion is restricted to the visible/infrared; Chapter 9 extends it to the microwave. Section 4.6 applies these results to the ideal instrument. Section 4.7 discusses the radiative transfer equation (RTE) and the atmospheric emission and scattering source terms. Section 4.8 derives two limiting solutions of the RTE, one for the infrared and microwave windows where absorption and emission dominate; the other for the visible wavelengths where absorption and scattering dominate. Section 4.9 concludes with a discussion of diffuse attenuation and skylight.

4.2 Description of the atmosphere

Figure 4.1 shows a characteristic vertical temperature profile of the atmosphere. The left-hand scale gives pressure in millibars; the right-hand scale gives the height above mean sea level; the horizontal scale gives the temperature in K. Proceeding upward from the surface, the atmosphere divides into the troposphere, stratosphere, mesosphere and thermosphere,

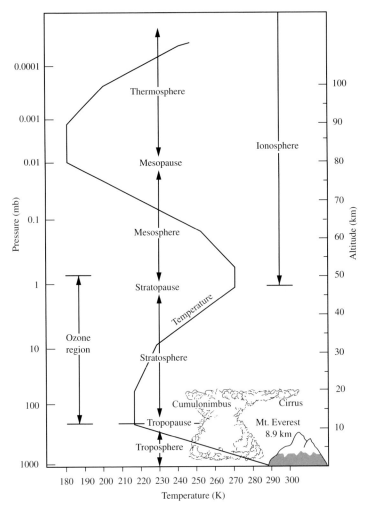

Figure 4.1. The one-dimensional structure of the atmosphere (Adapted from Eos Science, Steering Committee, 1989, not subject to US copyright).

which are respectively separated from one another by the tropopause, stratopause and mesopause. The exosphere, which is not shown, lies above the thermosphere. The figure also shows the approximate height of the cloud formation region in the troposphere and lower stratosphere, the stratospheric region of oxygen dissociation into ozone, and the ionosphere, which consists of the mesosphere, thermosphere and exosphere and contains the ions and electrons generated by solar dissociation of atoms and molecules. For remote sensing purposes, there are four important variable atmospheric constituents. These are water in the form of vapor, liquid and ice, aerosols, ozone and ionospheric free electrons. Each of these affects the atmospheric transmission and scattering properties at different wavelengths and on time scales ranging from hours to years.

Table 4.1. *Major components of the atmosphere, their molecular weight and content by volume*

Constituent	Molecular weight	Volume fraction
Nitrogen (N_2)	28.016	0.78
Oxygen (O_2)	32.00	0.21
Argon (Ar)	39.94	9.3×10^{-3}
Carbon dioxide (CO_2)	44.01	3.5×10^{-4}
Water vapor (H_2O)	18.02	variable
Ozone (O_3)	47.99	variable
Nitrous oxide (N_2O)	44.01	0.5×10^{-6}
Methane (CH_4)	16.04	2×10^{-6}
Carbon monoxide (CO)	28.01	trace, variable

Adapted from Weast 1976; Ulaby *et al.* 1981.

Examination of the atmospheric temperature profile shows that the air temperature oscillates vertically with height between 180 and 300 K, and illustrates the atmospheric stability. Proceeding upward from the surface, the troposphere is marginally stable and is characterized by strong vertical mixing up to the tropopause. Because of the mixing, the tropospheric variable components, which are primarily the different forms of water and a variety of aerosols, have time constants of order one day to one week. The stratosphere lies above the tropopause, and within it, the temperature increases with height up to the stratopause. The principal variable constituent of the stratosphere is ozone. The stable temperature profile means that the stratosphere is a region of weak vertical mixing, so that the time constant of ozone variability is of order months. Above the stratopause, the temperature continues to vary with height, but remains in the range 200–300 K.

The pressure scale on the left side of Figure 4.1 shows that approximately 90% of the atmospheric gases are in the troposphere, with an additional 9.9% in the stratosphere. In the troposphere, Ulaby, Moore and Fung (1981) show that the density of dry air ρ_a has the following dependence on height:

$$\rho_a = \rho_0 \exp(-z/H_a) \tag{4.1}$$

In (4.1), $\rho_0 = 1.225$ kg m^{-3} and the scale height H_a is about 9.5 km. This expression is accurate for $z \leq 10$ km; because of stratospheric ozone, at larger heights the observed densities deviate from Equation (4.1). Because the troposphere contains most of the atmospheric gases and almost all of the water vapor, it is where most of the scattering and absorption occurs.

Table 4.1 lists the constituents of the atmosphere and their relative weights and percentages. These divide into the well-mixed, constant constituents and the variable constituents. The constant constituents include oxygen, nitrogen, and a variety of trace gases; because the troposphere is well-mixed, their relative concentrations are constant regardless of location.

The variable constituents are next discussed in the order water, aerosols, stratospheric ozone and free electrons.

4.2.1 Atmospheric water

Atmospheric water occurs as water vapor, liquid water and ice crystals in clouds, rain and snow. Beginning with water vapor, from Ulaby *et al.* (1981) and Chahine *et al.* (1983), the vertical profile of atmospheric water vapor ρ_v is described by

$$\rho_v = \rho_{v0}\exp(-z/H_v) \tag{4.2}$$

In (4.2), ρ_{v0} is the surface density of water vapor and $H_v \simeq 2.5$ km is the water vapor scale height. Because $H_v \ll H_a$, the water vapor is concentrated in the lower part of the troposphere. From Equation (4.2), the total water vapor concentration is described either by ρ_{v0} with units of g m^{-3}, or by the *columnar water vapor V*, which is the total integrated amount of water vapor contained in a vertical column extending through the atmosphere. This book primarily uses V for water vapor; it is measured in units of g cm^{-2} or in terms of the height in mm of the columnar liquid water equivalent. Ulaby *et al.* (1981) and Chahine *et al.* (1983) show that ρ_{v0} varies from 10^{-2} g m^{-3} in the winter polar regions to values as large as 30 g m^{-3} in the tropics, with an average surface value of 10 g m^{-3}. From Equation (4.2) and for the same cases, V ranges from 0.03 to 75 mm of columnar water equivalent, with a global average of 25 mm. Since the total columnar mass of the atmosphere corresponds to 10 m of liquid water, vapor accounts for only about 0.3% of the atmospheric mass. In spite of its small contribution to atmospheric mass, Section 4.5 shows that water vapor is a major contributor to the atmospheric absorption. Chapter 9 describes the retrieval of V from passive microwave data, and Plates 10 and 11 give examples of its distribution, showing that most of the water vapor occurs in the tropics.

Liquid water and ice crystals occur in clouds and in the cloud-related phenomena of rain, hail and snow. Liquid water occurs in two forms: as non-raining cloud liquid water measured in terms of the amount of columnar liquid water L in mm, and as rain measured in terms of the rain rate R_R in mm h^{-1}. As Chapter 9 discusses in more detail, both L and R_R are retrievable from passive microwave observations. L ranges from 0 to 0.25 mm, so that non-raining clouds contain much less columnar water than the vapor, while R_R has a characteristic value of 2 mm h^{-1} and a maximum of about 20 mm h^{-1} (Wentz and Spencer, 1998).

4.2.2 Clouds

Clouds are transient atmospheric features that consist of small ice and liquid water particles with dimensions ranging from under a micrometer to a few millimeters (Baker, 1997). They participate in the vertical convection that mixes the troposphere, where this convection is in part driven by the evaporation, freezing and condensation of the cloud water droplets and ice crystals. This change of phase within clouds and especially at their edges contributes to cloud variability. Cloud properties vary with height; the lower troposphere contains marine stratus and cumulus consisting of water droplets, while the upper troposphere contains the high thin cirrus consisting of ice particles. Under certain conditions, convective cloud anvils extend into the lower stratosphere. In the VIR, the liquid water and ice crystals contained in clouds scatter and absorb radiation, so that thick clouds make it impossible to view the surface. Compared with the ocean surface, thick clouds are very reflective and generally colder. Because clouds occur in major weather systems, such as typhoons, cyclones, and atmospheric fronts, and because they stand out in both temperature and reflectivity against the ocean and land background, they are used by weather satellites to track storms.

At any time, clouds cover almost two-thirds of the globe. For example, Plate 2 shows a true color composite image of the Earth taken on April 19, 2000 by the Moderate Resolution Imaging Spectroradiometer (MODIS) aboard the TERRA spacecraft. Because the sun is north of the equator, the Antarctic continent is dark. Prominent land features include the North African desert and green vegetation in North America, southern Africa and Madagascar, India and northern Asia. The plate shows cyclonic swirls of cloud around Antarctica, a storm in the North Atlantic between Greenland and Europe, and another storm approaching the west coast of North America.

4.2.3 Atmospheric aerosols

Atmospheric aerosols consist of small liquid or solid particles from the Earth's surface and are another important source of atmospheric variability. Aerosols divide into three categories; marine- and land-generated aerosols in the troposphere and volcanic aerosols in the stratosphere. Marine aerosols occur in the marine boundary layer and are generated locally at the sea surface. They consist of water droplets with diameters of order 1–10 μm, with the addition of sea salt nuclei from breaking waves (Stewart, 1985). Land aerosols are generated over land, then advected over the ocean. Examples include desert dust, industrial and urban pollutants, and smoke and soot from biomass burning. Typical desert sources include dust from the North African Sahara Desert that is advected over the Atlantic Oceans to distances as far away as Florida, dust from the southwest African deserts advected over the South Atlantic, and dust from the East Asian Gobi Desert advected over the North Pacific. Industrial and urban pollutants are generated in Europe, Russia, North America, and southern Asia, where this material is respectively advected over the North Atlantic, the Arctic Ocean, and the North Pacific and Indian Oceans. Soot and particulate matter are seasonally generated by biomass burning in parts of Mexico, Central and South America, Africa and Asia, where these particulates are advected over their respective adjacent seas (Wang, Bailey and McClain, 2000). Typically, the scale height of the aerosol layer for land- and

marine-based aerosols is about 1 km, so that 90% of the aerosols are confined to within
2 km of the sea surface (Gordon and Castano, 1987). Volcanoes are another important
source of aerosols; volcanic emissions carry micrometer-scale diameter droplets of sulfuric
acid and other suspended particulates into the troposphere and stratosphere. As Chapter 7
shows, this material can cause changes in atmospheric absorption for periods of 1–3 years
after the initial eruptions.

4.2.4 Ozone

Ozone is the principal component of the stratosphere and forms from dissociation of oxygen
molecules by solar radiation. It is a stable chemical species with a residence time of order
months and a seasonal variability. Figure 4.2 gives the typical distribution of ozone with
height for mid-latitude summer and winter, and shows that its concentration is less in
summer than in winter. Stratospheric ozone is of great importance because it absorbs UV-
B, which occurs in the range 280–320 nm and causes skin cancer (Thomas and Stamnes,
1999). For ocean remote sensing, the importance of ozone is that it attenuates visible
radiation with a seasonal and latitudinal dependence, so that as Chapter 6 shows, it must
be considered in the ocean color retrieval. The ozone layer is also important because of the
austral summer ozone hole in the Southern Hemisphere and a less intense but similar hole in
the Northern Hemisphere. Other tropospheric gases that exhibit long-term variability are the
greenhouse gases such as methane and carbon dioxide. Although their long-term changes

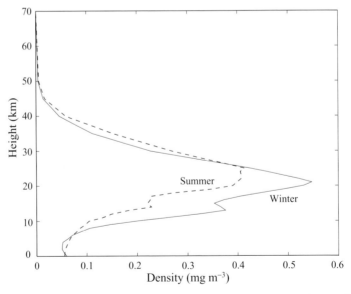

Figure 4.2. Comparison of ozone profiles derived from the standard MODTRAN cases of Mid-latitude
summer and Mid-latitude winter, where Section 4.5 describes the MODTRAN program and cases.

are important to atmosphere properties, they only affect satellite-observed radiances over decadal periods.

4.2.5 Ionospheric free electrons

Free electrons are generated in the ionosphere by solar-driven molecular dissociation (Chelton *et al.*, 2001b; Bird, 1988). These electrons occur at heights of 100–1000 km and because they form reflective layers at certain frequencies, have a great effect on terrestial radio communications. Because molecular dissociation occurs in sunlight with the molecules being restored at night, the densities have a strong diurnal cycle. Figure 4.3 compares the day and night profiles of electron density and shows the nighttime density decrease. The columnar concentration of the free electrons is given in units of TECU, the Total Electron Content Unit, where 1 TECU $= 10^{16}$ electrons m^{-2} (Chelton *et al.*, 2001b). The columnar concentration varies diurnally between 10 and 120 TECU and interannually with the 11-year solar cycle, which reached its minimum in 1997 and its maximum in 1990 and 2001. The importance of these diurnal and interannual changes is that the electron

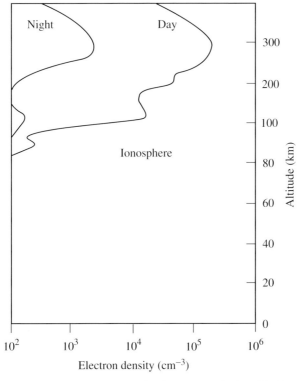

Figure 4.3. The day/night difference in electron density versus altitude (altitude scale as on original; redrawn from Bird, 1988, not subject to US copyright).

density affects the electromagnetic phase speed, which as Chapter 12 shows must be accounted for in the altimeter design.

4.3 Molecular absorption and emission

As Chapter 3 describes, atoms and molecules absorb and emit radiation in discrete quanta. An isolated molecule emits radiation by making a transition from a higher to a lower quantized energy state, which occurs when an electron moves to a lower orbit, or from changes in its rotational or vibrational molecular state. If $\Delta\hat{E}$ is the change in the internal molecular energy and $\hbar$ is the Planck constant, this energy change is governed by the following relation:

$$\hbar f = \Delta\hat{E} \tag{4.3}$$

The reverse occurs during a transition from a lower to higher energy state, when the molecule absorbs energy. Equation (4.3) shows that the frequency at which radiation is emitted or absorbed is determined by the magnitude of $\Delta\hat{E}$. Because these molecular state changes occur in discrete steps, each molecular species generates different line absorption and emission spectra.

In the atmosphere, emission and absorption do not occur in discrete lines, but because of processes called *line broadening*, they occur in spectral bands. Two of these processes are called Lorenz and Doppler broadening. Lorenz broadening occurs because, for molecular gas layers at a characteristic pressure and temperature, the molecular collisions perturb the energy level spacing of the individual molecules and broaden the spectral lines. As the gas pressure, density or temperature increases, Lorenz broadening increases. Doppler broadening occurs because the gas molecules are in motion. Each gas molecule with a velocity component toward or away from the viewer generates a Doppler shift in the line absorption or emission spectra, where the sum of these shifts generates line broadening. Because the peak and spread of this velocity distribution increase with increasing temperature, Doppler broadening increases with temperature. The Lorenz and Doppler broadening are the principal broadening mechanisms; together they generate what is called the Voigt line shape, shown approximately in the lower frame of Figure 4.4 (Ulaby *et al.*, 1981, Section 5.3; Liou, 1980).

4.3.1 Molecular extinction

The terminology used to describe the transmission of radiation in the atmosphere depends on the choice of wavelength window. In the visible/infrared, it is called the *transmittance*; in the microwave, the *transmissity* (Ulaby *et al.*, 1981, page 187, Table 4.1). In the visible/infrared, and depending on the source, the atmospheric attenuation of radiation divides into two parts: the attenuation of a narrow beam of radiation generated by a discrete source such as a laser or spotlight, and the attenuation of radiation generated by an extended source, such as that

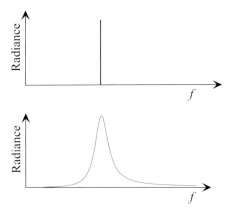

Figure 4.4. Comparison of a line spectrum and its broadened form (Modeled after the water absorption line at 22 GHz, Equation (5.19) from Ulaby *et al.*, 1981).

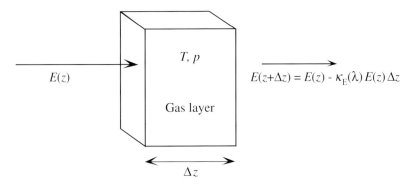

Figure 4.5. Attenuation by a gas layer, where the λ-dependence of E is omitted. See text for further description.

created by solar reflection from the ocean surface, or from any surface large compared with the instrument FOV. The narrow beam case yields what is called the *beam transmittance* or simply the *transmittance*. The extended source yields the *diffuse transmittance*, which because it depends on scattering, is only important in the visible. The symbol t with a variety of subscripts will be used for both terms. The discussion of the diffuse transmittance is delayed until Section 4.9.1 after the Rayleigh scattering discussion; the following discusses beam attenuation.

Consider a parallel beam with irradiance E propagating in the direction z and incident on a layer of gas with differential thickness Δz, where the gas consists of a single molecular species at a constant temperature T and pressure p (Figure 4.5). Within the gas layer, there are two kinds of attenuation: molecular absorption and scattering of radiation out of the beam. If the sources of radiation due to scattering and blackbody emission are neglected, Beer's law states that the change in the irradiance as it passes through the layer is proportional to the incident irradiance (Kidder and Vonder Haar, 1995), so that

$$\Delta E(\lambda) = -\kappa_E(\lambda)\, E(\lambda, z)\, \Delta z = -[\kappa_A(\lambda) + \kappa_S(\lambda)] E(\lambda, z)\, \Delta z \qquad (4.4)$$

In (4.4), $E(z)$ is the incident irradiance, ΔE is the change in irradiance across the gas layer, $\kappa_A(\lambda)$ is the volume absorption coefficient and $\kappa_S(\lambda)$ is the volume scattering coefficient. In the atmosphere these coefficients have units of m^{-1} or km^{-1}. If the extinction coefficient is defined as $\kappa_E(\lambda) = \kappa_A(\lambda) + \kappa_S(\lambda)$, then (4.4) becomes

$$\Delta E / E(z) = -\kappa_E(\lambda)\, \Delta z \qquad (4.5)$$

where κ_E is sometimes called the attenuation coefficient. Because Equation (4.5) is valid for narrow collimated beams, it also applies to the intensity and the radiance. Rewriting (4.5) in a differential form gives

$$\frac{dE}{dz} = -\kappa_E E \quad \text{or} \quad \frac{dE}{E} = -\kappa_E dz \qquad (4.6)$$

so that for κ_E constant, the decay is exponential. The magnitude of κ_E depends on T, p and the gas constituent. For identical processes in the ocean, the volume absorption coefficient is written as $a(\lambda)$, the volume scattering coefficient as $b(\lambda)$ and the attenuation coefficient as $c(\lambda)$, where a, b and c have units of m^{-1} (Mobley, 1994).

For the atmosphere, κ_E is sometimes given in units of *decibels* per km, or dB km^{-1}, where dB is a measure of the relative power or energy, defined as follows. If E_0 is a reference irradiance and E is the measured value, then from (4.5),

$$dB = 10 \log_{10}(E / E_0) \qquad (4.7)$$

Equation (4.7) shows that a factor of 10 decrease in the transmitted irradiance corresponds to -10 dB; a 50% reduction, to -3 dB.

4.3.2 Optical depth and transmittance

In the application of the extinction model to the atmosphere, for each molecular constituent, Equation (4.6) with E replaced by L is integrated across the atmosphere. The surface boundary conditions are that at $z = 0$, $L = L_0$. Integration of Equation (4.6) from the surface to an arbitrary height z gives

$$L(z) = L_0 \exp\left[-\int_0^z \kappa_E(z)dz \right] \qquad (4.8)$$

Equation (4.8) leads to the definition of two important terms, the *optical depth* or *thickness* $\tau(\lambda)$ and the *transmittance* $t(\lambda)$. Although both terms are functions of λ, in the following, the λ-dependence is omitted for brevity. The optical thickness τ can be defined relative to any reference height or path orientation. For a vertical path originating at the surface, the optical thickness $\tau(z)$ is written as

$$\tau(z) = \int_0^z \kappa_E(z)\,dz \tag{4.9}$$

If the top of the atmosphere (TOA) occurs at the height $z = z_H$, then the optical thickness τ of the atmosphere is

$$\tau = \int_0^{z_H} \kappa_E(z)\,dz \tag{4.10}$$

Given τ, the atmospheric beam transmittance or transmissivity t is defined from Equation (4.8) by $t = \exp(-\tau)$, so that $L(z_H) = L_0 \exp(-\tau) = L_0 t$. From these definitions, a transparent atmosphere has $\tau = 0$ and $t = 1$; an opaque atmosphere has $\tau = \infty$ and $t = 0$. The advantage of t and τ is that unlike κ, they are dimensionless.

4.3.3 Emission

For gases, Kirchoff's law described in Section 3.4.5 applies in a slightly different form that allows derivation of the relation between absorption and emission (Thomas and Stamnes, 1999, Section 5.3.1). Given a black box containing a small volume of gas, where the gas and its surrounding walls are in thermal equilibrium, Kirchoff's law also states that emission equals absorption on a per-unit-volume basis. For a gas volume of width Δz, where the z-direction is arbitrary, the radiance absorbed in the gas from the walls is

$$\Delta L_{\text{absorbed}} = -\kappa_A(T, p, \lambda) f_P(\lambda, T)\,\Delta z \tag{4.11}$$

Therefore, the thermal emission from the gas in the direction z must be

$$\Delta L_{\text{emitted}} = \kappa_A(T, p, \lambda) f_P(\lambda, T)\,\Delta z \tag{4.12}$$

so that the absorptance $\hat{a} = -\kappa_A \Delta z$, and the emissivity $e = \kappa_A \Delta z$. Since the direction z is arbitrary, Equation (4.12) shows that the emitted radiance is the product of the absorption coefficient and the Planck function and is isotropic.

Even though this relation is derived under restrictive conditions, Thomas and Stamnes (1999) state that it applies to the real atmosphere. For the case of a strongly directional solar radiation incident on an atmospheric layer, the absorption has a directional dependence, while the atmosphere radiates isotropically, so that even though the absorption and emission constants are equal and opposite, the incident and emitted radiances are not equal. Under these conditions, the atmosphere heats up during the day and cools at night. As the following shows, the radiative source term in (4.12) is particularly important at the infrared and microwave wavelengths.

4.4 Scattering

Scattering has at least two effects on a narrow beam of radiation. The first is the previously discussed case of energy loss from scattering out of the beam; the second occurs for a sensor viewing the atmosphere or ocean in a specific direction, where an energy gain can occur from scattering of solar or other external energy sources into the view direction. The first case is a loss from the beam; the second, an unwanted gain. Scattering divides into *single* and *multiple* scattering. In single scattering, a photon experiences at most one collision along its path. An example of single scattering is a searchlight beam viewed at a distance. On clear nights, the beam retains its pencil-like shape, but is visible because of photons that are single-scattered from the beam to the observer. In contrast, for an evening with rain or fog, multiple scattering occurs, so that the beam might be visible only as a diffuse glow around the beam source. In many cases, single scattering can be modeled analytically, while multiple scattering is more complex and is generally modeled numerically.

The previous section shows that absorption and emission are scalar processes. Because of its strong directional dependence, scattering is a more complicated vector process. Figure 4.6 defines the variables and coordinates used in the scattering discussion. Following Kirk (1996) and Thomas and Stamnes (1999), the scattering properties are derived from consideration of a plane wave of irradiance E propagating along the z-axis and incident on a small gas volume $dV = dA\ dz$, where $dA = dx\ dy$ is perpendicular to the incident irradiance. Within this volume, because a fraction of the incident flux is scattered into angles other than the propagation direction, the magnitude of E decreases with distance along the path. The scattered energy is assumed governed by the following assumptions. First, the volume $dA\ dz$ is sufficiently small that within it only single scattering takes place. Second, the scattered power has an axisymmetric distribution about the propagation direction so that it is only a function of the scattering angle α, and is described in terms of the angular distribution of the intensity $dI(\alpha)$ within a solid angle $d\Omega$. Third, there is no fluorescence

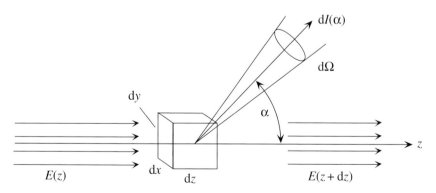

Figure 4.6. The geometry used in the definition of the volume scattering function.

or energy emitted at secondary wavelengths, so that the scattered radiation has the same wavelength as the incident.

The power incident on the volume is $\Phi = E\,dA$; from the definition of I in Section 3.3, the power scattered into any direction is $d^2\Phi = dI\,d\Omega$. The *volume scattering function* $\beta(\alpha, \lambda)$ is defined as the ratio of the power scattered per-unit-length and unit-solid-angle into a particular direction to the total power incident on the volume:

$$\beta(\alpha, \lambda) = \frac{dI(\alpha)}{E\,dA\,dz} = \frac{d^2\Phi}{\Phi\,dz\,d\Omega} \tag{4.13}$$

From (4.13), $\beta(\alpha, \lambda)$ has dimensions of $m^{-1}\,sr^{-1}$. This equation can be rewritten as

$$d^2\Phi(\alpha) = \beta(\alpha)\Phi\,d\Omega\,dz \tag{4.14}$$

where the λ-dependence of β is omitted.

To calculate the power loss per-unit-length, Equation (4.14) is integrated over all angles except for the forward direction, which is excluded since radiation propagating in this direction is not lost from the beam. The integration is from 0 to 2π in Φ, and from 0_+ to π in α, where the '+' subscript on 0 means that the forward direction is excluded. The result of this integration is

$$\frac{d\Phi}{\Phi_0\,dz} = 2\pi \int_{0_+}^{\pi} \beta(\alpha)\sin\alpha\,d\alpha \tag{4.15}$$

Equation (4.15) is the ratio of the flux scattered per-unit-length to the incident flux. It is independent of α and depends only on wavelength. Dividing the numerator and denominator of the left-hand side of (4.15) by dA and noting that the power lost from the beam is a negative number, transforms Equation (4.15) into a form similar to (4.5). Consequently, $\kappa_S(\lambda)$ can be written as

$$\kappa_S(\lambda) = 2\pi \int_{0_+}^{\pi} \beta(\alpha)\sin\alpha\,d\alpha \tag{4.16}$$

Equation (4.16) relates the volume scattering coefficient to the integral of the volume scattering function. Using (4.16), the following section discusses the limiting case of isotropic scattering and defines the *scattering phase function* $P(\alpha)$.

4.4.1 Isotropic scatter and the scattering phase function

For the isotropic case, the scattered radiation is the same in all directions, so that

$$\beta(\alpha) = \text{constant} = \beta_0 \tag{4.17}$$

Substitution of β_0 into (4.16) gives

$$\kappa_S(\lambda) = 4\pi\beta_0 \tag{4.18}$$

By analogy with (4.18), the scattering phase function $P(\alpha)$ is defined as

$$P(\alpha) = 4\pi\beta(\lambda, \alpha)/\kappa_S(\lambda) \qquad (4.19)$$

Equation (4.19) shows that $P(\alpha)$ is independent of λ and has units of sr^{-1}. For isotropic scatter, $P(\alpha) \equiv 1$. From (4.19), β can written as

$$\beta(\lambda, \alpha) = P(\alpha)\kappa_S(\lambda)/4\pi \qquad (4.20)$$

so that $P(\alpha)$ contains its angular dependence and $\kappa_S(\lambda)$ contains its λ-dependence. Given these definitions, the dependence of scattering on α, λ and the gas constituent is easily described.

4.4.2 Rayleigh and aerosol scattering

The description of scattering from molecules and particles in the atmosphere and ocean divides into two parts: *Rayleigh* or molecular scattering, and *Mie* (pronounced 'me') or aerosol scattering. The type of scattering depends on the size of the molecular or aerosol scatterer relative to the incident wavelength, or on the magnitude of the parameter q, defined as

$$q = 2\pi a/\lambda \qquad (4.21)$$

In (4.21), a is the radius of the molecule or particle and λ is the incident wavelength. For molecular scattering, $a \sim 0.1$ nm and for visible light with $\lambda \sim 500$ nm, $q \sim 10^{-3}$. For this case where $q \ll 1$, a simple closed form solution exists called Rayleigh scattering. As q approaches 1, diffraction of the incident radiation around the particle generates a strong forward scattering, which has a complicated mathematical form and is called Mie scattering. Examples in the atmosphere include scatter from water droplets and aerosols. In the visible, Rayleigh scattering dominates, with additional contributions from Mie scattering at small aerosol concentrations. Large aerosol concentrations or clouds obscure the surface. In the infrared, Rayleigh scattering is negligible, and Mie scattering is also neglected because infrared remote sensing depends on the absence of all heavy clouds and aerosols. At the microwave wavelengths, Rayleigh and Mie scattering occur from cloud water droplets and from large rain drops, with the scattering increasing at the shorter microwave wavelengths.

4.4.3 Molecular or Rayleigh scattering

From Stamnes and Thomas (1999), Rayleigh scattering can be written as follows:

$$P_R(\alpha) = (3/4)(1 + \cos^2\alpha), \qquad \kappa_R(\lambda) \sim 1/\lambda^4 \qquad (4.22)$$

In (4.22), the subscript R refers to Rayleigh scatter. Figure 4.7 compares the Rayleigh scattering phase function with the isotropic case, and shows that $P_R(\alpha)$ is symmetric about the fore and aft directions. Because $\kappa_R(\lambda)$ varies as λ^{-4}, scattering increases as λ decreases.

Table 4.2. *The λ-dependence of the*
Rayleigh scattering coefficient at
standard pressure from Equation (4.23)

λ (nm)	$\tau_{RO}(\lambda)$
400	0.390
500	0.152
600	0.072
700	0.038

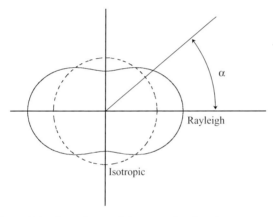

Figure 4.7. Comparison of the dependence of the scattering function on scattering angle α for the isotropic (dashed line) and Rayleigh scattering (solid line) cases.

For example, as λ decreases from 700 to 400 nm, or by about a factor of 2, the magnitude of the Rayleigh scattering increases by a factor of almost 16. This increase in the intensity of Rayleigh scattering with decreasing wavelength explains why the daytime clear sky is blue.

For the entire depth of the atmosphere and at the standard surface atmospheric pressure of $p_0 = 1013.25$ mb, the Rayleigh optical thickness at standard pressure τ_{RO} is

$$\tau_{RO}(\lambda) = 0.0089\lambda^{-4}(1 + 0.0113\lambda^{-2} + 0.000\,13\lambda^{-4}) \tag{4.23}$$

(Evans and Gordon, 1994, their Equation (10); Hansen and Travis, 1974). If τ_R is the Rayleigh optical thickness at an arbitrary surface pressure, then

$$\tau_R(\lambda) = \tau_{RO}(p/p_0) \tag{4.24}$$

For selected λ, Table 4.2 lists $\tau_{RO}(\lambda)$ and shows its wavelength dependence.

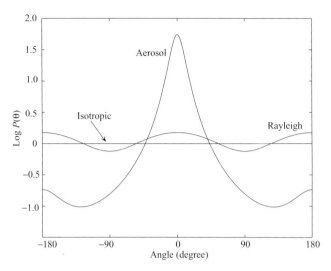

Figure 4.8. Comparison of the scattering function for isotropic and Rayleigh scattering with a strong forward scattering aerosol phase function that approximates a marine aerosol. See text for additional description.

4.4.4 Aerosol or Mie scattering

When the scatterers become comparable or larger in size relative to the incident wavelength, the scattering becomes strongly biased in the forward direction. For aerosol scattering in the marine boundary layer, Figure 4.8 compares the isotropic and Rayleigh phase function with an approximate solution called a Henyey–Greenstein function (Gordon and Castaño, 1987). As the figure shows, Mie scattering is characterized by a large forward-scattered component; it is also characterized by either a weak λ^{-1} or λ^0 wavelength dependence.

4.5 Atmospheric attenuation

This section discusses the atmospheric transmittance in the VIR; Chapter 9.2 discusses the microwave transmissivity. The total transmittance is given by the sum of the individual optical thicknesses or equivalently by the product of the individual transmittances for the different molecular species and processes. When only molecular absorption and scattering are important, the total optical thickness τ_{tot} and transmittance t_{tot} can be written

$$\tau_{tot} = \tau_R + \tau_{CO_2} + \tau_{O_3} + \tau_{H_2O} + \cdots \tag{4.25}$$

$$\text{and} \quad t_{tot} = t_R \cdot t_{CO_2} \cdot t_{O_3} \cdot t_{H_2O} \cdots \tag{4.26}$$

For a nadir-looking satellite, the radiance L_H received from a surface radiance L_0 is simply

$$L_H = t_{tot} L_0 \tag{4.27}$$

Table 4.3. *The dependence of atmospheric water vapor in the MODTRAN standard cases*

Case	Name	Surface density ρ_{v0} (g m^{-3})	Columnar density V (mm)
1	Tropical	17.3	48
2	Mid-latitude summer	13.0	35
3	Mid-latitude winter	3.5	11
4	Sub-arctic summer	8.7	25
5	Sub-arctic winter	1.3	5
6	1976 standard atmosphere	5.6	17

From this point on, the subscript tot will be dropped. Given these definitions, the next three figures examine the dependence of t on λ for several different molecular constituents and for atmospheric conditions ranging from tropical to sub-polar.

For the different constituents and atmospheric conditions, the transmittances are derived from the MODTRAN code (Anderson *et al.*, 1995). MODTRAN is part of a series of widely used computer codes (LOWTRAN, MODTRAN, HITRAN, FASCODE) that describe the radiative properties of the atmosphere at a variety of resolutions in wavelength. Table 4.3 lists the six default MODTRAN atmospheres with the values of their water vapor surface and columnar densities. In the table, Tropical refers to latitudes less than 30°; Mid-latitude to 30°–45°; Sub-arctic to 45°–60°. The table shows that for these cases, V lies between 5 and 50 mm.

For these atmospheres, three attenuation cases are considered: first, the contributions to t of several atmospheric molecular constituents and values of V for the wavelength band 0.2–15 μm; second, a detailed examination of the molecular and Rayleigh scattering contributions to t for the UV/VNIR wavelengths 0.2–1.0 μm; third, the variations in t associated with the seasonal ozone variability for 0.25–0.80 μm (250–800 nm).

For the six MODTRAN cases, a vertical path across the atmosphere and a wavelength range of 0.2–15 μm, Figure 4.9 shows the contributions to t from five molecular constituents, oxygen (O_2), nitrous oxide (N_2O), ozone (O_3), carbon dioxide (CO_2), water vapor (H_2O), as well as the total transmittance. Examination of the contributions from each molecule shows that although O_2 has an important absorption region in the near infrared, O_2 and N_2O are minor contributors to t. Between 9 and 10 μm, O_3 generates a major absorption region and as the next figure shows in more detail, blocks the transmission of ultraviolet radiation at wavelengths shorter than about 0.35 μm. For $\lambda > 3$ μm, the combination of CO_2 and water vapor primarily determines the transmittance. The CO_2 provides a long wavelength cutoff at about 14 μm and some major opaque regions at shorter wavelengths. Between 1 and 14 μm, water vapor determines much of the shape of the transmittance and is the dominant contributor to the variability. The total transmittance shows that for $\lambda > 3$ μm, or in the region where thermal emission is important, the three wavelength windows that are used in

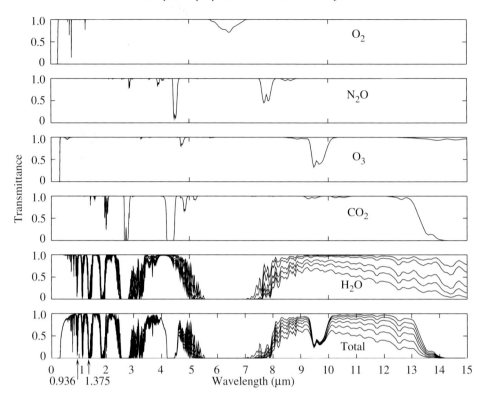

Figure 4.9. The wavelength dependence of the transmittance for the visible and infrared wavelengths for the five major molecular contributors to atmospheric absorption and for the values of the columnar water vapor represented by the six MODTRAN atmospheres in Table 4.3. The H_2O curves correspond in order of decreasing transmittance to Sub-arctic winter, Mid-latitude winter, 1976 standard, Sub-arctic summer, Mid-latitude summer, Tropical. The arrows at 0.936 μm and 1.375 μm mark the water vapor absorption bands used in the discrimination of high cirrus clouds discussed in Chapter 7. See text for further description.

the SST retrieval are 3–4 μm, 8–9 μm and 10–12 μm. As Chapter 7 discusses in detail, the 3–4 μm transmittance is least dependent on water vapor, while the 10–12 μm transmittance is most dependent.

For a more detailed examination of the range of $0.20 < \lambda < 1$ μm ($200 < \lambda < 1000$ nm), and for the two extreme Tropical and Sub-arctic winter MODTRAN cases, Figure 4.10 shows the contributions of O_2, O_3, water vapor and Rayleigh scattering to the total transmittance. The major difference between Figures 4.9 and 4.10 is that for $\lambda < 1$ μm, Rayleigh scattering becomes important. Examination of the curves also shows that O_3 provides a small but important transmittance change around 600 nm and attenuates the ultraviolet. O_2 generates two absorption regions in the near infrared, where the region at about 762 nm is called the oxygen-A band, and also completely attenuates the ultraviolet, although at shorter wavelengths than O_3. The water vapor contributions only affect a few specific

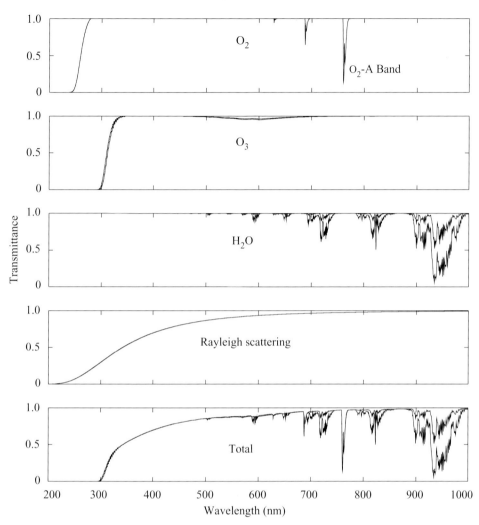

Figure 4.10. The wavelength dependence of the atmospheric transmittance for oxygen, ozone, water vapor, and Rayleigh scattering, for the two extreme MODTRAN cases of Tropical and Sub-arctic winter. For water vapor, the lower curve corresponds to the winter case.

bands occurring at wavelengths greater than about 600 nm. Consequently, over much of the visible spectrum, water vapor absorption can be ignored. Finally, at the short visible wavelengths or for $\lambda < 600$ nm, the combination of O_3 attenuation and Rayleigh scattering determines most of the variability in the absorption.

Because the effect of variable O_3 is difficult to see at the scale of Figure 4.10, Figure 4.11 shows the difference between the summer and winter transmittances associated with the Mid-latitude summer and winter MODTRAN ozone distributions, as shown by the profiles in Figure 4.2. The figure shows that the summer transmittance is greater in

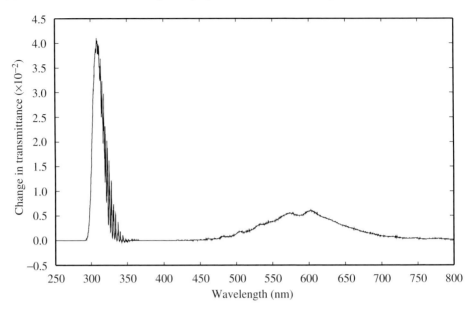

Figure 4.11. The difference between the MODTRAN Mid-latitude summer and winter transmittance associated with the decrease in summer ozone.

the ultraviolet for 300–350 nm, and increases slightly in the visible for 450–700 nm. As Chapter 6 discusses, the ocean color retrieval accounts for this ozone-induced change in the visible transmittance. In summary, Figures 4.9 and 4.10 show that in the visible/infrared, the atmosphere has a number of spectral windows or regions which permit observation of the surface from space. Most of the visible wavelengths are transparent, although strongly attenuated by Rayleigh scattering at the shorter wavelengths. Additional windows occur in the NIR between 0.8 and 0.9 μm, and in the TIR at 3–4 μm, 8–9 μm and 10–12 μm. Between these windows and as Section 4.8.1 describes, satellite instruments called sounders use the opaque regions to determine the temperatures at different depths in the atmosphere.

4.6 Application to the ideal instrument

For the ideal nadir-viewing telescope described in Section 3.5.2 that views the ocean through an attenuating atmosphere, Equation (3.35) can be written as

$$\Phi_{IN} = t\alpha_S A L \tag{4.28}$$

so that for a fixed λ, Φ varies linearly with t. Suppose that the satellite views the Earth at an off-nadir angle. If the curvature of the Earth can be neglected and the atmospheric variables p, T, and κ_E have a plane parallel distribution so that they are only functions of z, then the off-nadir case also has a simple solution.

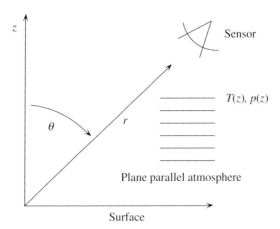

Figure 4.12. The coordinate system used in the discussion of a sensor viewing the surface through a plane parallel atmosphere.

For this case, Figure 4.12 shows that the sensor views the surface at an incidence angle θ and along a radial r. Because T and p are only functions of z, $\kappa_E(T, p, \lambda) = \kappa_E(z, \lambda)$, and with the explicit λ-dependence omitted, can be written as $\kappa_E(z) = \kappa_E(r \cos \theta)$. For constant θ, if κ_E is integrated along a slant path between the surface and the satellite, then $L_H = L_0 \exp(-\tau')$, where

$$\tau' = \int_0^{r_H} \kappa_E(r) \cos \theta \, dr \tag{4.29}$$

and r_H is the radial height of the TOA along θ. With the change of variable in Equation (4.29) from r to z, and noting that the secant is defined by $\sec \theta \equiv 1/\cos \theta$ so that $r = z \sec \theta$, Equation (4.29) can be written

$$\tau' = \int_0^{z_H} \kappa_E(z) \sec \theta \, dz = -\tau(z_H) \sec \theta \tag{4.30}$$

From Equation (4.30) and for the slant-looking case, the radiance received at the satellite is

$$L_H = L_0 \, e^{-\tau \sec \theta} = L_0 t^{\sec \theta} \tag{4.31}$$

From (4.31), the solutions for the off-nadir and nadir cases have the same form, except that τ is replaced by $\tau \sec \theta$, and t is replaced by $t^{\sec \theta}$. For the finite bandwidth instrument described in Section 3.5.4, Equation (3.36) becomes

$$\Phi_{IN} = L(\lambda_c) \Delta \lambda \, A \alpha_S t^{\sec \theta} \tag{4.32}$$

For an instrument looking at the surface through a purely attenuating, plane parallel atmosphere, Equation (4.32) gives the general form of the received radiant flux. As Chapter 7 shows, the $\sec \theta$ dependence in (4.32) becomes important in the infrared retrieval of SST.

The form of (4.32) also shows that in some cases, the θ-induced variability is not important. For example, suppose that the atmospheric transmittance is 0.8 and that θ varies from 0 to 45°. Although the increase in path length is 41%, the attenuation varies only by 10%, from 0.8 at nadir to 0.73 at 45°. This means that in some cases whiskbroom scanners can be used without correction for variable θ.

4.7 Radiative transfer equation

This section discusses the radiative transfer equation (RTE) and its attenuation and source terms. Sections 4.7.1 and 4.7.2 discuss the emission and scattering source terms, and for the beam transmittance case, Section 4.7.3 derives the solution for a radiance propagating across the entire atmosphere.

Consider a radiance at a location $x = x, y, z$ propagating along the path defined by the angles θ, ϕ. Along this path, the radiance loses energy by absorption and scattering out of the path, and gains energy from thermal emission and by scattering from external sources into the path. Combination of all these terms leads to the following form of the RTE (Kirk, 1996):

$$\frac{d}{dr}L(\lambda, x, \theta, \phi) = -\kappa_E(\lambda, x)L(\lambda, x, \theta, \phi) + \Pi(\lambda, x, \theta, \phi) \qquad (4.33)$$

In (4.33), r is in the direction specified by θ, ϕ, where, in this direction, the left-hand side is the change in radiance per-unit-distance. On the right-hand side, the first term describes the attenuation of the radiance by scattering and absorption; the second is the source term Π, given by

$$\Pi(x, \theta, \phi) = \Pi_{emit}(x, \theta, \phi) + \Pi_{scat}(x, \theta, \phi) \qquad (4.34)$$

In (4.34), and as the following sections discuss, $\Pi_{emit}(x, \theta, \phi)$ is the emission source term and $\Pi_{scat}(x, \theta, \phi)$ is the scattering source term generated by scattering into θ, ϕ from all directions other than the direction of propagation. To simplify these definitions, their λ-dependence is omitted.

4.7.1 Thermal emission source term

From Equation (4.12), the thermal emission source term is

$$\Pi_{emit} = \kappa_A(T, p, \lambda) f_P(\lambda, T(x)) \qquad (4.35)$$

Because both the atmosphere and ocean are at temperatures of about 300 K, their emissions are negligible in the visible, but must be considered in the infrared and microwave. Within the water column, because radiation at the infrared and microwave wavelengths cannot propagate more than a few millimeters, oceanic thermal emission is neglected in all observational bands.

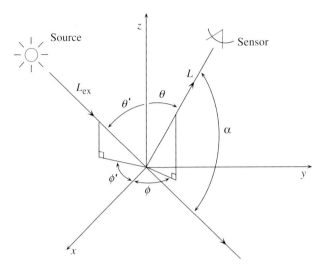

Figure 4.13. The coordinate system and geometry used in discussion of the scattering source term.

4.7.2 Scattering source term

The scattering source term $\Pi_{\text{scat}}(x, \theta, \phi)$ is more complicated than the emission term, and is derived from consideration at a location x of a volume element with a length dr in the view direction. At this location, evaluation of the source term consists of calculating the sum of the radiances propagating in the direction θ, ϕ that are generated by the scattering of all of the radiances incident on the volume element, except those propagating in the direction θ, ϕ. Figure 4.13 shows the scattering geometry. In this figure, an external radiance L_{ex} from a source such as the sun is incident on the origin at an angle θ', ϕ'. At the origin, a fraction of this radiance is scattered into the direction θ, ϕ toward the sensor, where α is the angle between the incident and scattered radiance.

For this geometry, Π_{scat} can be written in terms of the volume scattering function β. Equation (4.13) defines $\beta(\alpha, \lambda)$ in terms of the power scattered per-unit-length and unit-solid-angle out of an incident irradiance. Mobley (1994, Chapter 5.2) shows that β can be equivalently considered as the radiance scattered per-unit-path-length at an arbitrary point into the view direction from an external irradiance that is incident on the point at the relative angle α. Figure 4.14 shows an expanded view of this geometry, where the scattering occurs at the origin. The incident irradiance is $E_{\text{ex}} = L_{\text{ex}}\Delta\Omega'$, where L_{ex} is a source of external radiation such as the sun and $\Delta\Omega'$ is the angle subtended by the source. For this case, Mobley (1994) shows that the radiance per-unit-path-length received at the sensor from scattering at the origin is

$$\Pi_{\text{scat}} \equiv \frac{dL}{dr} = \beta(\alpha)L_{\text{ex}}\Delta\Omega' \tag{4.36}$$

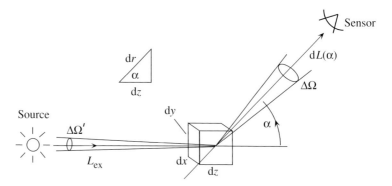

Figure 4.14. Coordinate system used in discussion of scattering into the beam.

For multiple sources of external radiation, integration of (4.36) over all solid angles gives

$$\Pi_{\text{scat}}(\theta, \phi) = \int_{4\pi} \beta(\theta, \phi; \theta', \phi') L_{\text{ex}}(\theta', \phi') \, d\Omega' \tag{4.37}$$

where L_{ex} now represents all sources of external radiance. As Section 4.8.2 shows, given β and L_{ex}, the right-hand side of (4.37) can be integrated to yield the scattering source term.

4.7.3 General solution for a radiance propagating across the atmosphere

This section derives the radiance received at the satellite for the beam attenuation case. The derivation proceeds from integration of Equation (4.33) across the atmosphere along a path inclined at an incidence angle θ. (This derivation follows the undated and unpublished course notes of J. L. Mueller and C. H. Wash, approximately 1984.)

For a plane parallel atmosphere where $dr = dz \sec \theta$, (4.33) becomes

$$\cos \theta \frac{d}{dz} L(z) + \kappa_E(z) L(z) = \Pi(z) \tag{4.38}$$

In (4.38), when the right-hand side or source term is set equal to zero, the solution of the homogenous equation has the form $\exp[-\tau(z) \sec \theta]$. Equation (4.38) is solved in the following way. First, the optical thickness τ is redefined for a path originating at the TOA, so that relative to z_H at the TOA, $\tau(z)$ becomes

$$\tau(z) = \int_z^{z_H} \kappa_E(z) dz \tag{4.39}$$

From this definition, $\tau(z_H) \equiv 0$.

Second, multiplying both sides of (4.38) by the term $\sec \theta \exp[-\tau(z) \sec \theta]$ and integrating from z to z_H yields

$$\int_z^{z_H} \exp[-\tau(z) \sec \theta] \left[\frac{dL}{dz} + \kappa_E(z) L(z) \sec \theta \right] dz$$

$$= \sec \theta \int_z^{z_H} \Pi(z) \exp[-\tau(z) \sec \theta] dz \tag{4.40}$$

If $u = L(z) \exp[-\tau(z) \sec\theta]$ and with the definition of τ from (4.39), the left-hand side of Equation (4.40) can be integrated to yield

$$\int_z^{z_H} du = L(z_H) - L(z) \exp[-\tau(z) \sec\theta] \qquad (4.41)$$

Therefore, for beam transmittance, the solution to Equation (4.40) is

$$L(z_H) = L(z) \exp[-\tau(z) \sec\theta] + \sec\theta \int_z^{z_H} \Pi(z) \exp[-\tau(z) \sec\theta] dz \qquad (4.42)$$

Physically, Equation (4.42) shows that the radiance received at the satellite from a source at height z consists of the exponential attenuation of the radiance at z plus the integral of the source term along the path between z and the TOA. In the infrared and microwave, Equation (4.42) is valid for both extended and discrete sources of radiance; in the visible, it is valid for discrete sources. For extended visible sources, Section 4.9 discusses the case where in the first term on the right, the surface radiance is attenuated by the diffuse transmittance.

A change in variables from z to τ, where $d\tau = -\kappa_E dz$, allows the second term to the right in (4.42) to be written as

$$\sec\theta \int_0^{\tau(z)} [\Pi \exp[-\tau(z) \sec\theta]/\kappa_E(z)] d\tau \qquad (4.43)$$

Equation (4.43) is the *path radiance*, which is the radiance generated along the path between the height z and the satellite by either scattering into the beam or molecular emission. As the next section shows, evaluation of (4.43) at $z = 0$ or equivalently at $\tau(0) \equiv \tau$ gives the path radiance generated across the entire atmosphere.

4.8 Specific solutions of the radiative transfer equation

This section discusses specific solutions of the RTE. Section 4.8.1 derives the case applicable to the infrared and microwave where scattering is negligible. For the visible case where molecular scattering predominates, Section 4.8.2 derives the single scattering Rayleigh path radiance. Section 4.8.3 then briefly discusses the single scattering aerosol path radiance.

Because the relative magnitudes of the scattering and emission terms vary greatly among the visible, infrared and microwave windows, it is easier to find an approximate solution to (4.42) that is applicable to each window than to find a general solution. The reason these approximate solutions are successful is that the ocean surface and atmospheric temperatures are both of order 300 K, and molecular scattering is only important in the visible. From Planck's equation, the temperature condition means that the ocean surface and atmosphere have their maximum radiance at a wavelength of about 10 μm. Thus, in the visible, atmospheric emission can be neglected and the RTE is dominated by Rayleigh and aerosol scattering, where the solution is called *scattering dominant*. In the infrared and microwave, because of their longer wavelengths, scattering can be neglected for cloud-free conditions in the infrared and for almost all conditions except heavy rain in the microwave. At these wavelengths, the RTE is primarily a balance between atmospheric absorption and emission, where the solution is called *absorption–emission dominant*. The following subsections first derive the RTE for the absorption–emission case, then derive the scattering-dominant case.

4.8.1 Absorption–emission dominant case

As Section 4.4.3 shows, molecular scattering can be neglected for wavelengths longer than the visible, so that if aerosol scattering can be similarly neglected, the RTE can be approximated as a balance between absorption and emission. This solution is particularly applicable to the infrared and microwave. Because cloud-free conditions are required to view the ocean surface in the infrared, only molecular scattering can occur, which is negligible so that the approximation holds. In the microwave, where the surface can be viewed through clouds, the approximation holds for long wavelengths, but as Chapter 9 describes in detail, breaks down in heavy rain and at the shorter microwave wavelengths.

For an absorption–emission balance with zero scattering, $\kappa_E = \kappa_A$. Because thermal emission is the only source term, substitution of Equation (4.43) and (4.35) into (4.42) gives the following solution for the RTE at the TOA:

$$L(z_H) = L_0 \exp[-\tau(z_H) \sec \theta] + \sec \theta \int_0^\tau f_P(\lambda, T) \exp[-\tau'(z) \sec \theta] d\tau' \qquad (4.44)$$

In (4.44), L_0 is the surface radiance, $L(z_H)$ is the TOA radiance and f_P is the Planck function defined in Equation (3.21). The first term on the right is the attenuated surface radiance, the second is the atmospheric emission source term. This important solution to the RTE is called the *Schwarzschild* equation (Kidder and Vonder Haar, 1995).

If the atmosphere can be characterized by a mean temperature $\overline{T}$ and mean τ, Equation (4.44) can be written

$$L(z_H) = L_0 \exp(-\tau \sec \theta) + f_P(\lambda, \overline{T})[1 - \exp(-\tau \sec \theta)] \qquad (4.45)$$

For this approximation, Equation (4.45) shows that the radiance received at the TOA divides into two parts: the surface radiance L_0 attenuated by atmospheric absorption, and an atmospheric emission term proportional to the Planck function. As Chapter 7 shows, the simplified equation in (4.45) is used in the SST retrieval algorithms.

Depending on the magnitude of the transmittance or transmissivity and thus on the choice of observational wavelength, the received radiance is dominated by one of the two terms of (4.45). If the window is highly transmissive, the ocean surface radiance dominates the received radiance; if the window is weakly transmissive, the path radiance dominates. For the weakly transmissive case and depending on wavelength, the retrieved radiance is proportional to the temperature at different levels in the atmosphere. Sounders use these regions to retrieve the atmospheric temperature profile (Kidder and Vonder Haar, 1995, Chapters 3 and 6).

To examine the relative magnitudes of the attenuated surface radiances and the atmospheric emission radiances for the wavelength range 3–15 μm, Figure 4.15 compares the TOA radiances for two MODTRAN atmospheres with different water vapor contents.

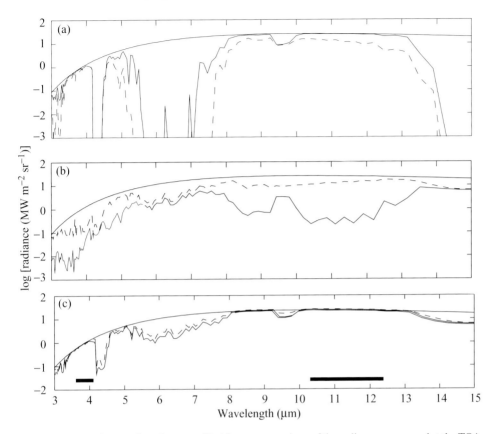

Figure 4.15. For the wavelength range of 3–15 μm, comparison of the radiances measured at the TOA for the MODTRAN Sub-arctic winter case (lower solid line) and the Tropical case (dashed line), each for a surface temperature of 285 K. On each figure, the upper solid line is the 285 K blackbody curve. (a) The attenuated surface radiance; (b) the atmospheric emission radiance; (c) the total radiances. The horizontal black bars in the lower panel show the location of the 3.5–4.1 μm and 10.3–12.4 μm windows used in the SST retrieval described in Chapter 7.

These are the Tropical and Sub-arctic winter cases, both at a surface temperature of 285 K. Figure 4.15a compares the attenuated surface radiances with the 285 K blackbody radiance; as expected, the tropical radiance is smaller because of the greater water vapor content and attenuation. Figure 4.15b shows the atmospheric emission radiances, here the situation is reversed because the tropical atmosphere is a stronger emitter than the winter. Figure 4.15c shows the total received radiance, in this case the tropical radiance is slightly larger.

The horizontal bars in the lower panel show the locations of the 3.5–4.1 μm and 10.3–12.4 μm windows used in SST retrieval. These are respectively called the 4- and 11-μm windows. For these, the blackbody curve shows that the radiance at 4 μm is at least an order smaller than at 11 μm. Also at 4 μm, Figure 4.15a shows that for both atmospheres the received radiance approximately equals the surface radiance. Consequently, at 4 μm, variations in the atmospheric water content have little effect on the attenuation and emission. At 11 μm, because of the water vapor contribution, the tropical surface radiance is more attenuated than the corresponding winter radiance, and, as Figure 4.15b shows, the tropical emissive radiance is much larger than the sub-arctic case. Therefore, any retrieval of the surface radiance at 11 μm depends on removal of the water vapor attenuation and emission. In summary, even though the surface radiance at 4 μm is an order smaller than at 11 μm, water vapor has a much smaller effect at the shorter wavelengths.

4.8.2 Single scattering approximation

In the visible, thermal emission can be neglected so that scattering dominates the RTE. The single scattering approximation divides into two parts. The present section derives the path radiance generated by an external source such as the sun; Section 4.9.1 discusses the attenuation of an extended surface radiance by the diffuse transmittance.

   This section specifically derives the Rayleigh path radiance for molecular single scattering and Section 4.8.3 states the result for aerosol single scattering, where both solutions are used to retrieve ocean color. The derivation assumes no tropospheric absorption of visible light, so that only scattering occurs. The atmosphere is divided into two layers: the stratosphere, where it is assumed that only ozone attenuation occurs with no scattering, and the troposphere, where Rayleigh and aerosol scattering occur. The marine troposphere is sometimes further divided into an upper layer where Rayleigh scattering dominates and a lower layer where aerosol scattering dominates.

   Before derivation of the Rayleigh path radiance, and following Mobley (1994), the *spectral single scattering albedo* $\omega_0(\lambda)$ is defined as

$$\omega_0(\lambda) = \kappa_S(\lambda)/\kappa_E(\lambda) \qquad (4.46)$$

The term $\omega_0(\lambda)$ is the ratio of the scattering coefficient to the extinction coefficient, and can be considered as the probability that as a photon travels a given distance, it will be scattered rather than absorbed. For pure scattering, $\omega_0 = 1$; for pure absorption, $\omega_0 = 0$. For Rayleigh single scatter, $\omega_0(\lambda) = \omega_R(\lambda) \equiv 1$. For aerosol single scatter, $\omega_0(\lambda)$ is replaced by single scattering aerosol albedo $\omega_A(\lambda)$, where

$$\omega_A(\lambda) = \kappa_{AS}(\lambda)/\kappa_{AE}(\lambda) \qquad (4.47)$$

In (4.47), κ_{AS} is the aerosol scattering coefficient, κ_{AE} is the aerosol extinction coefficient. In general, $\omega_A(\lambda) < 1$ (Gordon and Castaño, 1987).

   In the following, the sun is assumed to be the only source of external radiation, where the solar irradiance is approximated as a point source, located in Earth coordinates at the solar zenith and azimuth angles θ_S, ϕ_S. With this notation, the solar irradiance at the TOA is described by $F_S(\lambda)\delta(\theta - \theta_S, \phi - \phi_S)$, where $\delta(\theta - \theta_S, \phi - \phi_S)$ is the delta function. From Chapter 3, $F_S(\lambda) = \Delta\Omega\, L_S(\lambda)$, where $\Delta\Omega$ is the solid angle subtended by the sun at the

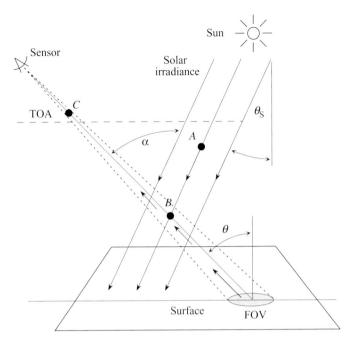

Figure 4.16. The coordinates and definitions used in the single scattering discussion. The gray ellipse shows the sensor FOV. See text for further description.

TOA and $L_S(\lambda)$ is the solar radiance. This means that the single scattering case approximates the sun as a bright point source in a black sky and neglects the additional skylight term associated with Rayleigh scattering of the incident solar radiation.

The task is to calculate the path radiance observed by a satellite sensor viewing the surface; with small changes, the same calculation applies to a sensor looking up from the surface in a direction away from the sun. Figure 4.16 shows the geometry of the source and sensor. For simplicity, the observing path and the solar irradiance are assumed to lie on the same azimuth, so that θ and θ_S lie in the same plane. In this calculation, the surface radiance is neglected. At each volume element along the observing path, the incident solar radiation is scattered toward the sensor. Because each photon is only scattered once, there is no further attenuation of the scattered radiance along the path, so that the integral along the path of the contribution from each path element equals the Rayleigh path radiance. Determination of the path radiance thus divides into two parts: calculation of the scattering source term at an arbitrary height z within the observing path, and integration of this term over the entire path.

Calculation of the scattering source term at a height z, shown on Figure 4.16 as point A, proceeds as follows. If the solar irradiance $F(\lambda, z)$ is only attenuated by Rayleigh scattering, then at A, the irradiance becomes

$$F(\lambda, z) = F_S(\lambda)\,\delta(\theta - \theta_S, \phi - \phi_S)\exp[-\tau_R(z)/\cos\theta_S] \qquad (4.48)$$

In (4.48), $\tau_R(z)$ is the Rayleigh optical thickness derived from Equation (4.39) and an unspecified distribution of $\kappa_R(z)$; in the final result, the τ_R term appears only as the optical thickness for the entire atmospheric depth.

Second, at the point B located at an arbitrary height z within the observing path, a portion of the radiance is scattered toward the sensor. Substitution of Equation (4.48) into the scattering source term in (4.37) and integration over all solid angles shows that at z the scattering source function can be written

$$\Pi_{\text{scat}}(\lambda, z; \theta, \phi) = \int_{4\pi} \beta(\theta, \phi, \theta', \phi') F(\lambda, z) \, d\Omega'$$
$$= F_S(\lambda) \, P_R(\alpha) \, \kappa_R(\lambda, z) \exp[-\tau_R(z) \sec \theta_S]/4\pi \qquad (4.49)$$

In (4.49), the second line is derived from substitution of β from (4.20). Equation (4.49) gives the Rayleigh scattering at a height z and in the direction α relative to the solar irradiance.

Substitution of (4.49) into the source term in Equation (4.43) and integration across the atmosphere along the path defined by θ gives the following for the Rayleigh path radiance $L_R(\theta)$:

$$L_R(\theta) = [F_S(\lambda) \, P_R(\alpha) \sec \theta / 4\pi] \int_0^{\tau_R} \exp[-\tau_R(z)(\sec \theta_S + \sec \theta)] \, d\tau_R \qquad (4.50)$$

On Figure 4.16, L_R is the Rayleigh path radiance evaluated at the TOA or at point C. Since Rayleigh scattering is the only source of attenuation, the τ in (4.43) are replaced by τ_R in (4.50). The single scattering approximation also means that there is no further attenuation by scattering of the radiance as it propagates toward the sensor. Therefore, within the integral in Equation (4.50), the $\sec \theta$ term is set equal to zero, so that the equation can be integrated to yield

$$L_R(\theta) = F_S(\lambda) P_R(\alpha) \cos \theta_S (1 - \exp[-\tau_R(\lambda) \sec \theta_S])/4\pi \cos \theta \qquad (4.51)$$

For the real atmosphere, F_S and L_R are further attenuated by stratospheric ozone. To decouple ozone attenuation and tropospheric scattering, it is assumed that attenuation takes place only in the stratosphere while scattering takes place only in the troposphere. For a satellite sensor, L_R is then determined by the downward attenuation of the solar irradiance through the ozone layer, the tropospheric Rayleigh scattering of this light toward the sensor, and the further attenuation of the scattered radiance as it passes upward through the ozone layer. Because both the sun and sensor lie above the atmosphere, this situation is modeled by replacement of $F_S(\lambda)$ in (4.51) with the solar irradiance $F_S'(\lambda)$ attenuated by two passes through the ozone layer. Given an ozone optical thickness of τ_{OZ}, $F_S'(\lambda)$ can be written as

$$F_S'(\lambda) = F_S(\lambda) \exp[-\tau_{OZ}(\sec \theta + \sec \theta_S)] \qquad (4.52)$$

Equation (4.51) can be further simplified if the solar incidence angle is restricted to $\theta_S \leq 45°$. For these angles, $\sec \theta_S \leq 1.4$, and given the visible wavelength values of $\tau_R(\lambda)$ from Equation (4.24) and Table 4.2, $\tau_R(\lambda) \sec \theta_S < 1$, so that $\exp[-\tau_R(\lambda) \sec \theta_S] \cong 1 - \tau_R(\lambda) \sec \theta_S$.

Substitution of this approximation and (4.52) into (4.51) yields

$$L_R(\theta) = F_S'(\lambda) P_R(\alpha) \tau_R(\lambda)/4\pi \cos\theta \qquad (4.53)$$

Equation (4.53) gives the single scattering Rayleigh path radiance that is generated across the entire atmosphere and received at the satellite (Gordon and Castaño, 1987). In Equation (4.42) and for $z = 0$, (4.53) is the solution for the second term on the right. For the first term on the right in (4.42), if the source area is small, beam attenuation is applicable and τ is replaced with the sum of the Rayleigh and ozone optical thicknesses. For an extended surface radiance in the visible, then as Section 4.9.1 discusses, the diffuse transmittance attenuates the surface radiance.

4.8.3 Aerosol single scattering

Unlike the Rayleigh scattering case where $\omega_R = 1$, a typical value for $\omega_A(\lambda)$ is about 0.8 (Gordon and Castaño, 1987). Given $\omega_A(\lambda)$, Gordon and Castaño derive the aerosol single scattering solution in a manner similar to (4.53). If $P_A(\alpha)$ is the aerosol phase function and $\tau_A(\lambda)$ is the aerosol optical depth, they find that the aerosol path radiance L_A received at the sensor is

$$L_A(\theta) = \omega_A(\lambda) \, F_S'(\lambda) \, P_A(\alpha) \, \tau_A(\lambda)/4\pi \cos\theta \qquad (4.54)$$

Both (4.53) and (4.54) will be used in the retrieval of ocean color described in Chapter 6.

4.9 Diffuse transmittance and skylight

In addition to generating the path radiances, scattering determines the magnitude of two additional quantities: the diffuse transmittance discussed below in Section 4.9.1 and the skylight term discussed in 4.9.2. The diffuse transmittance is smaller than the beam transmittance, is important only in the visible, and applies to the case of a radiance generated by an extended surface. For clear skies, skylight is generated from the blue sky Rayleigh scattering of the incident solar irradiance, which creates a downwelling irradiance incident from all directions.

4.9.1 Diffuse transmittance

As Gordon *et al.* (1983) and Wang (1999) describe, for radiances emitted from an extended ocean surface into a scattering atmosphere, the received radiance depends not only on the radiance emitted within the instrument FOV, but also on radiances scattered into the instrument solid angle from the surrounding area (Figure 4.17). This effect is most important at the shorter visible wavelengths. The contributions from outside the FOV have two effects: first, the source of the received radiance is larger than the FOV; second, the received radiance is greater than it would be for beam attenuation alone. This scattering creates

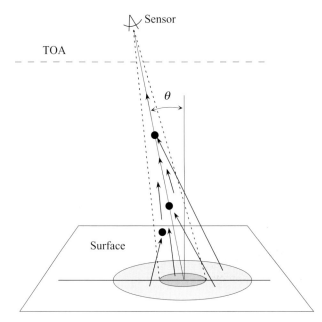

Figure 4.17. Generation of the diffuse transmittance. The inner dark gray ellipse is the instrument FOV; the larger light gray ellipse is the region from which single scattering of surface radiances into the beam contributes to the received radiance. The black dots show several examples of scattering sites. See text for additional description.

problems when an ocean FOV is close to land or adjacent to sea ice, icebergs or to any surface where the emitted or reflected radiance differs from the open ocean. Because of the contributions from these adjacent areas, the received radiance no longer represents the open ocean and is described as *contaminated*, as in land-contaminated. For this reason, ocean color observations can only be used if they are several pixels away from land.

With the additional scattering contribution, the received radiance from an extended surface depends on wavelength, the solar illumination angle, the optical properties of the atmosphere, the instrument look angle, and the angular distribution of the reflected radiation at the ocean surface. In most cases, the received radiance and its diffuse transmittance $t_D(\lambda, \theta)$ must be calculated numerically. Gordon *et al.* (1983) show from numerical calculations that for Lambertian reflection from a uniform extended surface and for Rayleigh and aerosol single scattering, t_D can be approximated by the following analytical expression:

$$t_D(\lambda, \theta) = \exp([-\tau_R(\lambda)/2 + \tau_{OZ}] \sec \theta) \qquad (4.55)$$

Equation (4.55) provides an approximate, but useful solution. Examination of this solution shows that scattering reduces the attenuation by a factor of 2, where this decrease is accompanied by an increase in the apparent FOV of the instrument.

Gordon *et al.* (1983) also show that although aerosol scattering also contributes to t_D, this contribution is neglected because for thin aerosols, the strong forward scattering means that the radiance is not appreciably attenuated, while for thick aerosols, the analysis breaks down. From comparison of Equation (4.55) with numerical solutions, Wang (1999) found that it is accurate to within 2–3% for nonabsorbing and weakly absorbing aerosols and for $\tau_A \leq 0.4$ and $\theta \leq 40°$. For $\tau_A \leq 0.6$ and $\theta \leq 60°$, Wang's modified numerical model of the diffuse transmittance is accurate to within about 1%. For single scattering, t_D is used in the first term of Equation (4.42) to describe the attenuation of the surface radiance.

4.9.2 Skylight

Skylight refers to the Rayleigh scattered solar radiances that generate the blue clear sky color. Because of skylight, the solar irradiance incident on the ocean surface divides into two parts: a direct solar term and a diffuse Rayleigh single scattering term. Gordon and Clark (1981) combine these terms into a model for the downward plane irradiance $E_d(0_+)$ evaluated at a height $z = 0_+$ just above the sea surface. They assume that the ozone attenuation occurs independently of the tropospheric Rayleigh scattering, then show from numerical calculations that $E_d(0_+)$ is given by

$$E_d(0_+) = F_S(\lambda) \cos \theta_S \exp([-\tau_R(\lambda)/2 + \tau_{OZ}] \sec \theta_S) = F_S(\lambda) \cos \theta_S t_D(\lambda, \theta_S) \quad (4.56)$$

Equation (4.56), which is used in the ocean color discussion, shows that the surface solar irradiance is also a function of t_D.

4.10 Further reading

Thomas and Stamnes (1999) present a more detailed and advanced approach to atmospheric radiative transfer, while Kidder and Vonder Haar (1995) give a more applied discussion. The Anderson *et al.* (1995) article can be found at http://imk-isys.fzk.de/isys-public/Software-tools/Modtran/science/fa-mo-lo.htm.

5

Reflection, transmission and absorption at the atmosphere/ocean interface

5.1 Introduction

At all wavelengths, the properties of a radiance received at a satellite depend on the small scale interaction of the radiation with the air/water interface. In the infrared however, the ocean is so highly absorbing that absorption and emission are confined to the top 1–100 μm of the ocean, and in the microwave, they are confined to the top 1–3 mm. For these bands and neglecting the atmosphere, the properties of the received radiance depend only on scattering and reflection at the ocean surface. In contrast, for the visible wavelengths, the received radiance also depends on the backscatter of solar radiation in the ocean interior.

Therefore, in the visible, two kinds of reflection take place (Figure 5.1). The first is the *direct* or *surface* reflection at the interface of the solar radiance and skylight. The second is the *diffuse* reflection associated with the *water-leaving radiance* that is generated by the propagation of the incident solar radiance across the interface into the water column, where a portion of this radiance is backscattered to recross the interface into the atmosphere. As Chapter 6 shows, the water-leaving radiance or *luminosity* of a water surface generated by the interior scattering is essential to remote sensing in the visible and makes possible the retrieval of such water column properties as the chlorophyll concentration.

To expand on luminosity, for daytime clear skies, Raman (1922) discusses how the sea surface color is determined not by surface reflection of skylight, but by scattering within the water column. This occurs because, as Section 5.4 shows, the volume scattering coefficient for water has a form similar to Rayleigh atmospheric scattering while being 160 times greater. For the sun at zenith and neglecting absorption, the magnitude of the scattering coefficient means that a 50-m deep water column scatters as much light as about 8 km of atmosphere, so that the water surface should be nearly as bright as the sky. Even including absorption, Raman shows that under direct sunlight, scattering in the water is the primary cause of the blue ocean color. For a dense cloud cover, this ceases to be the case. In this case, the water column scattering is reduced and the ocean color is determined by the direct surface reflection of the forward-scattered sunlight passing through the clouds, so that the surface appears gray.

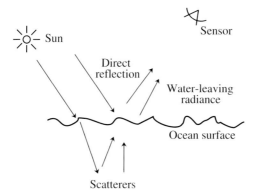

Figure 5.1. Comparison of the direct reflection of sunlight at the sea surface with the diffuse reflection associated with the water-leaving radiance generated by scattering in the water column.

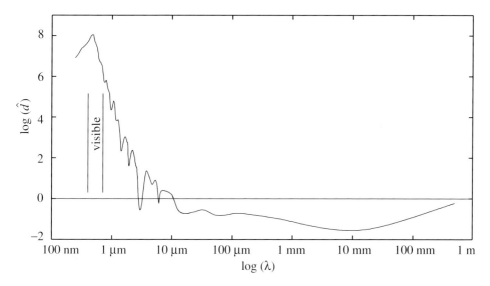

Figure 5.2. Wavelength dependence of the normalized attenuation depth $\hat{d}$ for 200 nm $< \lambda < 0.5$ m. The horizontal line shows where the absorption depth equals one wavelength.

For radiation propagating in the water column, and from Equation (3.13), the absorption depth d_a is

$$d_a = a(\lambda)^{-1} \qquad (5.1)$$

To give the magnitude of the absorption depth relative to the radiation wavelength, a non-dimensional depth is defined as $\hat{d} = d_a/\lambda$, and is plotted in Figure 5.2. The figure shows that $\hat{d}$ reaches its maximum in the blue at 475 nm and for $\lambda > 3$ μm, $\hat{d} < 1$. Because of the strong absorption at longer wavelengths outside of the VNIR, transmission in the water column is only important within a narrow window approximately centered on the visible wavelengths.

In the following, Section 5.2 discusses the reflection of an air-incident radiance from a flat ocean surface, then continues with reflection from an interface covered by waves. Reflection from these waves generates sun glint or glitter, which is the random reflection of sunlight into the sensor. For those wavelengths where radiation propagates to appreciable depths, Section 5.3 discusses the transmission of radiation across the interface. Section 5.4 describes the absorption and scattering properties of pure seawater, and the interaction of an incident radiance with the interface and the underlying water column. Section 5.5 describes the two kinds of remote sensing reflectances used in Chapter 6. Section 5.6 discusses reflection from foam.

5.2 The interface

For electromagnetic radiation incident on a surface, depending on the surface properties, some energy is reflected or scattered, some is absorbed and some is transmitted through the interface. For a radiance incident at a particular θ on a flat surface, the reflection is mirror-like or *specular*, meaning that the angle of incidence equals the angle of reflection. For a rough surface, the reflection process is more complicated. In the following, Section 5.2.1 defines roughness using a Rayleigh criterion, then describes reflection and scattering from smooth and rough surfaces. Three cases are considered. First, Section 5.2.2 discusses specular reflection from a flat surface and Section 5.2.3 gives the analytic solutions for specular reflection and transmission, which are called the Fresnel equations. Second, Section 5.2.4 describes reflection from an interface covered with capillary and short gravity waves. If this surface can be approximated as a mesh of small flat facets inclined at different angles to the horizontal where each facet serves as a specular reflector, then numerical solutions exist for the scattered radiance. Third, because the air bubbles associated with a foam-covered interface occur both at the surface and within the water column, Section 5.6 describes reflection from foam after discussion of the ocean interior.

5.2.1 General scattering considerations

For surface reflection and scattering, determination of whether the surface is rough or smooth depends on the Rayleigh roughness criterion. Following Rees (2001), Figure 5.3 shows a radiance incident on a surface, where σ_η is the rms surface height. In general, for a radiance incident at an angle θ and wavelength λ, the scattering is specular if

$$(\sigma_\eta \cos \theta)/\lambda < 1/8 \tag{5.2}$$

If (5.2) is satisfied, the surface is smooth, otherwise it is rough. Equation (5.2) shows that the scattering depends on three variables, σ_η, θ and λ. For σ_η and θ constant, as λ increases, the surface roughness becomes less important. For σ_η and λ constant, the roughness depends

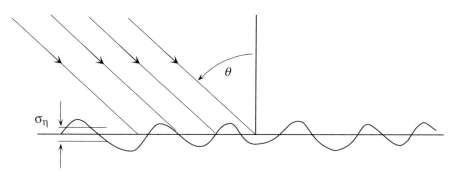

Figure 5.3. Geometry for discussion of the Rayleigh criterion for scattering and specular reflection from a surface (Adapted from Figure 3.10, Rees, 2001).

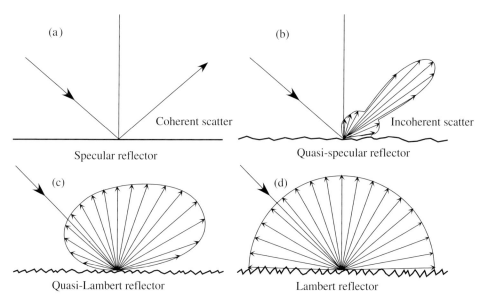

Figure 5.4. Limiting forms of reflection and scattering from a surface. (a) Specular reflection, (b) quasi-specular, (c) quasi-Lambertian, (d) Lambertian. The Lambertian example is only valid for an extended surface.

on θ. A surface that is rough at near vertical incidence angles can be smooth at near grazing angles. In the limiting case that Equation (5.2) is not satisfied at any angle, the reflection is Lambertian.

For four surfaces with increasing roughness, Figure 5.4 shows the reflection of an incident radiance. Figure 5.4a shows specular reflection from a perfectly flat surface, where the reflected energy propagates at an angle equal and opposite to the incidence angle. This is the case of pure *coherent* specular scattering or reflection, meaning that the reflected beam has specific phase relation with the incident radiance (Rees, 2001). For the small roughness case, Figure 5.4b shows that the reflection occurs partly by coherent scatter in the specular direction and partly by incoherent or diffuse scatter in all directions, where incoherent scatter has a random phase relation with the incident radiance. As the roughness increases, specular scatter decreases and incoherent scatter increases. For a rougher surface, Figure 5.4c shows that the scatter becomes quasi-Lambertian, meaning that most of the scatter is random, with only a small coherent component in the specular direction. Finally, Figure 5.4d shows the idealized case of reflection from a perfectly rough surface, where the reflection is completely Lambertian. In the visible/infrared, examples of surfaces that can be approximated as Lambertian include foam and clouds.

5.2.2 Specular reflection and transmission at a planar interface

Consider the ideal case of a flat planar interface as in Figure 5.4a, with air above and water below, where the properties in each medium vary only with distance upward or downward from the interface. This idealized physical situation applies to both a flat surface and a rough surface approximated as a large number of small facets. Following Mobley (1994), the air/water interface is assumed to be an infinitesimally thin slab, across which the real part of the index of refraction changes in a stepwise manner from its atmospheric value to its water value. Also from Mobley, the incident radiance is assumed to interact linearly with the interface, so that the magnitude of the reflected and transmitted radiances increase linearly with that of the incident radiance, and nonlinear effects such as frequency doubling do not occur. The discussion is restricted to the macroscopic, so that photon–atom interactions at the surface are not considered. Finally, the ocean is assumed to be sufficiently thick such that all of the transmitted radiation is absorbed before reaching the ocean bottom. For these conditions, the properties of the reflected and transmitted radiation are derived from the real and imaginary parts of the indices of refraction of air and water.

Figure 5.5 shows the familiar figure for the specular reflection and transmission of a narrow beam of radiation incident on a flat, planar interface, where the term "narrow beam" means that the incident radiance occupies a small solid angle. The upper half plane is air; the lower half is water. The task is to describe the reflection and refraction of this beam as it intersects the ocean surface. This description divides into two parts: the geometry of the interaction with the surface and the relative magnitudes of the reflected and refracted radiances given by the Fresnel relations.

On the figure, n_a and n_w are the real part of the index of refraction for air and water and θ_i, θ_r and θ_t are respectively the angles of incidence, reflection and transmission. Snell's law describes the geometry of the reflection, where the incidence and reflection angles are equal and opposite, so that $\theta_i = -\theta_r$. The transmitted radiance is refracted to an angle θ_t, given by

$$n_w/n_a \equiv n = \sin\theta_i/\sin\theta_t \qquad (5.3)$$

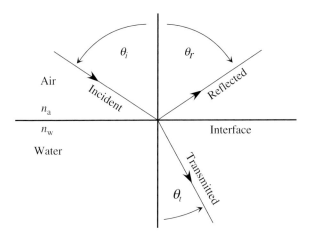

Figure 5.5. Reflection and refraction at a specular interface for an air-incident radiance.

In (5.3) and to simplify the following discussion, n is set equal to the ratio of the refraction indices. For the visible wavelengths where $n_a = 1$ and $n_w \cong 1.34$, the solution for θ_t is found by setting $n = 1.34$. Because $n > 1$ for an air-incident radiance, $\theta_t < \theta_i$.

5.2.3 Fresnel equations

The Fresnel equations give the magnitudes of the reflected and transmitted radiances relative to the magnitude of the incident radiance. The following discussion first considers an unpolarized radiance incident on a specular surface, then the case of V-pol and H-pol incident radiances. For the unpolarized case, the *radiance reflectance* $r(\lambda, \theta_r)$ is defined as the ratio of the reflected and incident radiances:

$$r(\lambda, \theta_r) = L_r(\lambda, \theta_r)/L_i(\lambda, \theta_i) \tag{5.4}$$

From the Fresnel equations, $r(\theta_i)$ can be written as the following function of θ_i and θ_t (Born and Wolf, 1999; Mobley, 1994):

$$r(\theta_i) = (1/2)([\sin(\theta_i - \theta_t)/\sin(\theta_i + \theta_t)]^2 + [\tan(\theta_i - \theta_t)/\tan(\theta_i + \theta_t)]^2) \tag{5.5}$$

In (5.5), $\theta_i \neq 0$ and θ_i and θ_t are related by Snell's law. When the incident radiance is normally incident, the reflectance becomes

$$r(0) = (n - 1)^2/(n + 1)^2 \tag{5.6}$$

For reflection of a polarized incident radiance, the Fresnel relations are given by the V-pol and H-pol reflection coefficients $\rho_V(\theta_i)$ and $\rho_H(\theta_i)$ defined in a similar manner to Equation (5.4). If the subscript on θ_i is omitted for brevity, then for $\theta > 0$, these coefficients are written as follows (Stewart, 1985; Born and Wolf, 1999):

$$\rho_H(\theta) = [(p - \cos\theta)^2 + q^2]/[(p + \cos\theta)^2 + q^2]$$
$$\rho_V(\theta) = [(\varepsilon'\cos\theta - p)^2 + (\varepsilon''\cos\theta + q)^2]/[(\varepsilon'\cos\theta + p)^2 + (\varepsilon''\cos\theta + q)^2] \tag{5.7}$$

In (5.7), $\varepsilon' = n^2 - \chi^2$ and $\varepsilon'' = 2n\chi$ are the real and imaginary parts of the complex dielectric constant defined in Equation (3.8). The terms p and q are given by

$$p = (1/\sqrt{2})[(\varepsilon' - \sin^2\theta)^2 + \varepsilon''^2]^{1/2} + [\varepsilon' - \sin^2\theta]^{1/2}$$
$$q = (1/\sqrt{2})[(\varepsilon' - \sin^2\theta)^2 + \varepsilon''^2]^{1/2} - [\varepsilon' - \sin^2\theta]^{1/2} \tag{5.8}$$

For normal incidence, $\theta = 0$ and the concept of V and H polarization loses its meaning, so that from (5.6), $\rho_H(0) = \rho_V(0) = r(0)$.

For $n = 1.34$ and for an air-incident radiance, Figure 5.6 shows the θ-dependence of r, ρ_V and ρ_H. The figure shows that the polarized reflectances lie above and below r, and that $\rho_V = 0$ at $\theta \cong 60°$, called the Brewster angle. A useful property of the reflectance is that for $\theta \leq 50°$, r is nearly constant at $r \cong 0.02$, so that for these angles, about 98% of the incident radiance is transmitted. For $\theta > 50°$, r increases rapidly.

5.2.4 Reflection from capillary waves

The Fresnel reflectance of radiation from a wind-roughened ocean surface generates sun glint, which refers to the scattering of the incident solar radiance from the rough surface into

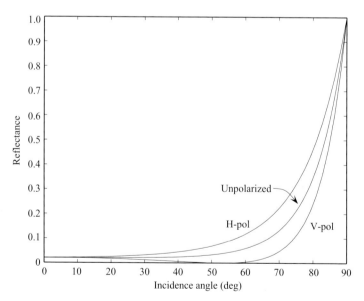

Figure 5.6. Radiance reflectance from a specular air/water interface as a function of incidence angle for the visible wavelengths, and for unpolarized, H-pol and V-pol incident radiance.

the sensor direction. Because at all observational wavelengths, sun glint can overwhelm the reflected or emitted ocean surface radiance, it must be avoided or masked. Accounting for sun glint is done as follows.

For a wind-roughened ocean surface, the concept of a unique reflection angle loses its meaning. To address this problem, Mobley (1995, 1999) assumes that the ocean surface is covered only by wind-driven capillary waves with their slopes described by a wind-speed dependence similar to that given in Equations (2.6). He further assumes that the wave surfaces can be approximated as a collection of congruent isosceles triangles, called facets, each of which serves as a specular reflector. This approximation is valid if the length scale of each facet is much greater than λ and if the deviation of the approximating facet from the wave surface is much less than λ (Rees, 2001). Equivalently, the approximation holds if the radius of curvature R_c of that part of the surface approximated by the facet satisfies

$$R_c \gg \lambda \tag{5.9}$$

For this case, (5.9) is satisfied, the radiation fields at the surface can be approximated by the fields that occur at a tangent plane (Valenzuela, 1978; Wu and Smith, 1997). Since the short water wavelengths λ_w and their curvatures are in the range 1–10 cm, while in the VIR the radiation wavelengths λ are less than 10 µm, this condition is easily satisfied. In contrast and as Sections 9.4 and 10.6.2 discuss for the passive and active microwave, because of the longer microwave wavelengths, this condition is not necessarily satisfied. Instead, scattering takes place from the short surface wavelengths that do not satisfy (5.9), and reflection takes place from the larger elements of surface area.

For a narrow beam of radiance incident on the surface at some θ in the VIR and assuming that Fresnel reflection occurs from each facet and that multiple reflections occur among the facets, Mobley (1999) numerically solves for the angular distribution of the reflected radiances. For a source radiance at $\theta = 40°$ in an area that Mobley calls a *quad*, measuring 10° in zenith by 15° in azimuth, and for wind speeds $U = 0, 2, 5$ and 10 m s^{-1}, Figure 5.7 shows the resultant distributions of the reflected radiances. For $U = 0$ or specular reflection, the figure shows the distribution on a hemisphere of the reflected radiances for 100 ray paths drawn from the source quad and reflected at the origin, where each dot shows the angular location of a single reflection. For this case, all of the incident radiances are reflected into the conjugate quad located at the equal and opposite angle from the incident radiance. Also, when the sun is the source radiance and for an observer looking at the surface along the conjugate angle, the reflection of the sun at the origin is sometimes called the sub-solar spot. For the other three cases, 5000 ray paths are used to construct the figures. These show that as the wind speed and capillary roughness increase, the angular extent of the reflected radiances increases, while the distribution of the reflected rays remains centered on the conjugate quad. For $U = 10$ m s^{-1}, the angular distribution of the reflected rays extends almost 120° in azimuth angle and slightly more than 90° in zenith angle.

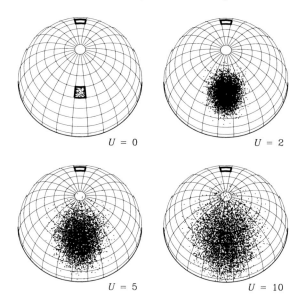

Figure 5.7. The angular distribution of the reflection of a radiance incident on a surface with its roughness proportional to wind speed. The source radiance is located on the far side of the hemisphere at $\theta = 40°$. Each box on the hemisphere represents an area 10° in zenith angle and 15° in azimuth (Figure 2 from Mobley, 1999, © Optical Society of America, used with permission).

By the principle of reciprocity, the opposite is also true, in that a sensor located in the solid angle defined by the source quad will, depending on wind speed, observe radiances from all of the darkened solid angles. For an instrument operating at any of the observing wavelengths, as Chapters 6, 7 and 9 respectively discuss for the visible, infrared and microwave, then depending on wind speed and the look and sun angles, the surface can directly reflect sunlight into the sensor. This means that either in the instrument design or in its data processing, care must be taken to avoid direct solar reflection and sun glint.

5.3 Transmission across an interface

For the visible spectrum only, the backscatter of sunlight within the water column creates an upward propagating irradiance incident on the interface from below. As this irradiance crosses the interface, it generates a water-leaving radiance. Because an understanding of these interfacial processes is critical to biological remote sensing, this section discusses the change in properties associated with a radiance incident on an interface both from above, called an *air-incident* radiance, and from below, called a *water-incident* radiance (Mobley, 1995). In the following, Section 5.3.1 discusses the case of a water-incident radiance on a specular surface and Section 5.3.2 discusses refractive convergence and divergence. Then with some approximations, Section 5.4 extends these concepts to the real ocean.

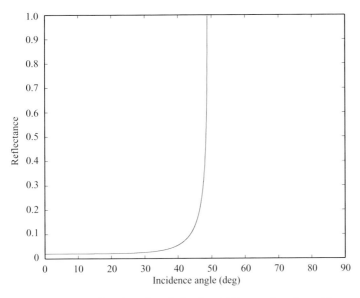

Figure 5.8. The reflectance as a function of angle for the visible wavelengths and for an unpolarized radiance incident on the water/air interface from below.

5.3.1 Radiance incident from below

For a radiance incident on a specular interface from below, the propagation direction in Figure 5.5 is reversed, and Snell's law becomes

$$\sin \theta_i / \sin \theta_t = 1/n = 0.75 \tag{5.10}$$

For an unpolarized radiance incident on the interface from below and from Equation (5.10), Figure 5.8 shows the dependence of r on incidence angle. For $\theta_i \leq 30°$, the figure shows that r is nearly constant at about 0.02, then at $\theta_i = 49°$ rises abruptly to $r = 1$, so that for $\theta_i > 49°$, total reflection occurs. Consequently, on the atmosphere side of the interface, a radiance incident on the interface from below at $\theta_i = 49°$ is refracted to $\theta_t = 90°$ so that the radiance is parallel to the interface. For $\theta_i > 49°$, the radiances are internally reflected and there is no transmission.

For another view of this phenomenon, Figure 5.9 shows the refraction of several ray paths incident at different angles on a flat ocean/atmosphere interface. The figure shows that the total reflection that occurs for $\theta_i > 49°$ gives rise to a *shadow zone*, which is the solid angle region within which incident radiances are totally reflected. Because radiances incident from below within the shadow zone do not propagate across the interface, if the irradiance incident from below has a Lambertian distribution, then almost half of the incident radiation does not cross the interface. The figure also suggests that light propagating upward within a narrow solid angle toward the interface is defocused into a large solid angle in the atmosphere and vice versa. This focusing and defocusing of radiances at the interface is

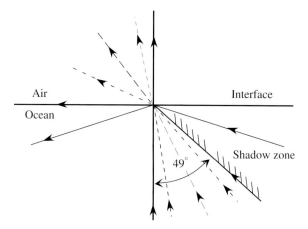

Figure 5.9. Radiation incident from below on the ocean surface. Because of the change in n across the interface, radiances at $\theta > 49°$ are internally reflected and do not propagate across the interface, giving rise to a shadow zone within which all incident radiances are reflected.

respectively called *refractive convergence* or *divergence*. Because of the importance of this phenomenon to the water-leaving radiance, the next section discusses it in detail.

5.3.2 Refractive convergence and divergence

For a narrow beam of unpolarized radiance incident from below on a planar interface separating two media with indices of refraction n_1 and n_2, this section derives the relation between the radiances on both sides of the interface.

Assume that a radiance is incident at an angle θ_1 on an area ΔA_S at the interface (Figure 5.10). A fraction of the incident radiant flux is transmitted at an angle θ_2. If $T(\theta_1) = 1 - r(\theta_1)$ is the unpolarized interface transmittance, where r is given in Figures 5.6 and 5.8, at the interface the relation between the radiant fluxes Φ_1 and Φ_2 is

$$\Phi_2 = T(\theta_1)\Phi_1 \tag{5.11}$$

For incidence angles less than about 40° and for an upward or downward radiance respectively incident on either air or water, $T \cong 0.98$. The corresponding radiances propagate within the associated solid angles $\Delta\Omega_i$, where $i = 1, 2$ indicates different sides of the interface. Using the form of the radiance from Equation (3.16) and referring to Figure 5.10, Equation (5.11) may be written as

$$L_2 \cos\theta_2 \, \Delta\Omega_2 = T(\theta_1)L_1 \cos\theta_1 \, \Delta\Omega_1 \tag{5.12}$$

By definition,

$$\Delta\Omega_i = \sin\theta_i \, \Delta\theta_i \, \Delta\phi_i \tag{5.13}$$

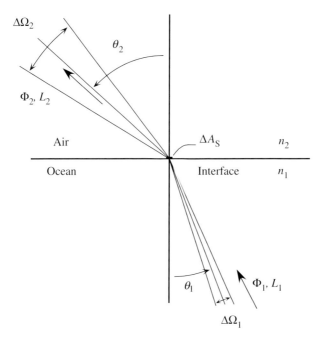

Figure 5.10. An upward propagating radiance focused on an element of surface area at the ocean/air interface, and the transformation of this radiance through refractive divergence as it crosses the interface.

For both sides of the interface, calculation of $\Delta\Omega_i$ and the radiances proceeds as follows. Because the azimuthal angles ϕ_i lie in the plane of the interface, they are independent of Snell's law and $\Delta\phi_1 = \Delta\phi_2$. The relation between θ_i is determined from Snell's law in Equation (5.3), so that

$$\sin\theta_1 = (n_2/n_1)\sin\theta_2 \qquad (5.14)$$

Squaring both sides of (5.14), differentiating, and substituting $\Delta\Omega_i$ from Equation (5.13) into the result yields

$$\Delta\Omega_1 \cos\theta_1 = (n_2/n_1)^2 \Delta\Omega_2 \cos\theta_2 \qquad (5.15)$$

Equation (5.15) gives the relation between the solid angles, incidence angles and the refraction indices on both sides of the interface and is called Staubel's invariant (Mobley, 1994, p. 160).

Substitution of Staubel's invariant in (5.15) into (5.12) gives

$$L_2 = (n_2/n_1)^2 T L_1 \qquad (5.16)$$

For $T = 1$, Equation (5.16) is called the fundamental theorem of radiometry (Mobley, 1994, p. 161). For the visible wavelengths and the case shown in Figure 5.10, where L_1 is the upward radiance just beneath the surface, L_2 is the water-leaving radiance, $n = n_1/n_2 = 1.34$, and for $\theta_1 < 40°$ or $T \cong 0.98$, Equation (5.16) becomes

$$L_2 = TL_1/n^2 \cong 0.55L_1 \tag{5.17}$$

For the opposite case of an air-incident radiance at $\theta < 50°$, Equation (5.16) can be rederived to give

$$L_1 = n^2 TL_2 \cong 1.76L_2 \tag{5.18}$$

where L_1 and L_2 are again respectively the radiances in the water and air. Examination of (5.17) and (5.18) shows that for air-incident radiance, the transmitted radiance is reduced almost by half, whereas for the water-incident radiance, it is nearly doubled. This illustrates an important difference between the transmitted radiant flux and the radiance, in that for the water-incident case and from Equation (5.11), the transmitted flux is reduced only by T or by a factor of 0.98, but because this energy propagates within a larger solid angle on the atmosphere side of the interface, L_2 from Equation (5.17) is reduced by 0.55. For the opposite case of an air-incident radiance, the radiant flux is again reduced by 0.98, while from Equation (5.18), the transmitted radiance is nearly doubled. This enhancement is one reason never to look at the sun from underwater, since its radiance is nearly doubled.

5.4 Absorption and scattering properties of seawater

For the visible and near visible wavelengths, this section describes the absorption and scattering of light within the water column. Specifically, Section 5.4.1 discusses the absorption and scattering properties of pure seawater. Section 5.4.2 describes how the absorption and scattering interact with the downwelling solar irradiance to generate an upwelling irradiance. Section 5.4.3 describes how the interface modifies this irradiance to produce in the atmosphere a water-leaving radiance $L_w(\lambda, \theta)$.

The oceanic optical and remote sensing properties divide into *inherent optical properties* (IOP) and *apparent optical properties* (AOP) (Mobley, 1994, 1995). The inherent properties depend only on the nature of the medium and include the absorption, scattering and attenuation coefficients and the Fresnel reflectances. The apparent optical properties depend on the medium and on the directional structure of the ambient light field, and include the irradiance reflectance and the diffuse transmittance. Even though the oceanic terminology used to describe extinction, absorption, and scattering differs from that used in the atmosphere, the mathematical formulation remains the same. In the ocean, $a(\lambda)$ is the volume absorption coefficient, $b(\lambda)$ is the scattering coefficient and $c(\lambda)$ is the attenuation coefficient, corresponding to the atmospheric extinction coefficient. The coefficients a, b and c have units of m^{-1}. For the volume scattering function, the terminology recommended by the International Association for Physical Sciences of the Ocean (IAPSO) is $\beta(\alpha, \lambda)$, with units of $m^{-1}\,sr^{-1}$, which is the same symbol used in the

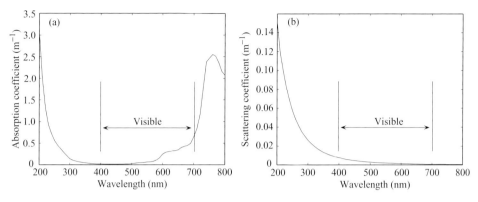

Figure 5.11. The wavelength dependence of (a) the absorption and (b) the scattering coefficient for pure seawater. The vertical lines mark the visible spectrum (Data from Smith and Baker, 1981, courtesy Curt Mobley).

atmosphere (Mobley, 1995). Further, the atmospheric scattering phase function $\rho(\alpha)$ is written for the ocean as $\tilde{\beta}(\alpha)$, also with units of sr^{-1} (Mobley, 1994). In the next chapter, subscripts will be added to these variables to distinguish between the case of absorption and scattering from pure seawater and from seawater containing suspended and dissolved material.

5.4.1 Properties of pure seawater

For pure seawater, Figure 5.11 shows the dependence on wavelength of the absorption and scattering coefficients for the wavelength range 200–800 nm, where the vertical lines show the boundaries of the visible spectrum. Figure 5.11a shows that the absorption has a minimum between 300 and 600 nm, and increases rapidly at shorter and longer wavelengths. Within the visible wavelengths, the absorption minimum is offset toward the UV, and absorption increases within the red (600–700 nm) and at the very short UV wavelengths. For comparison, Figure 5.11b shows the volume scattering coefficient, where in contrast to absorption, scattering rapidly decreases with increasing wavelength.

Mobley (1994, p. 103) gives the volume scattering function, the phase function and the scattering coefficient of seawater as

$$\beta(\alpha, \lambda) = 4.72 \times 10^{-4}(\lambda_0/\lambda)^{4.32}(1 + 0.835 \cos^2 \alpha) \text{ m}^{-1}\text{sr}^{-1} \tag{5.19}$$

$$\tilde{\beta}(\alpha) = 0.06225(1 + 0.835 \cos^2 \alpha) \text{ sr}^{-1} \tag{5.20}$$

$$\text{and} \quad b(\lambda) = 16.06(\lambda_0/\lambda)^{4.32}4.72 \times 10^{-4} \text{ m}^{-1} \tag{5.21}$$

In the above, $\lambda_0 = 400$ nm. Comparison of Equations (5.20) and (5.21) with the Rayleigh scattering described in Equation (4.22) shows that for both the atmosphere and ocean, molecular scattering strongly increases with decreasing wavelength. Given these

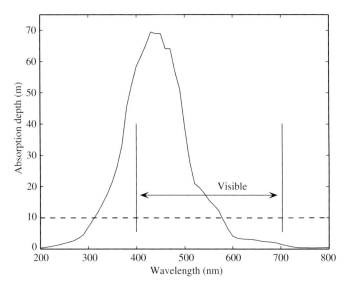

Figure 5.12. The absorption depth for seawater; the horizontal dashed line shows the 10-m absorption depth (Data from Smith and Baker, 1981, courtesy Curtis Mobley).

similarities, oceanic scattering is sometimes incorrectly described as Rayleigh scattering. But because seawater is a thousand times denser than air, its scattering properties are derived from the Einstein–Smoluchowski consideration of small-scale fluctuations in the liquid, or very differently from Rayleigh scattering (Mobley, 1994).

Within the water column, the attenuation depth d_a has a strong λ-dependence. For 200 nm $\leq \lambda \leq 800$ nm, Figure 5.12 shows from (5.1), the dependence of d_a on λ, where the horizontal dashed line marks the 10-m depth. From the figure, d_a reaches its maximum in the blue at about 430 nm, or just above the short wavelength edge of the visible spectrum, then decreases rapidly for both longer and shorter wavelengths. For $320 < \lambda < 570$ nm, d_a is greater than 10 m, so that light penetration into the ocean is strongly biased toward the blue-green wavelengths. At 640 nm, $d_a = 3$ m, and at 750 nm, $d_a = 0.4$ m, so that d_a decreases rapidly with increasing wavelength.

The preceding equations allow the verification of Raman's (1922) luminosity argument in Section 5.1. Equation (4.56) shows that the skylight due to Rayleigh scatter is proportional to $\tau_R(\lambda)$, which as Equation (4.9) shows, equals the integration of the volume scattering coefficient across the atmosphere. For the ocean, $b(\lambda)$ from Equation (5.21) is now used to define an optical thickness analogous to τ_R. For $\lambda = 400$ nm and assuming that $b(400)$ is independent of depth, if b is integrated over the top 50 m of the water column, then from Equation (4.9), $\tau_w(400) = 0.379$. Comparison of this number with the Rayleigh optical depths in Table 4.2 shows that the τ_w derived from a 50-m deep water column approximately equals the τ_R generated from the entire height of the atmosphere. Thus for a cloud-free sky, the water surface and sky should be equally bright.

5.4.2 Irradiance reflectance

This and the following section first define the plane irradiance reflectance $R(\lambda, z)$, then use this reflectance to relate the water-leaving radiance to the incident solar irradiance and the seawater scattering and absorption properties. Although the concept of irradiance is much simpler than radiance, because $R(\lambda, z)$ is an apparent optical property that depends on the directional dependence of the incident light field, it is more difficult to calculate. Following Zaneveld (1995, Equation 27), $R(\lambda, z)$ is defined as the ratio of the upwelling plane irradiance $E_u(z)$ to the downwelling irradiance $E_d(z)$:

$$R(\lambda, z) = E_u(\lambda, z)/E_d(\lambda, z) \tag{5.22}$$

Just below the interface, the irradiance reflectance is given by $R(\lambda, 0_-)$, where $z = 0_-$ refers to the water side of the interface, and in practical terms, to a sufficient distance below the interface so that water waves do not cause the optical measuring device to emerge through the surface.

The reflectance $R(\lambda, 0_-)$ can be thought of as a hypothetical reflector located just below the ocean surface that represents all of the scattering and absorption processes occurring in the water column. Its location separates the reflector from the problem of transmission through the interface (Figure 5.13). This reflectance is measured directly with spectral radiometers and as the next chapter shows, is a function of such water properties as the concentrations of chlorophyll and suspended sediments. Of equal importance, $R(\lambda, 0_-)$ is directly related to the water-leaving radiance, which can be measured by aircraft or satellite. To derive the properties of this reflector, the *spectral backward scattering coefficient* $b_b(\lambda)$ is next defined (Mobley, 1995).

Similar to the definition of the atmospheric scattering coefficient in Equation (4.16), $b_b(\lambda)$ is the integral of $\beta(\alpha, \lambda)$ in (5.19) over the upper half plane, or over $\pi/2 \leq \alpha \leq \pi$, where

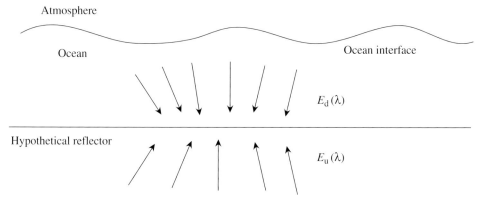

Figure 5.13. The location of and terminology for a hypothetical irradiance reflector located just beneath the interface.

π is the backscatter direction, so that

$$b_{\mathrm{b}}(\lambda) = 2\pi \int_{\pi/2}^{\pi} \beta(\alpha, \lambda) \sin \alpha \, d\alpha \qquad (5.23)$$

The reason for introducing b_{b} is that in combination with $a(\lambda)$, it leads to a conceptually simple model for $R(\lambda, 0_-)$. In the water column, radiative processes are to first order a balance between absorption and scattering. If a downwelling photon is absorbed, it cannot be scattered, but if a photon is backscattered by suspended material or water molecules, it becomes an upwelling photon (Mobley, 1994). The simplest model of this process assumes that $R(\lambda, 0_-)$ is directly proportional to b_{b} and inversely proportional to a, since for the case of large backscatter and small absorption, a strong upwelling irradiance is more likely than for the opposite case (Zaneveld, 1995; Mobley, 1994, pp. 493–496; Roesler and Perry, 1995). Thus to first order,

$$R(\lambda, 0_-) \equiv R(\lambda) \sim b_{\mathrm{b}}(\lambda)/a(\lambda) = Gb_{\mathrm{b}}(\lambda)/a(\lambda) \qquad (5.24)$$

In (5.24), G is a constant that depends on the nature of the incident light field and the volume scattering function, and in the following discussion, $R(\lambda)$ replaces $R(\lambda, 0_-)$. For a flat sea surface and the sun at zenith, $G = 0.33$ (Gordon, Brown and Jacobs, 1975). The above authors discuss various refinements of this simple model; Carder *et al.* (1999) use Equation (5.24) in the chlorophyll retrieval algorithms described in Chapter 6.

Calculation of $R(\lambda)$ for pure seawater proceeds as follows. For this case, the scattering coefficient is symmetric in the forward and backward directions, so that $b_{\mathrm{b}} = 0.5b$. Therefore, given the absorption a from Figure 5.11, $R(\lambda)$ is easily calculated from (5.24) and is shown in Figure 5.14. The figure shows that for λ greater than about 550 nm, the reflectance is near

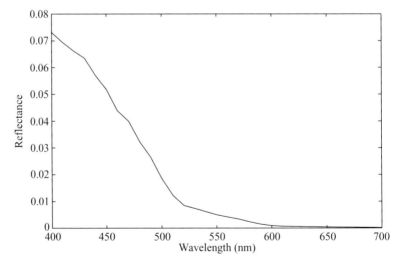

Figure 5.14. The sub-surface reflectance of pure seawater, calculated from Equation (5.24) and from data in Smith and Baker (1981).

zero; as λ decreases, R rises to its peak at 400 nm. Because the peak in the solar radiance occurs at about 490 nm, the upwelling radiance generated by the product of the solar irradiance with $R(\lambda)$ lies between 400 and 490 nm and has the color blue.

5.4.3 Water-leaving radiance

As defined in Equation (4.56), $E_d(0_+)$ is the solar irradiance just above the interface. Therefore, the solar irradiance just below the interface $E_d(0_-)$ is approximately given by

$$E_d(0_-) = T E_d(0_+) \tag{5.25}$$

For the solar zenith angle $\theta_S < 50°$ and $U < 16$ m s^{-1}, Kirk (1996, Figure 2.10) shows that T is nearly constant at about 0.98, so that $E_d(0_-)$ is a linear function of $E_d(0_+)$. For larger θ_S, Kirk shows that T is a function of θ_S and U.

Given $E_d(0_+)$ and $R(\lambda)$, Equations (5.22) and (5.25) can be solved for $E_d(0_-)$. If the radiance distribution within $E_d(0_-)$ is assumed to be Lambertian, then from Equation (3.17) the upwelling radiance just beneath the surface is $L_{up}(\lambda) = E_u(0_-)/\pi$, so that

$$L_{up}(\lambda) = R(\lambda)T E_d(\lambda, 0_+)/\pi \tag{5.26}$$

Substitution of Equation (5.26) into (5.17) shows that just above the surface, the water-leaving radiance $L_w(\lambda)$ can be written as

$$L_w(\lambda) = T^2 R(\lambda)E_d(\lambda, 0_+)/n^2 Q \cong 0.54 R(\lambda)E_d(\lambda, 0_+)/Q \tag{5.27}$$

In (5.27), the factor T^2 occurs because the solar radiation crosses the interface in the downward direction and the upwelled radiation crosses it in the upward direction. In the right-hand term, the quantity 0.54 is derived from $T = 0.98$ and $n = 1.34$, again for the visible wavelengths. The factor of Q replaces π because the presence of the shadow zone means that $E_u(0_-)$ is not necessarily Lambertian. For a range of θ_S and sea states, Q ranges from 3 to 6 (Mobley, 1994, p. 495).

Equation (5.27) gives the water-leaving radiance in terms of the downwelling radiance and the reflectance. $L_w(\lambda)$ is further refined by definition of the *normalized water-leaving radiance* $[L_w(\lambda)]_N$, which is sometimes written as nL_w. In the derivation of $[L_w(\lambda)]_N$, $L_w(\lambda)$ is rewritten from substitution of $E_d(0_+)$ from (4.56) into (5.27), yielding

$$L_w(\lambda) = [T^2 R(\lambda)F_S(\lambda)/n^2 Q]\exp[(-\tau_R(\lambda)/2 + \tau_{OZ})\sec\theta_S]\cos\theta_S \tag{5.28}$$

In (5.28), the terms in the left-hand square brackets are nearly independent of θ_S; the other terms are the product of $\cos\theta_S$ and the exponential term, which is the single scattering diffuse transmittance $t_D(\lambda, \theta_S)$ defined in Equation (4.55). Following Gordon and Clark (1981), $[L_w(\lambda)]_N$ is derived by division of $L_w(\lambda)$ in (5.28) by $t_D(\lambda, \theta_S)\cos\theta_S$ so that

$$[L_w(\lambda)]_N = L_w(\lambda)/t_D(\lambda, \theta_S)\cos\theta_S = T^2 R(\lambda)F_S(\lambda)/n^2 Q \tag{5.29}$$

Comparison of Equations (5.28) and (5.29) shows that $[L_w(\lambda)]_N$ is independent of θ_S and can be thought of as the radiance that exits the ocean for the case of a non-attenuating atmosphere and the sun at zenith (Gordon and Voss, 1999). The importance of $[L_w(\lambda)]_N$ is that it removes the θ_S term from L_w and permits comparison of water-leaving radiance measurements made at different sun angles. For this reason, $[L_w(\lambda)]_N$ is frequently used in the literature (Gordon and Voss, 1999; Gordon and Clark, 1981).

5.5 Two kinds of remote sensing reflectance

From the above, the following defines two kinds of remote sensing reflectances. The first is the *remote sensing reflectance R_{rs}*, which is the ratio of $L_w(\lambda)$ to the solar irradiance at the surface and is a linear function of $R(\lambda)$:

$$R_{rs}(\lambda) = \frac{L_w(\lambda)}{E_d(\lambda, 0_+)} \cong \frac{[L_w(\lambda)]_N}{F_S(\lambda)} = \frac{T^2 R(\lambda)}{n^2 Q} = \frac{GT^2 b_b(\lambda)}{n^2 Q a(\lambda)} \tag{5.30}$$

In (5.30), the fourth term is derived by substitution of (4.56) and (5.24) into (5.28). $E_d(\lambda, 0_+)$ is either directly measured or taken from Equation (4.56); the right-hand term is exact for single scattering. R_{rs} is calculated just above the surface and because of its application to the analysis of shipboard observations, is also frequently used in the literature.

Second, from Gordon and Voss (1999), the *extraterrestrial reflectance $\rho_w(\lambda)$* is the ratio of $L_w(\lambda)$ to the solar irradiance at the TOA, as in

$$\rho_w(\lambda) = \pi L_w(\lambda)/F_S(\lambda) \cos \theta_S = \pi T^2 R(\lambda) t_D(\lambda, \theta_S)/n^2 Q \tag{5.31}$$

where the factor of π converts the solar irradiance to units of radiance and the third term is derived from substitution of $L_w(\lambda)$ from (5.28). The term ρ_w is the ratio of the water-leaving radiance to the extraterrestrial solar radiance; its advantages are that it is dimensionless and that future ocean color instruments may be calibrated in terms of reflectance (Gordon and Wang, 1994a).

The analogous normalized reflectance $[\rho_w(\lambda)]_N$ is defined as

$$[\rho_w(\lambda)]_N = \rho_w(\lambda)/t_D(\lambda, \theta_S) = \pi [L_w(\lambda)]_N/F_S(\lambda) = \pi T^2 R(\lambda)/n^2 Q \tag{5.32}$$

which is the reflectance for the sun at zenith. The right-hand side of the equation shows that $[\rho_w(\lambda)]_N$ is a function of only the surface parameters and the irradiance reflectance. Finally, from Equations (5.30) and (5.32) and to a good approximation, the relation between the remote sensing reflectance R_{rs} and $[\rho_w]_N$ is $[\rho_w]_N = \pi R_{rs}$ (Gordon and Voss, 1999). Both ρ_w and R_{rs} will be used in the discussion of ocean color retrieval in Chapter 6, and with a variety of subscripts, ρ will be used as the ratio of any surface or atmospheric radiance to the solar irradiance.

5.6 Reflection from foam

The problem of reflection from the sporadic foam patches generated by breaking waves is a current topic of field and laboratory research. From Gordon and Wang (1994b), foam is generally assumed to be a Lambertian reflector with an irradiance reflectance $R_F(\lambda)$. Within an instrument pixel, the irradiance reflected from foam is the product of $R_F(\lambda)$ and the area covered by whitecaps, so that determination of the reflectance requires knowledge of both the areal foam extent and $R_F(\lambda)$.

Frouin *et al.* (1996) and Moore *et al.* (2000) show that $R_F(\lambda)$ decreases with increasing wavelength. This occurs because foam consists of small volumes of air contained within a fine lattice of seawater on the surface and of bubbles entrained into the near surface water column (Section 2.2.2), so that reflection from foam has both surface and subsurface components. In the visible, both contribute to the reflectance, while in the NIR and because of increased seawater absorption, the subsurface contribution is greatly reduced. From field observations in the surf zone, Frouin *et al.* (1996) find that $R_F(\lambda) = 0.40$ for $400 \leq \lambda \leq 650$ nm, while R_F is reduced by 40% to 0.25 at $\lambda = 850$ nm, by 50% at 1.02 μm, and by 85% to 0.06 at 1.65 μm. From instruments deployed on a 6000-km ship traverse across the equatorial Pacific Ocean, Moore *et al.* (2000) confirm this result. They also show that on the scale of a satellite pixel and for a wind speed range of 9–12 m s^{-1}, the additional satellite reflectance associated with foam in the visible is in the range 0.001–0.002, with no apparent dependence on wind speed. These foam properties are incorporated into the ocean color algorithms described in the next chapter.

5.7 Further reading and acknowledgements

This chapter draws heavily on the first four chapters of Mobley (1994) and on the Mobley (1995) article in the *Handbook of Optics*. The chapter also uses some of the definitions presented in Gordon and Voss (1999). I thank Curtis Mobley for his assistance and especially for provision of several of the data sets used in the figures.

6

Ocean color

6.1 Introduction

Satellite observations in the visible and near infrared allow the measurement of the oceanic chlorophyll a, the principal photosynthetic pigment associated with land surface and oceanic plants. Most oceanic plants are microscopic single- or multi-celled free-floating plants called algae or *phytoplankton*, from the Greek *phyton*, meaning plant, and *planktos*, meaning wandering (Jeffrey and Mantoura, 1997). Phytoplankton use photosynthesis to fix inorganic carbon into organic forms of carbon such as carbohydrates. They reproduce asexually, are globally distributed, consist of tens of thousands of species and make up about 25% of the total planetary vegetation (Jeffrey and Mantoura, 1997). Jeffrey and Vesk (1997) and Lalli and Parsons (1993) provide an introduction to the kinds and variety of phytoplankton species.

Globally, phytoplankton play at least two roles. First, following Jeffrey and Mantoura (1997), phytoplankton make up the base of the marine food web. Small oceanic animals called zooplankton derive their energy by grazing on the phytoplankton. In turn, larger species of fish and mammals consume the zooplankton. Second, because they can fix inorganic carbon and convert solar to chemical energy, phytoplankton contribute to the global carbon cycle. As the phytoplankton increase in number and mass, they convert the CO_2 in the upper water column to organic carbon. Their rate of growth and of carbon fixation is called *primary production* and is measured using radiocarbon techniques, with typical units of μg of carbon m^{-3} s^{-1}.

As the phytoplankton die, they sink into the abyss and sequester carbon in the deep ocean, in a process called the biological pump. Because of fossil fuel consumption, the carbon cycle is out of balance, with the excess CO_2 transported into the ocean and atmosphere. In the atmosphere, CO_2 increases the opacity of the thermal-infrared windows and thereby contributes to global warming. Carbon fixation by phytoplankton transfers some of this excess atmospheric carbon from the upper to deep ocean. Given the concerns about the imbalance of the carbon cycle, and additional concerns about feeding the growing human population and determining the carrying capacity of the planet, there is an immediate need to determine the oceanic global and regional distribution of chlorophyll and primary productivity.

Measurements of ocean color from space depend on the small scale nature of photosynthesis. Most oceanic carbon is inorganic; the photosynthetic pigments within each phytoplankton cell make possible the reduction or *fixation* of carbon dioxide into organic carbon, so that solar energy is converted to chemical energy, with oxygen as a by-product. These pigments consist of the ubiquitous chlorophyll *a* (Chl-*a*), the accessory pigments chlorophyll *b* and *c*, and the photosynthetic carotenoids. The annual production of oceanic chlorophyll is about 10^{12} kg (Jeffrey and Mantoura, 1997). For cells growing in high light environments, additional *photoprotectant* carotenoids protect the cell from photo-oxidation (Trees *et al.*, 2000). All of the above pigments account for about 95% of the light absorbed by phytoplankton (Aiken *et al.*, 1995). Because chlorophyll *a* is the only photosynthetic pigment that occurs in all phytoplankton, it provides a measure of phytoplankton abundance and biomass.

Jeffrey and Vesk (1997) summarize the species of phytoplankton, which include diatoms, dinoflagellates and cyanobacteria. At temperate and high latitudes, diatoms are generally the dominant class of phytoplankton (Lalli and Parsons, 1993). Plate 3 shows several diatoms; the central diatom belongs to the *chaetoceros* species, which consists of a chain of silica-shelled single cells with spines protruding from each cell and from both ends of the chain. The plate shows that within each cell the pigments are not distributed uniformly, rather they are located within small packages called chloroplasts. These pigments are a mixture of the green chlorophyll and the brownish yellow carotenoids. Because the pigment packaging differs from species to species, the response to incident light can differ by species even for the same chlorophyll concentrations.

A goal of biological remote sensing is to use observations of ocean color in models of the global distribution of primary production. As the following sections show, instead of primary production, ocean color is used to derive chlorophyll concentrations that are proportional to biomass or the standing phytoplankton stock. Whereas chlorophyll is a measure of biomass, primary production is a measure of phytoplankton growth and the two are not necessarily related. It is impossible to say from a satellite image alone if a change in observed chlorophyll occurs because of increased growth, or because the phytoplankton are being grazed less or are closer to the surface (Balch and Byrne, 1994). For example, in the North Pacific, when the primary productivity increases, zooplankton can graze the phytoplankton at such a rate that the phytoplankton standing stock is unchanged, and the productivity increase is represented by an increase in zooplankton (Lalli and Parsons, 1993). In contrast, in the North Atlantic, the productivity increase is accompanied by an increase in both phytoplankton biomass and observed chlorophyll.

Because primary production varies with the availability of nutrients and sunlight, growth occurs in regions of upwelling, or in regions with wind mixing that bring nutrients to the surface. Such regions occur along the west coast of continents, in the equatorial Atlantic and Pacific during La Niña and in the western Indian Ocean. In contrast, in the equatorial regions away from the immediate vicinity of the equator, the radiation-induced warm surface layer yields a stable upper ocean with small upwelling, so that the productivity is small. Behrenfeld and Falkowski (1997) review primary productivity models that relate productivity to the observed chlorophyll. In addition to chlorophyll, inputs to these models include such variables as the cloud-corrected estimates of daily surface solar irradiance, the oceanic optical depth and the physiological variables governing the ability of the organisms to take up carbon. Surface temperature also provides an indirect measure of nitrate and other nutrients, where because of upwelling and mixing at high latitudes, the nutrient concentrations decrease with increasing SST up to about 15 °C (Balch and Byrne, 1994).

This chapter concentrates primarily on ocean color retrieval and leaves estimation of primary productivity to the above references. In modeling of the absorption and scattering of light within the ocean, the following substances must be considered: phytoplankton pigments, dissolved organic material, and suspended organic and inorganic particulate matter. As the previous chapter discusses, the scattering and absorption of sunlight by pure seawater yields a blue upwelled light. In contrast, dissolved and particulate material yield a brownish yellow color. Because chlorophyll *a* preferentially absorbs in the blue, as its concentration within the water column increases from zero, the water appears less blue and more green. Historically, these easily viewed color changes suggested that visible sensors on aircraft and satellites could be used to survey large oceanic regions for biological activity. Retrieval of ocean color is a complex task. Unlike observations in the infrared, where the radiation is emitted from the top 10–100 μm of the sea surface, ocean color radiances in the blue-green can be upwelled from depths as great as 50 m. Also, because in the visible, aerosol and molecular scattering dominate atmospheric attenuation, the water-leaving radiances only comprise about 10% of the total received radiance. This means that determination of the water-leaving radiances requires precise determination of all other radiances.

The following discussion divides into six parts. Section 6.2 summarizes how the phytoplankton and the suspended and dissolved material alter the seawater scattering, absorption and reflectance properties from their pure seawater values. Section 6.3 describes the ocean color satellite instruments and their choice of wavelength bands with particular emphasis on CZCS, SeaWiFS and MODIS. Section 6.4 discusses the atmospheric correction algorithms. Section 6.5 describes the reflectance and fluorescence properties of the dissolved and suspended material. Section 6.6 describes the different kinds of ocean color algorithms, and Section 6.7 gives examples.

6.2 Scattering and absorption by phytoplankton, particles and dissolved material

The sources of color change in seawater include phytoplankton and its pigments, dissolved organic material and suspended particulate matter. The dissolved organic material, called *colored dissolved organic material* (CDOM) or *yellow matter* or *gelbstoff*, is derived from both terrestrial and oceanic sources. Terrestrial CDOM consists of dissolved humic and fulvic acid, which are primarily derived from land-based runoff containing decaying vegetable matter. In the open ocean, CDOM is produced when the phytoplankton are degraded by grazing or photolysis (Carder *et al.*, 1999). The organic particulates, called detritus, consist of phytoplankton and zooplankton cell fragments and zooplankton fecal pellets (Roesler,

Perry and Carder, 1989). The inorganic particulates consist of sand and dust created by erosion of land-based rocks and soils. These enter the ocean through river runoff, by deposition of wind-blown dust on the ocean surface or by wave or current suspension of bottom sediments (Mobley, 1994). Both CDOM and the particulates absorb strongly in the blue, yielding a brownish yellow color to the water (Hoepffner and Sathyendranath, 1993).

Given the diversity of this dissolved and suspended material, Morel and Prieur (1977) divide the ocean into *case 1* and *case 2* waters. In case 1 waters, phytoplankton pigments and their covarying detrital pigments dominate the seawater optical properties. In case 2 waters, other substances that do not covary with Chl-*a* such as suspended sediments, organic particles and CDOM are dominant. Even though case 2 waters occupy a smaller area of the world ocean than case 1 waters, because they occur in coastal regions with large river runoffs and high densities of human activities such as fisheries, recreation and shipping, they are equally important.

Scattering in the water column depends in part on the size distribution of the suspended living and inert particulate matter. Following Stramski and Kiefer (1991) and Mobley (1995), the smallest living organisms are viruses, with diameters of 10–100 nm and with oceanic concentrations of 10^{12}–10^{15} m^{-3}. Because of their small size, viruses tend to be Rayleigh scatterers. Next are bacteria, with diameters of 0.1–1 µm and concentrations as large as 10^{13} m^{-3}; these can be significant absorbers of light in the blue. Third, phytoplankton range in size from 2–200 µm, where the larger sizes consist of collections of cells. Because phytoplankton are larger than the visible wavelengths, they tend to be Mie scatterers. Fourth, the zooplankton that graze on phytoplankton have length scales of 100 µm to 20 mm.

The relative concentrations of these organisms depend on their size, where large organisms occur less frequently than small ones. The concentrations of organisms with diameters in the range 30 nm to 100 µm have an inverse fourth-power law dependence on diameter (Stramski and Kiefer, 1991). This relation approximately holds at larger scales, so that even though the ocean contains fish and marine mammals with characteristic sizes of 0.1–10 m, they occur so infrequently that, at the satellite observational scales, they do not affect scattering or absorption. Inert organic particles are comparable in size to phytoplankton. Because any photosynthetic pigments in this material are rapidly oxidized, the organic particles lose their characteristic Chl-*a* absorption properties. The inorganic particulates consist of fine sands, mineral dust, clay, and metal oxides, and have scales ranging from much less than 1 µm to of order 10 µm.

The importance of these scattering and absorption properties to remote sensing follows from Equation (5.27), which shows that the water-leaving radiance $L_w(\lambda)$ directly depends on the subsurface reflectance $R(\lambda)$. Rewriting $R(\lambda)$ from Equation (5.24) to account for scattering and absorption from suspended and dissolved material gives

$$R(\lambda) = G b_{bT}(\lambda)/a_T(\lambda) \qquad (6.1)$$

where G is a constant that depends on the incident light field. In (6.1), the addition of the subscript T to the absorption and backscattering coefficients means that they now take into account dissolved and suspended material, so that b_{bT} is the total backscatter coefficient defined similarly to Equation (5.23), and a_T is the total absorption coefficient. Given that the water-leaving radiance depends on the wavelength dependence of b_{bT} and a_T, the next section discusses the scattering and absorption properties of phytoplankton and particles and CDOM absorption. For completeness, it also briefly discusses chlorophyll fluorescence.

6.2.1 Scattering

This section discusses the scattering properties of living and inert material beginning with the total scatter, then concentrating on backscatter. Because measurements of the scattering properties of particles in seawater require observations at many different angles and because seawater absorption interferes with these observations, the scattering properties are difficult to determine and are a subject of current research.

In general, the presence in the water column of even a small amount of particulate matter generates a strong forward scatter and increases the scattering coefficient by an order of magnitude (Mobley, 1995, p. 43.33). Scattering from small particles tends toward the Rayleigh solution with a smaller forward scattering peak and a strong wavelength dependence; scattering from larger particles tends toward the Mie case with a large forward scatter and a weak wavelength dependence. The total volume scattering function $\beta_T(\alpha, \lambda)$ can be written as

$$\beta_T(\alpha, \lambda) = \beta_w(\alpha, \lambda) + \beta_p(\alpha, \lambda) \tag{6.2}$$

where the subscript w refers to pure seawater and p to organic and inorganic particulate matter (Mobley, 1994, Section 3.8).

For three different water masses, Figure 6.1 compares some of the very few measurements of the angular dependence of $\beta_T(\alpha, \lambda)$ with the pure seawater case (Petzold, 1972, from data in Mobley, 1994). Each scattering function was measured at a single wavelength of 514 nm; for the same wavelength, the pure seawater values are from Equation (5.19). The measurements are from turbid water in San Diego Harbor, coastal water in the Santa Barbara Channel, California, and clear water from the Tongue of the Ocean, Bahama Islands. Even though these scattering functions are derived from different waters and locations, they have similar shapes. Comparison of the ocean and pure seawater curves shows that the addition of suspended materials increases the forward scattering by four to five orders of magnitude, and the backscatter by up to one order of magnitude. Because of this strong forward scatter, the particulate backscatter is relatively small, being only about 2% of the total scatter (Carder, 2002).

The wavelength dependence of these scattering functions is as follows. From Equation (5.21), the backscatter coefficient for pure seawater is given by $b_{bw}(\lambda) \sim \lambda^{-4.32}$. For suspended particulate matter, this strong power law dependence disappears. The model of Kopelevich (1983, from Mobley, 1994) divides into scattering from large particles, which have a strong forward peak and a small wavelength dependence ($\lambda^{-0.3}$), and scattering from small particles, which have a more nearly symmetric scattering function and a stronger

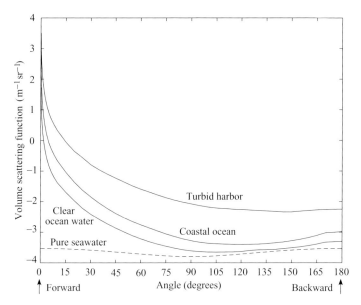

Figure 6.1. The dependence of the volume scattering function on angle for pure seawater (dashed line) and three different natural waters at 514 nm. The arrows at the bottom of the figure mark the directions of forward and backward scatter (Data from Petzold, 1972, as listed in Mobley, 1994, Table 3.10, courtesy Curtis Mobley).

wavelength dependence ($\lambda^{-1.7}$). In a phytoplankton example Gordon and Morel (1983, described in Mobley, 1994) show that if C_a is the phytoplankton concentration measured in mg m^{-3}, the particulate backscatter coefficient $b_{bp}(\lambda)$ can be written as

$$b_{bp}(\lambda) \sim \lambda^{-1} C_a^{0.62} \, \text{m}^{-1} \tag{6.3}$$

In their more general modeling of the backscatter from an arbitrary collection of particles, Carder *et al.* (1999) assume that $b_{bp}(\lambda)$ has the following form:

$$b_{bp}(\lambda) = X\lambda^{-Y} \tag{6.4}$$

where X is proportional to the particle concentration and Y depends on the particle size distribution. For large particles and Mie scatter, $Y \cong 0$; for small particles, $Y > 0$ (Carder *et al.*, 1999). As Section 6.6.4 discusses, this model is used in ocean color algorithms for retrieval of the constituent concentrations. Equation (6.4) is an approximation; in related work, Zaneveld and Kitchen (1995) investigate the behavior of the scattering function in the vicinity of the Chl-*a* absorption peak and show that it varies strongly across the peak. With new instrumentation, these and other measurements of the detailed wavelength dependence of scattering are a subject of current research (C. Roesler, private communication, 2000).

6.2.2 Absorption

The total absorption coefficient $a_T(\lambda)$ can be written as

$$a_T(\lambda) = a_w(\lambda) + a_p(\lambda) + a_\phi(\lambda) + a_{CDOM}(\lambda) \qquad (6.5)$$

where the subscript w refers to pure seawater, p to particles, ϕ to phytoplankton pigments and CDOM to colored dissolved organic material (Hoepffner and Sathyendranath, 1993; Roesler and Perry, 1995; Roesler *et al.*, 1989). For phytoplankton pigments, the absorption in units of m^{-1} is frequently normalized on the concentration of the pigment of interest, such as Chl-*a* or the carotenoids, where this *normalized* or *specific absorption* has units of m^2 (mg pigment)$^{-1}$. The following begins with the absorption properties of CDOM and particulate matter, then continues with the phytoplankton properties.

CDOM and particulates   For three different concentrations of CDOM and particulates, Figure 6.2 shows the wavelength dependence of $a_T(\lambda)$. For each curve, the absorption is greatest in the blue, then decreases exponentially toward longer wavelengths. Roesler *et al.* (1989) and Hoepffner and Sathyendranath (1993) show that for 350 nm $< \lambda < 700$ nm, $a_p(\lambda)$ and $a_{CDOM}(\lambda)$ can be expressed as follows:

$$a_i(\lambda) = A_i(400) \exp[-q_i(\lambda - 400)] \qquad (6.6)$$

In (6.6), the subscript *i* equals p or CDOM, $A_i(400)$ is the concentration-dependent reference absorption with values of order 10^{-1}–10^{-2} m^{-1}, and q_i is a species-specific constant

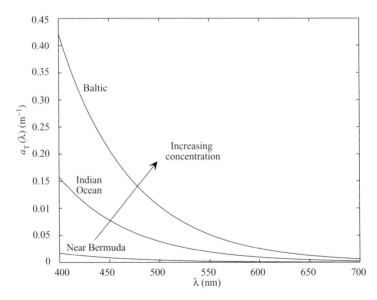

Figure 6.2. Dependence of the total absorption coefficient on λ as observed in three locations with different concentrations of CDOM and particulate matter (Data from Mobley, 1995, Table 7, courtesy Curtis Mobley).

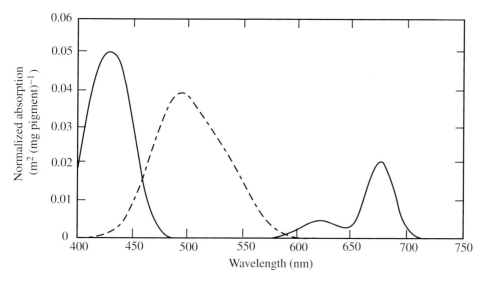

Figure 6.3. The normalized absorption for Chl-*a* (*solid line*) and the carotenoids (*dashed line*). Each curve is normalized by division of the measured absorption by the respective pigment density measured in mg of pigment m^{-3} (Figure 9 from Hoepffner and Sathyendranath, 1993, © 1993 American Geophysical Union, reproduced/modified by permission of AGU).

(Mobley, 1994). For CDOM, q_i ranges from 1.1 to 1.8 $\times 10^{-2}$; for organic particles, from 0.6 to 1.4 $\times 10^{-2}$, where inorganic particles have a similar range (Roesler *et al.*, 1989; Hoepffner and Sathyendranath, 1993).

Phytoplankton Phytoplankton absorption has a more complicated wavelength dependence. From case 1 waters in the summer North Atlantic, Figure 6.3 shows the absorption curves for chlorophyll *a* and the carotenoids, where each curve is normalized on its respective pigment concentration (redrawn from Hoepffner and Sathyendranath, 1993, Figure 9). The carotenoid curve includes contributions from both the photosynthetic and photoprotective carotenoids. Because the concentrations of chlorophyll *b* and *c* are generally much smaller than Chl-*a*, they are omitted. Examination of the Chl-*a* curve shows that it has two major absorption peaks, a blue maximum near 440 nm, called the Soret band (Trees *et al.*, 2000) and a red maximum centered at 665 nm, where in most cases, the blue peak is about three times greater than the red (Mobley, 1994). Between 550 and 650 nm, the absorption approaches zero, giving chlorophyll-rich water its characteristic green color (Kirk, 1996). The dashed curve shows the carotenoid absorption, with its peak shifted toward 500 nm and its bandwidth extending from about 450 to 550 nm.

From samples taken in the North Atlantic during September, Figure 6.4 shows the normalized total absorption with the pure seawater absorption removed, the absorption due to particles and that due to phytoplankton, which consists of the total minus the particulate absorption, where each curve is normalized on the Chl-*a* concentration (Hoepffner and

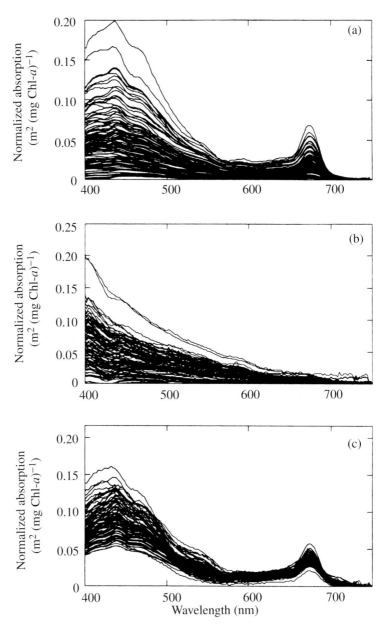

Figure 6.4. The dependence of absorption on wavelength from measurements made in the western North Atlantic. In each case, the pure seawater absorption is removed. (a) Total absorption; (b) particulate and CDOM absorption; (c) phytoplankton absorption (Figure 3 from Hoepffner and Sathyendranath, 1993, © 1993 American Geophysical Union, reproduced/modified by permission of AGU).

Sathyendranath, 1993, Figure 3). The figure shows the characteristic Chl-*a* peaks at 440 and 665 nm, the exponential decay with increasing λ of the particulate absorption, and the variance in the normalized phytoplankton absorption associated with differences in species, packaging and accessory pigments.

Finally, as Section 6.5.2 discusses further, a feature that does not occur in the absorption spectra but is important in the reflectance is the presence of a Chl-*a* fluorescence peak at 683 nm, adjacent to the 665-nm absorption peak. At the fluorescence peak, the phytoplankton emit radiation that is detectable by satellite.

The above discussion shows that to observe CDOM, phytoplankton and fluorescence, ocean color instruments need to employ the following wavelengths. Determination of the chlorophyll and CDOM concentrations involves observations at the chlorophyll absorption peak of 443 nm and at a CDOM-dominated wavelength such as 410 nm. Measurements must also be made in the 500–550 nm range where the chlorophyll absorption is zero and carotenoid absorption dominates. Fluorescence requires observations in the vicinity of the 683-nm peak. These absorption and emission properties provide the basis for the choice of instrument wavelengths described in the next section.

6.3 Ocean color satellite instruments

Satellite observations of ocean color began in 1978 with the launch of the Coastal Zone Color Scanner (CZCS) instrument on the NIMBUS-7 satellite (Mitchell, 1994). CZCS observations continued through about June 1986, although with problems caused by sensor degradation in its later years (Evans and Gordon, 1994). The next satellite instruments were the Japanese Ocean Color and Temperature Sensor (OCTS) on the ADEOS-1 satellite that operated from August 1996 to June 1997, and the German Modular Optical Scanner (MOS) launched in 1996 on the Indian Remote Sensing Satellite IRS-P3. Other US instruments include the Sea-viewing Wide Field-of-view Sensor (SeaWiFS) launched in August 1997, which has a five-year design life and continues to operate in 2004, and the Moderate Resolution Imaging Spectroradiometer (MODIS) launched on TERRA in December 1999 and on AQUA in May 2002. As Chapter 14 discusses, the future US ocean color instrument will be the NPOESS Visible/Infrared Imager/Radiometer Suite (VIIRS).

Other ocean color satellites and instruments include the following. In 1999, the German Ocean Color Monitor (OCM) was launched on the sun-synchronous Indian IRS-P4 OCEANSAT and the Ocean Color Imager (OCI) was launched on the Republic-of-China Satellite (ROCSAT). Instead of being in a sun-synchronous orbit, ROCSAT is in a near-equatorial low-inclination orbit. In March 2002, the European Medium Resolution Imaging Spectrometer (MERIS) was launched on ENVISAT, where MERIS is a pushbroom instrument in a 1000 descending sun-synchronous orbit with 15 observing bands between 400 and 900 nm. In May 2002, the Chinese Ocean Color and Temperature

Table 6.1. *Comparison of the ocean color bands for the MODIS,*
SeaWiFS and CZCS instruments

MODIS/SeaWiFS band number	Center wavelength (nm)	Wavelength range (nm)		
		MODIS	SeaWiFS	CZCS
8/1	412	405–420	402–422	—
9/2	443	438–448	433–453	433–453
10/3	490	483–493	480–500	—
—/4	510	—	500–520	510–530
11/—	531	526–536	—	—
12/5	555	546–556	545–565	540–560
13/6	670	662–672	660–680	660–680
14/—	678	673–683	—	—
15/7	765	743–753	745–785	—
16/8	865	862–877	845–885	700–800

With the exception of MODIS bands 11 and 14, the center wavelengths correspond
to the SeaWiFS bands.

Sensor (COCTS) was launched on the Haiyang-1 satellite (Ocean-1, or HY-1) into descending sun-synchronous orbit with an approximate crossing time of 1000 local time. HY-1 will be followed by the HY-2 and HY-3 satellites, which are scheduled at approximately two-year intervals. In December 2002, Japan launched the Global Imager (GLI) on the ADEOS-2 satellite into a 1030 descending sun-synchronous orbit, where ADEOS-2 failed in October 2003.

For all these instruments, the International Ocean Colour Coordinating Group (IOCCG) publication IOCCG (1999) describes their wavelength bands and properties. Given the biological, oceanographic and atmospheric constraints discussed in this chapter, each of these instruments uses similar wavelength bands. Because SeaWiFS and MODIS have generated a program of research cruises and surface observations as well as extensive series of papers, reports and conferences concerning the instruments and their algorithms, the following concentrates on these two instruments.

For MODIS, SeaWiFS and CZCS, Table 6.1 lists the wavelength bands used for ocean color observations. An important difference between MODIS and SeaWiFS is that the MODIS bands are narrower by factors of one-half to one-quarter, the MODIS data are 12-bit digitized as opposed to 10-bit for SeaWiFS, and as Table 6.2 shows below, the MODIS bands have about twice the signal-to-noise ratio. The largest shift in band locations between SeaWiFS and MODIS is that the 510-nm SeaWiFS band was moved to 531 nm for MODIS. The purpose of this move was to improve the instrument response to accessory pigments and to match the 531-nm laser wavelength used in aircraft remote sensing (Esaias *et al.*, 1998). The other change is the addition of a narrow fluorescence band at 678 nm. The locations

Table 6.2. *Comparison of the measured and derived radiances and their uncertainties for MODIS, SeaWiFS and CZCS, a solar zenith angle of $\theta_S = 60°$ and for measurements near the scan edge*

MODIS/SeaWiFS bands	λ_0 (nm)	L_{Tmax} ($\mu W\,cm^{-2}\,nm^{-1}\,sr^{-1}$)	L_T	$[L_w]_N$	$NE\Delta L$ ($\mu W\,cm^{-2}\,nm^{-1}\,sr^{-1}$)		
					MODIS	SeaWiFS	CZCS
8/1	412	13.6	9.3	1.1	0.005	0.019	—
9/2	443	13.8	8.7	1.1	0.005	0.013	0.033
10/3	490	11.1	7.1	0.7	0.004	0.010	—
—/4	510	8.9	5.6	0.3	—	0.0109	0.017
11/—	531	8.9	5.6	0.3	0.004	—	—
12/5	555	7.4	4.5	0.12	0.003	0.008	0.019
13/6	670	4.1	2.6	0.10	0.001	0.006	0.012
14/—	678	4.1	2.5	0.01	0.001	—	—
15/7	765	2.9	1.6	—	0.002	0.004	—
16/8	865	2.0	1.1	—	0.001	0.002	—

See text for additional description.
Adapted from Gordon and Voss, 1999, Table 1, and O'Reilly *et al.*, 1998, Table 4

of all these bands depend on two constraints: the optical properties of phytoplankton and suspended and dissolved oceanic material discussed in Section 6.2, and the locations of the atmospheric and solar absorption bands discussed next.

To illustrate the constraints imposed by atmospheric and solar absorption, Figure 6.5 compares the locations of the MODIS and SeaWiFS bands with the wavelength dependence of the solar irradiance at the TOA, the atmospheric transmittance and the solar irradiance at the Earth's surface. Figure 6.5a shows the TOA solar irradiance from Figure 3.9, where the chemical symbols mark the location of the major Fraunhofer absorption lines generated in the solar corona (Phillips, 1992). Figure 6.5b shows the atmospheric transmittance for the MODTRAN 1976 standard atmosphere, where the oxygen, oxygen-A and water vapor absorption bands are marked; Figure 6.5c shows the solar irradiance at the surface for normal incidence, where the upper horizontal bars give the location of the SeaWiFS bands; the lower bars, the MODIS bands. The figure shows that the SeaWiFS 745–785 nm band overlaps the oxygen-A band, while the MODIS bands are located so that they avoid all of the major Fraunhofer and atmospheric absorption bands.

The table and figure show that SeaWiFS has a similar set of bands to MODIS, although with one less band between 650 and 700 nm. The MODIS visible bands have characteristic widths of 10 nm, compared with 20 nm for SeaWiFS. For comparison, the table shows that CZCS had only three bands in the blue-green, one band in the red, and a single band in the NIR (not listed), which had insufficient gain for aerosol removal. The MODIS and

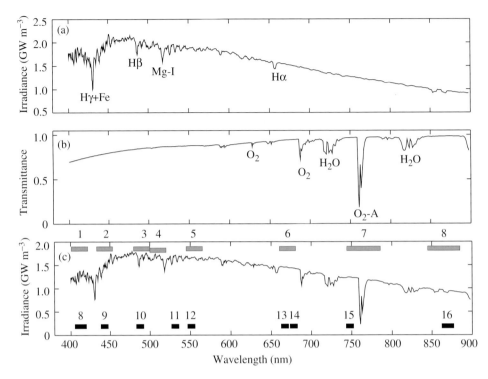

Figure 6.5. The solar irradiance, the atmosphere transmittance, and the surface irradiance shown with the locations of the SeaWiFS and MODIS bands. (a) The solar irradiance at the TOA with labels on the Fraunhofer lines, where Hα, Hβ, Hγ are the different hydrogen lines (locations from Phillips, 1992, Table 3.2); (b) the transmittance from the MODTRAN 1976 standard atmosphere with the absorption lines labeled; (c) the surface irradiance for the sun at zenith, where the gray bars and the numbers above the frame show the SeaWiFS bands; the lower numbered black bars, the MODIS bands. For clarity, SeaWiFS band 4 is slightly offset in the vertical.

SeaWiFS bands have the following purposes: the 412 nm band is used for detection of CDOM and suspended sediments; the 443, 490, 510 and 555 nm bands are used to determine chlorophyll concentrations. For MODIS, band 13 at 670 nm, band 14 at 678 nm and band 15 at 765 nm are used for measurement of the chlorophyll *a* fluorescence peak at 683 nm; the reason that MODIS band 14 and SeaWiFS band 6 are located at a slightly shorter wavelength than the fluorescence peak is to avoid the oxygen absorption band at 687 nm. For SeaWiFS, the 765 and 865 nm bands, and for MODIS, the equivalent bands are used for atmospheric aerosol correction. For the CZCS, the 670 nm band alone was used for this purpose.

For MODIS, SeaWiFS and CZCS, Table 6.2 gives an example of the magnitudes of the received radiances and their instrument-associated uncertainties. The radiances are from measurements made in a region of very low pigment concentration in the summer Sargasso Sea and are taken near the scan edge to maximize their uncertainty (Gordon and Voss, 1999).

The first five columns in the table list the MODIS band number, the center wavelength λ_0, the maximum total radiance L_{Tmax} received at the satellite, a more typical total received radiance L_T, and the corresponding values of the water-leaving radiances $[L_w]_N$. The last three columns list the values of the noise-equivalent delta-radiance $NE\Delta L$ defined in Section 3.5.4, where the MODIS and SeaWiFS values are from preflight specifications; the CZCS values, from in-orbit measurements.

Examination of the table shows that MODIS is typically two to three times more sensitive than SeaWiFS, which is about twice as sensitive as CZCS. The exceptions are MODIS bands 13 and 14, which are six times more sensitive than SeaWiFS and about ten times more sensitive than CZCS. At 443 nm, the table also shows that $[L_w]_N$ makes up only 13% of L_T, so that about 90% of L_T consists of contributions from other sources. Since the goal of SeaWiFS and MODIS is to determine $[L_w]_N$ within 5% (Hooker and McClain, 2000), to achieve this goal, the sum of the other radiances must be determined to within 1%, or within $\pm 0.06\ \mu\text{W cm}^{-2}\ \text{nm}^{-1}\ \text{sr}^{-1}$. Because for MODIS or SeaWiFS, the desired accuracy is 5–10 times the $NE\Delta L$, the 5% goal should be achievable.

6.3.1 SeaWiFS

The SeaWiFS sensor and its OrbView-2 (formally SeaStar) spacecraft were constructed and launched by the private Orbital Science Corporation (Figure 6.6). In August 1997, the instrument was launched from an altitude of about 15 km using an L-1011 aircraft as the first stage. SeaWiFS occupies a sun-synchronous orbit at an altitude of 705 km with a 1200 descending crossing time. The instrument is a cross-track scanner with a swathwidth of 2800 km, corresponding to a scan-angle range of $\pm 58.3°$. Its resolution of 1.6 mrads yields a surface resolution at nadir of 1.1 km. This design provides global coverage at two-day intervals. A detailed description of the satellite and sensor, and a large collection of images can be found at http://seawifs.gsfc.nasa.gov/SEAWIFS.html.

Figure 6.7 and 6.8 respectively show a drawing and photograph of the SeaWiFS cross-track optical scanner and electronics module. Relative to the satellite, the scanner consists of a rotating folded off-axis telescope and a non-rotating optical bench. The telescope rotates at six revolutions per second in the cross-track direction, so that continuous coverage is provided at nadir. This rotation rate means that the SeaWiFS output is compatible with the existing AVHRR direct broadcast format described in Chapter 7. From Figure 6.7, the primary mirror collects the surface radiance and reflects it from a polarization scrambler into the half-angle mirror, which focuses the radiance into the non-rotating Aft Optics Bench. The half-angle mirror rotates at half the rate of the telescope, and uses alternate sides on successive telescope scans. To avoid interference with sun glint, the entire instrument can be tilted in the along-track direction to angles of $+20°$, $0°$ and $-20°$.

Because the entire instrument tilts instead of only the telescope, the SeaWiFS optical paths are independent of tilt angle. The sensor is calibrated in three ways. For periods of order days, the instrument observes the sun at approximately daily intervals. The solar

Figure 6.6. The OrbView-2 satellite and the SeaWiFS instrument, where the instrument is mounted at the bottom of the satellite between the vertical rod-shaped antennas (Courtesy of SeaWiFS Project NASA/GSFC and ORBIMAGE, image courtesy Gene Feldman).

calibration takes place when the satellite passes over the South Pole, at which time the instrument is tilted 20° aft so that it views the solar reflection in the Solar Diffuser Plate mounted inside the Solar Calibrator. Because this plate deteriorates slowly, this technique cannot be used for long-term calibrations and is intended only to detect abrupt calibration changes. So far, none have been observed.

Monthly observations of the full moon provide a longer term calibration. During the full moon and on the nighttime segment of its orbit, the spacecraft rolls over 180° along its flight axis from its normal Earth-oriented position, so that the instrument points at the moon. Because the nighttime lunar radiances are similar in magnitude to the daytime upwelled ocean radiances, this provides a calibration for similar radiances along the same optical path. For the first 850 days of the mission, Figure 6.9 shows the results of the lunar calibration in a time plot of the normalized radiances for each band. On the figure, the

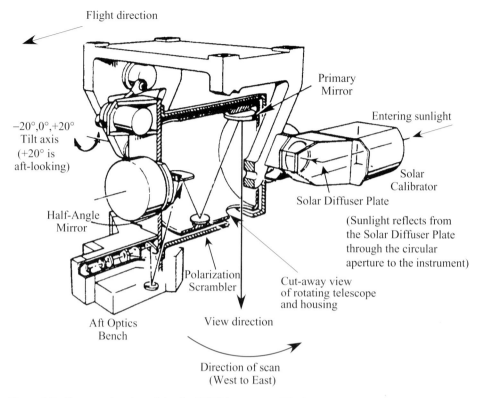

Figure 6.7. Cutaway drawing of the SeaWiFS instrument. The instrument attaches to the bottom of the spacecraft by the four top mounting points (Figure 7 from Hooker *et al.*, 1992, not subject to US copyright, courtesy William Barnes, Orbital Science Corporation and the NASA SeaWiFS program).

radiances are normalized in two ways. Each individual radiance is divided by its initial value and by the mean of the band 3 and 4 radiances, the properties of which changed least over the observational period. The figure shows that the sensitivity of the NIR bands 7 and 8 degraded with time, while the calibration of the other bands remained roughly constant. All these changes were incorporated into the SeaWiFS algorithms. A third calibration is provided by a moored buoy near Hawaii called MOBY (Marine Optical Buoy), which measures the water-leaving radiance in an ocean region of both low cloudiness and low marine productivity. Comparison of the buoy measurements with the satellite observations provides a check on the chlorophyll algorithm and the aerosol corrections.

6.3.2 MODIS

MODIS is the principal visible/infrared instrument on the TERRA and AQUA satellites, and is described in Barnes and Salomonson (1993) and Barnes, Pagano and Salomonson (1998).

Figure 6.8. Photograph of the SeaWiFS instrument with a caliper for scale. The view direction is upward, the solar calibrator is to the back right, the Aft Optics Bench is not visible. See text and Figure 6.7 for further information (Courtesy Raytheon Co., Santa Barbara Remote Sensing, used with permission).

MODIS is a hybrid cross-track scanner with a swathwidth of 2300 km, corresponding to a scan-angle range of ±55°, where the swaths are nearly contiguous at the equator and provide global coverage every 1–2 days (see Plate 2 for an example of single day coverage). The instrument has 36 bands with a spectral range of 0.4–14.4 μm. Table A.2 in the Appendix lists the bands and their resolutions. The instrument makes simultaneous observations of ocean color and SST, where the ocean color and thermal bands have a 1-km nadir resolution. Because of the bowtie effect described in Section 1.6.4, a pixel with a nadir resolution of 1 km has a resolution at the swath edge of approximately 6 km in the along-scan direction and 2 km in the along-track direction. The land/cloud properties bands (bands 3–7) have a 500-m nadir resolution and the land/cloud boundary bands (bands 1 and 2) have a 250-m resolution.

MODIS operates differently than SeaWiFS. First, instead of the SeaWiFS rotating tele-scope, MODIS uses a fixed telescope focused on a double-sided rotating paddle wheel

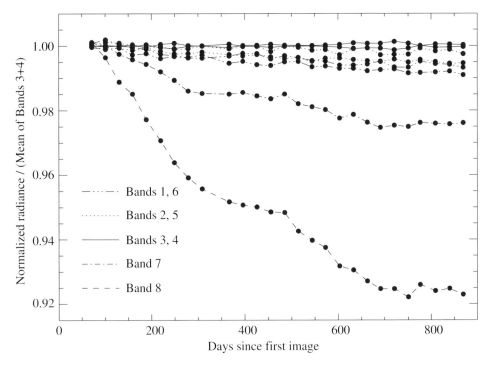

Figure 6.9. The change in the sensitivity of the SeaWiFS bands derived from the lunar calibration during the first 850 days of the mission (Figure 10 of Eplee and Barnes, 2000, not subject to US copyright).

mirror. This mirror continuously rotates at 20.3 rpm, where the surface scans are collected alternately by the two sides referred to as A and B below (Figure 6.10). The scan mirror reflects the surface radiances onto another mirror that is not visible in the figure, then into a telescope that transmits the radiances to an optics bench. Second, at the optical bench, the radiance for each band is focused onto linear strips of sensors that subdivide the along-track scan into multiple pixels. For the 1-km bands, 10 sensors subdivide the 10-km wide swath into 1-km pixels; for the 500-m bands, 20 sensors are used; for the 250-m bands, 40 sensors are used. The use of these sensor strips instead of the single sensors used with SeaWiFS greatly reduces the mirror rotation speed and increases the dwell time, yielding a better signal-to-noise ratio. Third, to reduce costs, and because of the additional MODIS on AQUA, the instrument does not tilt. Instead, the expectation is that despite sun glint, two non-tilting instruments will provide better coverage than one tilting instrument.

A problem that occurs for MODIS on TERRA is that mirror sides A and B have slightly different reflectivities, so that without careful calibration, striping occurs on the images. The calibration includes a solar diffuser, a solar diffuser stability monitor, a spectral radiometric calibration assembly, and a blackbody. Similar to SeaWiFS, TERRA also performs a lunar calibration, which involves a monthly spacecraft roll to look at the full moon,

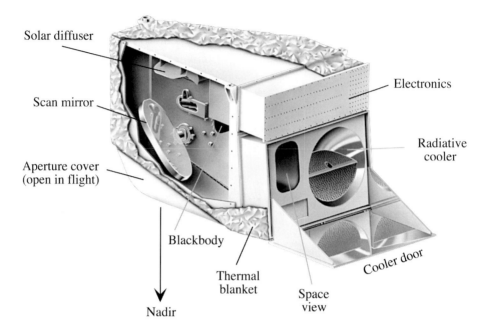

Figure 6.10. The MODIS instrument. The instrument measures approximately 1 m × 1 m × 1.6 m and weighs 250 kg; the length of the scan mirror is 0.58 m (Courtesy NASA and William Barnes).

followed by a return roll. The timing and results of this calibration are described at the MODIS Characterization Support Team website (http://mcstweb.gsfc.nasa.gov).

6.4 Atmospheric correction and retrieval of the water-leaving radiance

Given the amount of analysis and algorithm development invested in the SeaWiFS instrument and observations since its launch in 1997, and because SeaWiFS, MODIS, MERIS and the other existing and proposed instruments use similar bands for atmospheric correction, the discussion of the retrieval of the water-leaving radiance follows the SeaWiFS algorithm. The first step in this procedure is to determine if every oceanic pixel in the image under investigation is cloud-free. The SeaWiFS cloud detection test uses the 870-nm band. For this band, because the water-leaving radiance is near zero and clouds are reflective, pixels with a reflectance greater than a preset threshold are classified as cloud. Because the MODIS cloud algorithms depend on both the visible and infrared bands, their description will be delayed until Section 7.7, after discussion of the infrared SST algorithms.

Assuming cloud-free conditions, the retrieval of $L_w(\lambda)$ depends on the removal of all other radiances from the total radiance $L_T(\lambda)$ received at the satellite and on the calculation of the beam and diffuse transmittances. In the following, Section 6.4.1 discusses on a term by term basis, the contributions to the total radiance and their evaluation. For SeaWiFS and MODIS, Sections 6.4.2 and 6.4.3 discuss the aerosol retrieval; Section 6.4.4 describes the special case of the CZCS aerosol correction.

6.4.1 Contributions to the total radiance

Retrieval of the water-leaving radiance is described both in terms of radiances and in terms of the extraterrestrial reflectances defined in Equation (5.31) (Esaias *et al.,* 1998). Because they are the quantities measured by the field and satellite instruments, the following discussion primarily uses radiances. Given that for any λ, the reflectances and radiances differ by only a multiplicative constant, the equations for the reflectances have a similar form.

The corrections include the determination and removal from $L_T(\lambda)$ of the ozone attenuation, the radiances associated with sun glint and foam, the Rayleigh path radiances and, most difficult computationally, the aerosol path radiances. Because, as Table 6.2 shows, these radiances make up about 90% of the retrieved signal, their removal is critical to the $L_w(\lambda)$ retrieval. Figure 6.11 illustrates the contributions to the total radiance received at the sensor and shows the terms included in the CZCS and SeaWiFS/MODIS algorithms. Each algorithm corrects for ozone attenuation and the Rayleigh and aerosol path radiances. The CZCS algorithm assumes a flat surface and single molecular and aerosol scattering; the CZCS instrument avoided sun glitter by tilting at angles of ±10° and ±20°. The SeaWiFS/MODIS algorithm assumes a rough ocean surface and multiple molecular and aerosol scattering, accounts for radiance from foam, and avoids sun glint by tilting the SeaWiFS instrument and by using a sun glint mask for SeaWiFS and MODIS.

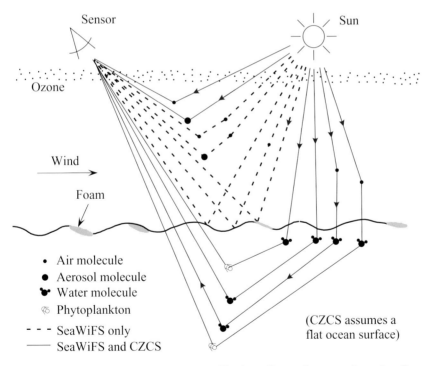

Figure 6.11. The contributions received at the satellite from direct reflectance, the path radiance, the reflected path radiance and the water-leaving radiance for CZCS and SeaWiFS (Adapted from Figure 4, McClain *et al.,* 1992).

From Hooker and McClain (2000), the total radiance $L_T(\lambda)$ received at the satellite can be written as

$$L_T(\lambda) = t_D(\lambda)L_w(\lambda) + t_D(\lambda)L_F(\lambda) + t(\lambda)L_G(\lambda)$$
$$+ L_R(\lambda) + L_A(\lambda) + L_{RA}(\lambda) \tag{6.7}$$

In (6.7), $t_D(\lambda)$ is the diffuse transmittance, $L_F(\lambda)$ is the radiance reflected by foam, $L_G(\lambda)$ is sun glint, and $L_R(\lambda)$, $L_A(\lambda)$, and $L_{RA}(\lambda)$ are respectively the atmospheric Rayleigh, aerosol and mixed Rayleigh–aerosol scattering terms. The θ-dependence of all variables is omitted for simplicity. As described below, each of these terms is corrected for ozone attenuation using Equation (4.52). Because $L_G(\lambda)$ is generated by Fresnel reflection from wave facets, it is attenuated by the beam transmittance $t(\lambda)$. In contrast, because $L_F(\lambda)$ and $L_w(\lambda)$ are generated by Lambertian reflection, they are attenuated by the diffuse transmittance $t_D(\lambda)$. An alternative way to look at these radiances follows Hooker and McClain (2000), who show that the contributions to $L_T(\lambda)$ divide into the path radiances generated in the atmosphere $[L_R(\lambda) + L_A(\lambda) + L_{RA}(\lambda)]$, the sun glint and foam radiances generated at or immediately below the ocean surface $[t_D(\lambda)L_F(\lambda) + t(\lambda)L_G(\lambda)]$, and the diffusely attenuated water-leaving radiance generated within the water column $[t_D(\lambda)L_w(\lambda)]$. These terms are evaluated as follows.

Ozone For wavelengths between 500 and 700 nm, Figure 4.11 shows that the attenuation has a small but non-negligible seasonal dependence on ozone. From Gordon and Voss (1999) and for the SeaWiFS bands, the ozone attenuation $\tau_{OZ}(\lambda) \leq 0.035$, where ozone is assumed to be strictly absorbing and non-scattering. Because all terms in $L_T(\lambda)$ depend on the solar irradiance, they are each reduced by a seasonally and latitudinally dependent downward and upward passage through the ozone layer. For MODIS and SeaWiFS, the spatial and temporal distribution of ozone and τ_{OZ} is determined by observations from the Total Ozone Mapping Spectrometer (TOMS) instrument on the Earthprobe satellite, and by a similar instrument on the EOS-AURA satellite launched in 2004.

Sun glint As Section 5.2.4 and Figure 5.7 show, the angular distribution of the solar radiances generated by Fresnel reflectance from wave facets is a function of sun angle and vector wind speed. As an example, the white arrows in Plate 2 mark the stripes of sun glint along the individual MODIS swaths, where because of the northerly position of the sun, the glint is north of the equator. For each SeaWiFS and MODIS image, the combination of the wave facet model described in Section 2.2.4 with vector wind speeds derived from NOAA numerical weather prediction models allows calculation of a sun glint mask. Although the data from the forecast models are available at intervals of 3–6 hours, this procedure unavoidably neglects the effect of local wind gusts. An additional check on sun glint is provided by examination of the NIR radiances, where if these radiances exceed a preset threshold, sun glint is assumed and the pixel is masked.

Foam   There is an important difference between the spatial and angular distribution of the reflected radiances due to foam and sun glint. The sun glint radiances are distributed around the solar conjugate angle, so that depending on wind velocity these radiances may affect only a fraction of the image, which can be masked. The foam coverage also depends on wind speed, but because the foam reflectance is more nearly Lambertian, it has a much weaker dependence on solar angle, so that $L_F(\lambda)$ can be nearly uniform across an image. In the processing, $L_F(\lambda)$ is estimated, then either subtracted from $L_T(\lambda)$ or if $L_F(\lambda)$ is too large, the image is discarded. Estimation of $L_F(\lambda)$ follows the model of Frouin *et al.* (1996) and Moore *et al.* (2000) described in Section 5.6. In almost all cases, the correction for foam is small, perhaps because strong winds are often accompanied by clouds.

Rayleigh path radiances   At the shorter wavelengths, the Rayleigh path radiance is generally the largest term in the received radiance. The single scattering Rayleigh path radiance $L_R(\lambda)$ is from Equation (4.53), where the Rayleigh optical thickness $\tau_R(\lambda)$ is derived from Equation (4.24) with the surface pressure p taken from a numerical weather prediction model. In addition to the direct path radiance and for both Rayleigh and aerosol scattering, there are two additional smaller path terms, so that the total path radiance divides into the following three parts (Figure 6.12):

a. the dominant path radiance generated by the scattering of the downward solar irradiance into the sensor look direction, from Equation (4.53);
b. the path radiance generated along the conjugate path to the sensor look direction that is then reflected at the surface into the sensor direction;
c. the path radiance in the sensor direction generated by the reflected solar radiance.

Because the Fresnel surface reflectivities are small, the second and third terms are smaller than the first. Inclusion of multiple scattering further complicates these terms, but does not affect their relative magnitudes. Also, because wind waves alter the magnitudes of the reflected radiances in 'b' and 'c', these radiances are functions of wind speed (Wang, 2000). Finally, the angles and reflectances in the above three terms are generally incorporated into an expanded phase function (Gordon and Wang, 1992).

Aerosol path radiances   The retrieval of the aerosol path radiances and diffuse transmittance are complicated and at the heart of the $L_w(\lambda)$ retrieval. As the following shows, the NIR observing bands at 765 and 865 nm permit retrieval of the aerosol radiances and their dependence on wavelength. Given the complexity of this retrieval, it is discussed twice, first in a brief summary, then more extensively in the next two sub-sections.

Although the magnitudes of the aerosol path radiances $L_A(\lambda)$ strongly depend on the aerosol type and concentration, their determination and removal is conceptually straightforward. As Figure 5.14 shows, because for $\lambda > 700$ nm the seawater absorption is much larger than in the visible, the NIR subsurface reflectances and water-leaving radiances approach zero. For the 765 and 865 nm bands, this means that under most conditions, $L_w(\lambda)$ is set equal to zero. This is the *black pixel* assumption (Siegel *et al.*, 2000). For the

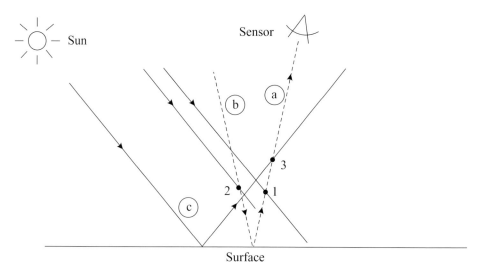

Figure 6.12. The sources of the solar generated single scattering path radiance. (a) The path between the sensor and surface, with a scattering location shown at 1; (b) the atmospheric path radiance generated at locations such as 2, which is reflected into the sensor; (c) the contribution to the path radiance along (a) generated at locations such as 3 by the reflected solar radiance. Path radiances are shown as dashed lines; solar irradiances as solid lines. The points 1, 2, 3 are illustrative only and are symbolic of an integration across the atmosphere.

NIR bands, this assumption holds except in coastal environments with large concentrations of surface sediments or biological blooms, or as discussed in Section 6.6.5, in regions of coccolithophore blooms.

Estimation of the aerosol radiances proceeds as follows. First, the ozone attenuation, sun glint, foam reflection and Rayleigh scattering terms are removed from all bands. For the two NIR bands where $L_w(\lambda)$ is assumed equal to zero, this yields values of $L_A(865)$ and $L_A(765)$. These observed radiances are compared with radiances that are numerically calculated from many different aerosol models. If the observed and calculated radiances agree, this comparison provides an estimate of the aerosol type and concentration, which allows the observed NIR aerosol radiances to be extrapolated to the visible. These extrapolated radiances are then removed from $L_T(\lambda)$, leaving only the attenuated water-leaving reflectance $t_D(\lambda)L_w(\lambda)$. The final step in the recovery of $L_w(\lambda)$ is the estimation and removal of $t_D(\lambda)$.

Diffuse transmittance   As Section 4.9.1 discusses, $t_D(\lambda)$ describes the attenuation of a radiance that is generated at an extended surface and propagates through a scattering atmosphere. In the treatment of $t_D(\lambda)$, two factors are considered: its contribution to land contamination and its method of calculation. First, for a scattering atmosphere and an extended surface, the received radiance has contributions not only from the instrument

FOV, but also from the surrounding area. Thus when the FOV in question is close to land, the received radiance becomes land-contaminated, so that the ocean color algorithms break down within a few pixels of the coast. For example in Puget Sound, the color retrieval breaks down within about three pixels of the coast (B. Sackmann, 2000, private communication). This contamination can also occur adjacent to an ice edge or to any location where the surface reflectances change abruptly. Second, for single scattering and the assumption of a Lambertian distribution of radiance at the surface, Equation (4.55) describes $t_D(\lambda)$. For multiple scattering, $t_D(\lambda)$ is numerically determined from the choice of aerosol model. Improvements to the $t_D(\lambda)$ retrieval will come from more exact measurements of the aerosol scatter and from refinements of the assumption of a Lambertian water-leaving radiance (Wang, 1999).

6.4.2 Determination of the aerosol path radiances

Determination of the aerosol and mixed Rayleigh–aerosol scattering terms in Equation (6.7) divides into the single and multiple scattering cases. The single scattering approximation is valid for thin layers of marine aerosols that occur in oceanic regions far from land, but breaks down for thick aerosols where multiple scattering becomes important. Such aerosols include those generated by African and Asian dust storms, by urban pollution advected over the ocean or by a thick marine atmospheric boundary layer. This section reviews the methods used to retrieve the aerosol properties and radiances, defines the terminology used in single and multiple scattering, and discusses the global distribution of oceanic aerosols.

For single scatter and in the visible, when the sun glint, foam and Rayleigh path radiances are removed from each band, the remaining terms are the aerosol path and water-leaving radiances. In the NIR with the assumption of a zero water-leaving radiance, the only remaining term is the single scattering aerosol radiance from Equation (4.54):

$$L_A(\lambda) = \omega_A(\lambda)\tau_A(\lambda)F'_S(\lambda)P_A(\lambda, \theta, \theta_S)/4\pi \cos\theta \tag{6.8}$$

In (6.8), θ_S is the solar zenith angle, θ is the look angle, $\omega_A(\lambda)$ is the single scattering aerosol albedo defined in Equation (4.47) and $P_A(\lambda, \theta, \theta_S)$ is the aerosol phase function expanded to include the contributions from the reflected path radiances shown in Figure 6.12.

Estimation of the aerosol type and concentration proceeds by division of the $L_A(\lambda)$ at 765 and 865 nm by their respective $F'_S(\lambda)$ from Equation (4.52), then taking their ratio. Following Gordon and Castaño (1987), this ratio becomes

$$\varepsilon(\lambda, \lambda_0) = \frac{L_A(\lambda)F'_S(\lambda_0)}{L_A(\lambda_0)F'_S(\lambda)} = \frac{\omega_A(\lambda)\tau_A(\lambda)P_A(\lambda, \theta, \theta_S)}{\omega_A(\lambda_0)\tau_A(\lambda_0)P_A(\lambda_0, \theta, \theta_S)} \tag{6.9}$$

In (6.9), $\lambda_0 = 865$ nm and the term $\varepsilon(\lambda, \lambda_0)$ is called the *single scattering color ratio* (King *et al.*, 1999).

Gordon and Wang (1994a) and Gordon and Voss (1999) describe the atmospheric correction procedure. For each pixel, $\varepsilon(765, 865)$ is calculated, then compared with values of ε

derived from known aerosols and listed in lookup tables. For SeaWiFS, the aerosol lookup tables are based on approximately 25 000 radiative transfer simulations derived from twelve different MODTRAN aerosol models. Comparison of $\varepsilon(765, 865)$ with the lookup table values allows the observed NIR aerosol radiances to be extrapolated into the visible and permits calculation of the diffuse transmittance. If the observed value of $\varepsilon(765, 865)$ equals that of a model aerosol, then ε is assumed to equal the model result in the visible; if ε lies midway between two model aerosols, it is also assumed to lie midway between the same two model results in the visible.

Consequently, once the aerosol type or ε is determined, L_A can be solved for in the visible, so that for example at 443 nm,

$$L_A(443) = \varepsilon(443, 865)L_A(870)[F_S'(443)/F_S'(865)] \tag{6.10}$$

The use of (6.10) for each band allows removal of the aerosol radiances from $L_T(\lambda)$, leaving the term $t_D(\lambda)L_w(\lambda)$. For single scatter, $t_D(\lambda)$ is from Equation (4.55); for multiple scatter, $t_D(\lambda)$ is from lookup tables.

From Gordon and Voss (1999) and for the SeaWiFS range of wavelengths, Figure 6.13 shows the dependence of ε on three relative humidities and three aerosols, called Maritime, Coastal and Tropospheric. The Tropospheric aerosol occurs above the marine boundary layer and consists of 70% water-soluble material and 30% atmospheric dust. The Maritime and Coastal aerosols are made up of different proportions of the Tropospheric and Oceanic aerosol, where the Oceanic aerosol occurs in the marine boundary layer and has a large sea-salt component. Specifically, the Maritime aerosol consists of 99% Tropospheric and 1% Oceanic aerosol; the Coastal consists of 99.5% Tropospheric and 0.5% Oceanic aerosol. Because the particles making up these aerosols are hygroscopic, as the humidity increases, the particle sizes also increase so that the scattering becomes more Mie-like and less dependent on wavelength. For these aerosols,

$$\varepsilon(\lambda, 865) = \exp[K_A(\lambda - 865)] \tag{6.11}$$

where the constant K_A depends on aerosol type. The figure shows for the Maritime aerosol at 98% humidity, $K_A \cong 0$, for which case the aerosol radiances at all wavelengths approximately equal those in the NIR. This behavior is characteristic of thin maritime aerosols. For the thickest aerosol, the Tropospheric at 50% humidity, $L_A(440)$ is approximately twice $L_A(865)$. For comparison, the Rayleigh path radiances give $L_R(440) \cong 15L_R(865)$, so that the Rayleigh dependence on wavelength is much greater than the aerosol dependence.

6.4.3 Distribution of oceanic aerosols

The global oceanic distribution of aerosols can be derived from the NIR bands. Aerosols are described in terms of two variables, the aerosol optical thickness τ_A and the Ångström exponent α, determined as follows. Given the aerosol type from the procedure described in the previous section, the scattering albedo ω_A and the aerosol phase function P_A can be

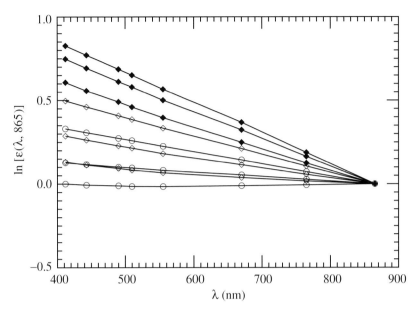

Figure 6.13. The wavelength dependence of $\varepsilon(\lambda, 865)$ for a nadir-view, a sun angle of $\theta_0 = 60°$, and for the Maritime $\circ$, Coastal $\diamond$ and Tropospheric $\blacklozenge$ model aerosols. For each aerosol, the relative humidities are 50, 80 and 98%, and decrease from the lower to upper curves (Figure 3a, Gordon and Voss, 1999, not subject to US copyright).

estimated in the NIR, which allows retrieval of $\tau_A(\lambda)$ at 765 and 865 nm (King *et al.*, 1999). The ratio of these optical thicknesses can be written as follows:

$$\tau_A(765)/\tau_A(865) = (765/865)^{-\alpha} \qquad (6.12)$$

For particles that are large relative to λ, Mie scatter occurs so that τ_A is nearly constant and α is approximately zero, while for small particles, the scattering tends toward Rayleigh scattering and α is larger. Thus the optical thickness $\tau_A(865)$ is proportional to the aerosol concentration, while α characterizes the size of the aerosol particles.

For April and October 1998, Plate 4 shows the global oceanic distribution of α and τ_A (Wang *et al.*, 2000). On the figure, land is black and regions with no data are gray. To the north and south, the gray regions correspond to sea ice; to the west of Africa, to optically thick dust blowing off the Sahara desert (April), or to biomass burning in southern Africa (October). The upper images show that most of the global oceanic aerosols have a small optical thickness ($\tau \sim 0.1, 0.2$). The lower images show that in the Southern Hemisphere and away from land, α is small, implying large marine particles. In contrast, off the east coast of North America, around Europe and on the east coast of Asia, $\alpha \cong 1$, implying the presence of what are probably the small sulfate particles characteristic of industrial pollution.

6.4.4 CZCS atmospheric correction algorithm

Because CZCS had no bands in the NIR, for each image an operator determined the values of ε from the 670-nm band using one of two different methods. In the first, the operator estimated the aerosol type by guessing the values of $\varepsilon(\lambda, 670)$. These were generally assumed to be close to unity and were based on the particular geographic region. The aerosol concentrations and resultant water-leaving radiances varied with this guess. In the second, the operator searched each image for a *clear water pixel*, defined as a pixel containing only pure seawater for which $L_w(\lambda)$ could be calculated. For this pixel, the $L_w(\lambda)$ and the Rayleigh radiances were removed from $L_T(\lambda)$, yielding the aerosol radiances at all wavelengths. These aerosol radiances were then used to calculate the values of $\varepsilon(\lambda, 670)$, which were assumed constant across the entire image. For both procedures, the additional assumption was made that for all pixels in the image, $L_w(670)$ equaled its pure seawater value. As long as the aerosol composition remained constant, then even though its concentration varied from pixel to pixel, both procedures removed the aerosol radiances. This assignment of ε or search for a clear water pixel meant that each image had to be individually processed, which was both time-consuming and a source of uncertainty. In contrast, the presence of NIR bands on both SeaWiFS and MODIS means that the images can be machine processed.

6.5 Chlorophyll reflectance and fluorescence

From the scattering and absorption behavior described in Section 6.2 and for seawater samples taken off the Oregon coast, in the Gulf of Maine and from Puget Sound, Figure 6.14 shows the dependence on wavelength and chlorophyll concentration C_a of the subsurface reflectance $R(\lambda)$ defined in Equation (6.1). On the figure, the horizontal bars show the location of the SeaWiFS and MODIS bands, where, as Section 6.6 discusses, the black bars identify the bands used in the empirical Chl-*a* algorithms discussed below. Examination of Figure 6.14 shows that as C_a increases, the reflectances exhibit the following behavior. First, for $\lambda < 550$ nm, R decreases as C_a increases; for $\lambda > 550$ nm, R increases as C_a increases; while at $\lambda = 550$ nm, R is approximately independent of C_a. Second, the radiance emitted at the 683-nm fluorescence peak increases with increasing C_a. Based on this behavior, there are at least two ways to retrieve the phytoplankton properties.

6.5.1 Reflectance

Many algorithms make use of the reflectance behavior shown for $\lambda \leq 550$ nm in Figure 6.14. At the 443-nm absorption peak and for $C_a \ll 1$ or nearly clean water, the figure shows that $R(443) \cong 0.08$. As C_a increases, $R(443)$ decreases dramatically and the maximum reflectance shifts toward 500 nm, so that as C_a increases, the water-leaving radiances become more green and less blue. For C_a greater than about 1–2 mg m^{-3}, $R(443)$ becomes so small that it approaches the noise floor of the instruments, while at 490 and 530 nm, the presence of accessory pigments with the weaker dependence of their absorption on

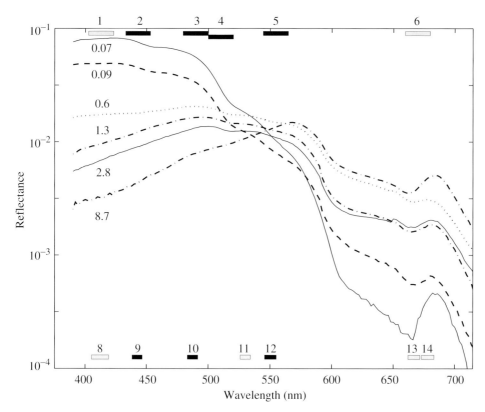

Figure 6.14. The subsurface reflectance $R(\lambda)$ plotted versus wavelength, for several values of C_a shown to the left and adjacent to each curve in units of mg m^{-3}. The lower horizontal bars show the MODIS bands; the upper, the SeaWiFS bands. SeaWiFS band 4 is offset vertically for clarity. For each set of bands, the black bars identify those used in the SeaWiFS and MODIS empirical Chl-a algorithms discussed in Section 6.6 (Data from Roesler and Perry, 1995, courtesy Collin Roesler).

C_a means that R decreases more slowly with increasing concentration. Consequently, the algorithms discussed below depend not only on the radiances measured at the 443-nm Chl-a absorption peak, but also on radiances in the 490–550 nm range that are dominated by carotenoid absorption.

For the algorithms to be successful, the concentrations of carotenoids and accessory pigments must covary with Chl-a in all parts of the ocean. Trees *et al.* (2000) show that even though the ratios of the accessory pigments to the Chl-a concentrations vary locally, globally the concentrations covary. They find that a log-regression of the measured accessory pigments against Chl-a yields a correlation coefficient of 0.934 with a root-mean-square error of 28%. This strong correlation in part explains the success of the algorithms. Finally, although at small C_a, $R(412)$ is very sensitive to changes in C_a, its companion sensitivity to CDOM and suspended particles means that the 412-nm band cannot be used in a chlorophyll algorithm without an accompanying CDOM algorithm.

6.5.2 Fluorescence

Figure 6.14 also shows that the magnitude of the 683-nm chlorophyll fluorescence peak increases with C_a. Fluorescence is generated from re-emission in the red of a portion of the solar radiation absorbed at all visible wavelengths by the phytoplankton chlorophyll. Of the absorbed radiation, about 85% is lost as heat, up to 12% is converted to chemical energy through photosynthesis and about 3% is re-emitted as fluorescence (Esaias *et al.*, 1998). Measurement of the fluorescence emission provides an alternative method for estimation of C_a as well as for determination of other phytoplankton properties. Letelier and Abbott (1996) and Esaias *et al.* (1998) show that determination of the fluorescence magnitude requires radiance measurements at the triplet of 10-nm wide bands centered at 667, 678 and 748 nm, where the 678-nm measurement determines the fluorescence and the 667- and 748-nm measurements allow removal of the background contribution. Measurements at this triplet allow determination of the chlorophyll-generated fluorescence line height (FLH).

Although SeaWiFS band 6 is too wide to observe the details of the fluorescence peak, ocean color instruments that carry this observing triplet include MODIS, MERIS and GLI. The advantage of this measurement is that the fluorescence is only produced by chlorophyll and is independent of CDOM and particles. From measurement of FLH and by making assumptions about the absorbed light and the species-specific fluorescence quantum efficiency, C_a can be calculated. Alternatively, knowledge of the species and its concentration yields the fluorescence quantum efficiency. In summary, there are at least two kinds of Chl-*a* algorithms: those that function in the blue-green (400–550 nm), where, as C_a increases, the reflectances at 440, 490 and 530 nm decrease relative to 550 nm, and those that function in the vicinity of the 683-nm fluorescence peak where the emitted light increases with C_a.

6.6 Ocean color algorithms

This section discusses a variety of bio-optical algorithms and their dependence on the reflectance ratios, then gives examples of several SeaWiFS and MODIS algorithms. Specifically, Section 6.6.1 describes the difference between the semianalytic and empirical algorithms, then briefly discusses the CZCS algorithm. Sections 6.6.2 and 6.6.3 respectively describe the SeaWiFS and MODIS empirical algorithms, Section 6.6.4 discusses the semi-analytic algorithms and Section 6.6.5 describes a species-specific algorithm for coccolithophores.

6.6.1 Two kinds of algorithms and their reflectance basis

For wavelengths of 400–550 nm, the two kinds of operational bio-optical algorithms are called *empirical* and *semianalytic* (O'Reilly *et al.*, 1998; Carder *et al.*, 1999). The empirical algorithms are derived from regression of coincident ship and satellite observations of $L_w(\lambda)$ against shipboard observations of C_a. The input to these algorithms is satellite observations of $L_w(\lambda)$ or equivalently R_{rs} at several wavelengths; the output is chlorophyll concentration. Because these algorithms yield only chlorophyll, their use is restricted to case 1 waters. The value of these algorithms is that they provide for continuity within the almost 25-year record of CZCS, SeaWiFS and MODIS observations.

In contrast, the semianalytic algorithms combine theoretical models of the relation of R_{rs} to the backscatter/absorption ratio with empirical models of the dependence of absorption and backscatter on the oceanic constituents, where the empirical relations change with season, geography and SST. As Section 6.6.4 discusses, these algorithms, combined with the $R_{rs}(\lambda)$ measured at several wavelengths, yield a set of simultaneous equations for quantities such as chlorophyll concentration and CDOM absorption. This allows chlorophyll to be retrieved in the presence of CDOM, or from both case 1 and case 2 waters.

The empirical and semianalytic algorithms are written in terms of any of the following three variables: the normalized water-leaving radiances $[L_w(\lambda)]_N$ defined in Equation (5.29), the remote sensing reflectances $R_{rs}(\lambda)$ in (5.30), or the extraterrestrial reflectances $[\rho_w(\lambda)]_N$ in (5.32) (O'Reilly *et al.*, 1998). The empirical algorithms primarily use R_{rs}, where from Equations (5.30) and (6.1),

$$R_{rs}(\lambda) = T^2 R(\lambda)/n^2 Q = [GT^2 b_{bT}(\lambda)]/[n^2 Q a_T(\lambda)] \qquad (6.13)$$

From (6.13), $R_{rs}(\lambda)$ is a linear function of subsurface reflectance $R(\lambda)$. Because as Figure 6.14 shows, $R(555)$ is approximately independent of concentration, the radiances or reflectances used in the algorithms are expressed as ratios relative to their value at 555 nm. The relationships among these ratios are

$$\frac{R_{rs}(\lambda)}{R_{rs}(555)} \equiv \frac{[\rho_w(\lambda)]_N}{[\rho_w(555)]_N} \equiv \frac{[L_w(\lambda)]_N F_S(555)}{[L_w(555)]_N F_S(\lambda)}$$
$$= \frac{R(\lambda)}{R(555)} = \frac{b_{bT}(\lambda)}{b_{bT}(555)} \frac{a_T(555)}{a_T(\lambda)} \qquad (6.14)$$

As examination of (6.13) and (6.14) shows, an advantage of working with these ratios instead of with the individual radiances or reflectances is that in the ratios the uncertainties associated with light propagation across the interface represented by G, T^2 and Q cancel.

The empirical algorithms use ratios of $[L_w(\lambda)]_N$ or R_{rs} based on the wavelength pairs 443/555, 490/555 and 510/555, where the CZCS algorithm uses the first and third pair, SeaWiFS uses all three, and MODIS uses the first two. From field data, Figure 6.15 shows the dependence of the R_{rs} ratios or C_a (adapted from Aiken *et al.*, 1995, Figure 6). The figure shows that as C_a increases, the 443-ratio decreases most rapidly, and the 490- and 510-ratios respectively decrease more slowly. This means that the 443-ratio is largest for small C_a, then as C_a increases, the 490-ratio becomes the largest, followed by the 510-ratio. This behavior provides the basis for the CZCS, SeaWiFS and MODIS empirical algorithms.

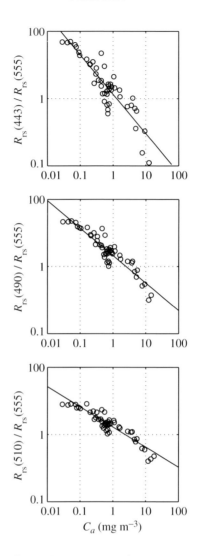

Figure 6.15. The dependence of $R_{rs}(\lambda)/R_{rs}(555)$ on C_a for $\lambda = 443$, 490 and 510 nm. The straight lines are linear least-square fits to the data (Adapted from Aiken *et al.*, 1995).

The CZCS algorithm was written in terms of $[L_w(\lambda)]_N/[L_w(550)]_N$, where $\lambda = 433$ and 520 nm, and was called a *switching* algorithm for the following reason. For $C_a < 1.5$ mg m^{-3}, the chlorophyll concentrations were derived from the 443/550 ratio; for $C_a > 1.5$ mg m^{-3}, they were derived from the 520/550 ratio (O'Reilly *et al.*, 1998). The difficulty with this algorithm was that there could be an abrupt shift in properties at the 1.5 mg m^{-3} switch-over point between bands. Because of the problems induced by this shift, the Sea-WiFS and MODIS empirical algorithms use a different, *maximum band ratio* approach.

6.6.2 SeaWiFS maximum band ratio empirical algorithm

This section first discusses the field data sets used in the derivation of the SeaWiFS and MODIS algorithms, then describes the maximum band ratio algorithms. Beginning in 1997, the NASA-sponsored SeaWiFS Bio-optical Algorithm Mini-Workshop (SeaBAM) began collection of a global surface radiance and chlorophyll data set. This data set contains simultaneous *in situ* measurements of radiance and chlorophyll, and the less-frequent simultaneous *in situ* observations of chlorophyll and satellite-observed radiances. In 1997, the data set consisted of 919 different stations, with $0.019 < C_a < 32.79$ mg m^{-3} (O'Reilly *et al.*, 1998).

From these data, O'Reilly *et al.* tested two semianalytic and fifteen empirical regional and global algorithms, where *global* refers to a single algorithm that provides reasonable results for tropical, subtropical and temperate waters. The SeaWiFS global algorithm that provided the best fit to the SeaBAM data was the maximum band ratio Ocean Chlorophyll-4 (OC4) empirical algorithm, which is version 4 of the currently recommended SeaWiFS algorithm (see http://seawifs.gsfc.nasa.gov/SEAWIFS/RECAL/Repro3/ and http://seawifs.gsfc.nasa.gov/SEAWIFS/RECAL/Repro4/). As Section 6.6.3 and Chapter 14 describe, a similar maximum band ratio algorithm is used with MODIS and is proposed for use with VIIRS.

The OC4 algorithm is written in terms of R_{rs}-ratios. The reason it is called a maximum band ratio algorithm is that unlike the CZCS, there is no fixed value of C_a at which the algorithm switches ratios. Instead, the algorithm uses whichever R_{rs}-ratio (443/555, 490/555, 510/555) is largest. Consequently, as C_a increases, the OC4 algorithm first uses the 443-ratio, then when the 490-ratio is greater than that derived from the 443-ratio, OC4 switches to the 490-ratio, and finally to the 510. The advantage of this approach is that over a broad range of C_a, the signal-to-noise ratio remains as large as possible. For SeaWiFS, OC4 consists of the following fourth-order polynomial (http://seawifs.gsfc.nasa.gov/SEAWIFS/RECAL/Repro3/OC4_reprocess.html):

$$R_{MAX} = \text{Maximum of } [R_{rs}\text{-ratio}(443/555, 490/555, 510/555)]$$
$$R_L = \log_{10}(R_{MAX}) \tag{6.15}$$
$$\log_{10}(C_a) = 0.366 - 3.067R_L + 1.930R_L^2 + 0.649R_L^3 - 1.532R_L^4$$

Figure 6.16 compares the current SeaBAM data set with the OC4 algorithm, where in each sub-figure, C_a is plotted against R_{MAX}. The four sub-figures are identical, except that on the different plots, the darkened data points show the range of C_a to which each ratio contributes. Figure 6.16a shows that as R_{MAX} increases, C_a decreases. As expected, at small values of C_a, the 443-ratio dominates; at mid values, 490 dominates; at large values, 510 dominates. Because the range of the dominant bands overlaps by 10–30%, the algorithm experiences smooth transitions as R_{MAX} decreases.

For the same SeaBAM data points shown in Figure 6.16, Figure 6.17 compares the surface and OC4 satellite-retrieved values of C_a. On the figure, the central $45°$ straight line is the line of perfect agreement; the closest irregular line is the regressive fit to the data;

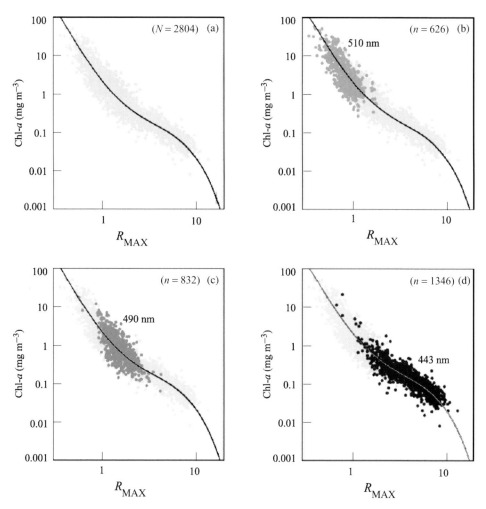

Figure 6.16. A log-log comparison of the OC4 algorithm with the SeaBAM data. In each panel, the solid line shows C_a from Equation (6.15); the data points show the dependence of the chlorophyll concentration on the R_{rs}-ratios. Each panel compares all data with the OC4 curve; on the sub-figures b, c and d, the R_{rs}-ratios of 510, 490 and 443 nm to 555 nm are respectively darkened. Also, N is the total number of points, n is the number of points used in each respective ratio (Courtesy of the NASA SeaWiFS project, http://seawifs.gsfc.nasa.gov/SEAWIFS/RECAL).

the two solid lines above and below these lines show the 35% range of agreement; the two dashed lines above and below the regression line are at values of 5 and 1/5 times the perfect agreement line. For the logarithms of the *in situ* and derived chlorophyll shown in the figure, the correlation coefficient is 0.892; the root-mean-square agreement is 0.222. The figure shows that the agreement between the two data sets is best for $C_a < 1$ gm m^{-3}, while for larger values of C_a, the disagreement is greater. This suggests that the desired

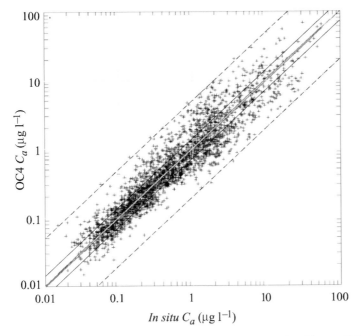

Figure 6.17. A comparison of the *in situ* and satellite Chl-*a* for the OC4 algorithm and 2804 observations. The inner diagonal solid lines mark the ±35% agreement relative to the 1:1 central solid line; the dashed lines are the 1:5 and 5:1 lines bordering the data set (Courtesy of the NASA SeaWiFS project, http://seawifs.gsfc.nasa.gov/SEAWIFS/RECAL).

35% uncertainty can be met for small values of C_a. Hooker and McClain (2000) further discuss the algorithm validity.

6.6.3 MODIS empirical algorithm

Because MODIS lacks the 510-nm band used in the OC4 SeaWiFS algorithm, the MODIS empirical maximum band ratio algorithm is based on three bands instead of four. The algorithm, called OC3M for MODIS, is the successor to the CZCS and SeaWiFS empirical algorithms and is written as follows:

$$R_L = \log_{10}(\max[R_{rs}\text{-ratio}(443/551, 488/551)]) \tag{6.16}$$
$$\log_{10}(C_a) = 0.283 - 2.753R_L + 0.659R_L^2 + 0.649R_L^3 - 1.403R_L^4$$

For OC3M, the relation between R_{rs} and chlorophyll was parameterized with the same SeaBAM data used with OC4. The statistics of the OC3M algorithm are about the same as the SeaWiFS OC4 statistics. Comparison with additional coastal and mid-ocean data shows that, as expected, the algorithm overestimates chlorophyll in case 2 waters. OC3M also underestimates Chl-*a* at concentrations below 1 mg m^{-3}, and overestimates them

at larger values (information on OC3M from http://modis-ocean.gsfc.nasa.gov/qual.html/
dataqualsum/chlor_a_2_qualsum.html).

6.6.4 Semianalytic algorithms

The semi-analytic algorithms combine theoretical models of the R_{rs} dependence on the
backscatter and absorption with empirical formulas that describe the absorption and
backscatter dependence on particles and on the CDOM and phytoplankton pigments. The R_{rs}
models are similar to those described by Equations (6.13) and (6.14). The empirical formulas that give the dependence of the absorption and backscatter on the distribution of particle
size and on the concentrations of the various constituents are similar to those described in
Section 6.2 and change with season, geography and SST. As Carder *et al.* (1999) show, the
combination of these theoretical and empirical relations with the satellite-observed $L_w(\lambda)$
yields a set of simultaneous equations for quantities such as C_a, and absorption by CDOM
and phytoplankton.

   In their algorithm, Carder *et al.* (1999) use the following form of the R_{rs}-ratio, where
$R_{rs}(\lambda)$ is expanded to show the absorption and scattering contributions from the different
constituents,

$$\frac{R_{rs}(\lambda)}{R_{rs}(555)} = \frac{b_{bT}(555)}{a_T(555)} \frac{[a_w(\lambda) + a_\phi(\lambda) + a_{CDOM}(\lambda)]}{b_{bw}(\lambda) + b_{bp}(\lambda)} \tag{6.17}$$

In (6.17), the absorption divides into contributions from pure seawater (a_w), phytoplankton
(a_ϕ) and CDOM (a_{CDOM}); the backscatter divides into contributions from pure seawater
(b_{bw}) and particulates (b_{bp}). The reason for the absence of the particulate absorption (a_p) is
discussed below. For terms in the numerator, Section 5.4.1 describes the absorption of pure
seawater; an analytic function based on the curves in Figure 6.3 gives the phytoplankton
absorption. Because as Section 6.2.2 describes, a_p and a_{CDOM} have identical analytic forms,
in the numerator they are combined into the single a_{CDOM} term given by Equation (6.6).
In the denominator, b_{bw} is derived from Equations (5.19) and (5.23), and b_{bp} is given by
Equation (6.4), where the magnitude of X and Y respectively depend on concentration and
the particle size distribution. Carder *et al.* (1999) show that the particle size distribution
varies according to season and region, being largest adjacent to the mouths of major rivers.

   The semianalytic algorithm has two additional inputs: the SST and the Nitrate-Depletion
Temperature (NDT). As Carder *et al.* (1999) describe, NDT is the near surface temperature
at which nitrate in the surface waters can no longer be detected, as determined regionally
from ship surveys. When the SST exceeds the NDT, the water is nutrient-poor and is characterized by the growth of small diameter phytoplankton cells with unpackaged pigments;
for the opposite case, the water is nutrient-rich and characterized by fast-growing cells
with packaged pigments. This relation between the SST and NDT is used to classify the
ocean into three regimes: (1) a warm regime consisting of tropical, subtropical and summer
waters with unpackaged chlorophyll within each cell; (2) a transitional regime; (3) a cold

regime, consisting of high latitude and upwelling regions where the phytoplankton consist of fast growing diatoms with packaged chlorophyll. As shown below and especially at high latitudes, this classification provides an important correction to the empirical estimates of Chl-*a*.

From the SST and NDT classification, and from the regional differences in the constituent scattering and absorption properties, the ocean is divided into different geographic regions. For the measured R_{rs}, the constituent concentrations for each region are described by multiple equations and multiple unknowns. The inputs include the MODIS-derived SST, the regional NDT and the radiances at 412, 443, 488, 531 and 551 nm; the outputs include Chl-*a*, the absorbed radiation by phytoplankton (ARP), the 400-nm CDOM absorption $a_{CDOM}(400)$ and the 675-nm phytoplankton absorption a_ϕ (675). For these last two, formulas exist to extend a_{CDOM} and a_ϕ across the entire visible spectrum. The sum of these yields the total absorption $a_T(\lambda)$, from which the visible radiation absorbed in the water column can be derived. The various components of the absorption can then be inverted to yield concentrations of phytoplankton and CDOM.

6.6.5 Species-specific algorithms

With some scale differences associated with chlorophyll packaging, most phytoplankton species share a common absorption curve and cannot be individually identified from space (M. J. Perry, private communication, 2002). A few species, however, have unique absorption curves that permit their identification. These include coccolithophores, the toxic red tide dinoflagellate *Karenia brevis* (*K. brevis*), and phycoerythrin-containing species such as *Trichodesmium*, which is the major ocean nitrogen-fixing phytoplankton. Esaias *et al.* (1998) describe identification of the phycoerythrin species; Cannizzaro *et al.* (2004) discuss *K. brevis*; coccolithophores are discussed next.

Coccolithophores are an unusual class of phytoplankton. When they bloom, they grow and shed external plates or shells, consisting of micrometer-scale white platelets made up of calcium carbonate (calcite). Holligan *et al.* (1993) describe a field study of a bloom of the most widespread coccolithophore, *Emiliania huxleyi* (*E. huxleyi*) shown in Figure 6.18. *E. huxleyi* occurs in the mid to high latitudes, where its blooms take place in one-month periods over areas of order 10^5 km^2 and produce more calcite than any other single organism. The importance of these blooms is two fold. First, they generate large fluxes of dimethyl sulphide (DMS) to the atmosphere, which can trigger cloud formation. Second, because each bloom generates of order 10^9 kg of calcite, they enhance the absorption of atmospheric CO_2. The presence of the white shells in the surface waters changes the water color to a milky blue over large areas. In contrast to the phytoplankton pigments, which strongly decrease the radiance in the blue while only slightly increasing it in the green, coccolithophores increase the water-leaving radiance uniformly in the blue and green (Balch *et al.*, 1999). Because of this uniform increase, the *E. huxleyi* algorithm is written in terms of the normalized

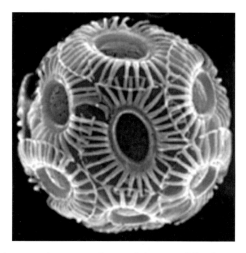

Figure 6.18. Scanning electron microscope picture of the coccolithophore species *Emiliania huxleyi*. The sphere diameter is about 3 μm. Each plate is a separate lith, which remains in the ocean after the organism has died (Micrograph courtesy Jeremy R. Young, used with permission).

water-leaving radiances $[L_w(\lambda)]_N$ at 443 and 555 nm, instead of in terms of the R_{rs}-ratios (Gordon and Balch, 1999).

6.7 Examples of satellite images

This section gives examples of SeaWiFS and MODIS imagery processed by the empirical and semianalytic algorithms. First, Section 6.7.1 examines a SeaWiFS image band by band to show how the atmospheric correction procedure and variable chlorophyll concentrations affect the received and water-leaving radiances and the OC4 chlorophyll distribution. Second, Section 6.7.2 uses 2001 MODIS imagery to compare the annual averaged Chl-*a* distributions derived from the empirical and semianalytic algorithms and to display near global distributions of CDOM, FLH and calcite.

6.7.1 An OC4 processed SeaWiFS image

From a SeaWiFS image of the coastal waters of British Columbia, Canada and Washington and Oregon, United States taken on September 1, 1999, this section examines the wavelength dependence of the total received radiances, the OC4 Chl-*a* distribution and the water-leaving radiances. Plate 8 discussed in Chapter 7 shows an additional OC4 SeaWiFS image that compares the 1997–98 transition from El Niño to La Niña and the effect of this change on the surface temperature and biology of the equatorial Pacific.

Beginning with the total radiances, Figure 6.19 shows $L_T(\lambda)$ for the 865, 765, 670, 555, 443 and 412 nm SeaWiFS wavelengths, and illustrates the effect of Rayleigh scattering.

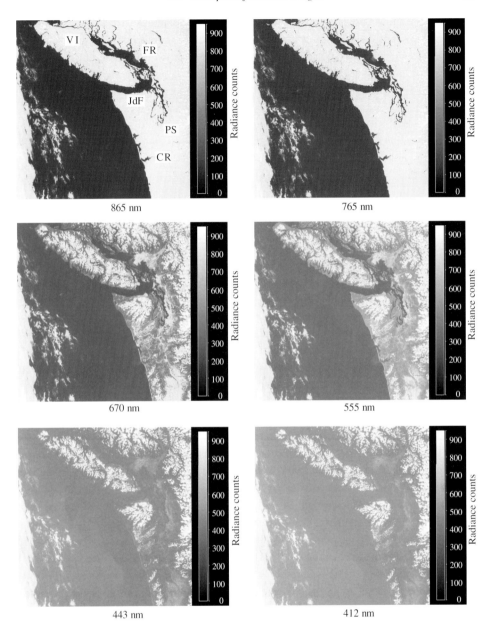

Figure 6.19. The total radiances received at the SeaWiFS instrument for the northeast Pacific adjacent to British Columbia, Canada and Washington and Oregon, United States on September 1, 1999. The channels include the two NIR channels at 865 and 765 nm, the 670-nm channel spanning the fluorescence peak, and three channels in the blue-green at 555, 443 and 412 nm. On the figure, VI is Vancouver Island; FR, Fraser River; JdF, Strait of Juan de Fuca; PS, Puget Sound; CR, Columbia River (OrbView-2 Imagery provided by ORBIMAGE, the SeaWiFS Project and NASA/Goddard Spaceflight Center, processing courtesy Brandon Sackmann and Miles Logsdon).

The gray scales to the right of each image are expressed in radiance counts. For the images at 765 and 865 nm where L_w is near zero, land and clouds are bright, and except for the radiances associated with the sediment plumes at the mouths of the Columbia and Fraser Rivers, the ocean is dark. The 865-nm image particularly shows how the clouds stand out against the ocean background. Both the Columbia and Fraser Rivers carry large sediment loads, which from personal observation generate a brownish yellow color in the surface waters. At 670 nm, the river plumes remain visible and there is a suggestion of a large vortex or jet off the Strait of Juan de Fuca. For shorter wavelengths, examination of the remaining images shows that Rayleigh scatter progressively obscures the land and ocean surface.

Plate 5a shows a true color image of the region, obtained by a red–green–blue (RGB) mixing of the 670, 555 and 410 nm bands, with Rayleigh scattering removed and the colors enhanced. The image shows green land, snow in the mountains and white clouds off the coast. Away from the coast, the water is blue; closer to the coast, it is darker blue with hints of green. The image also shows the sediment plumes off the mouths of the Columbia and Fraser Rivers. For comparison, Plate 5b shows the OC4 Chl-*a* concentrations in units of mg m^{-3}. The land and clouds are masked in black, the coast is outlined in red. Because the processing of this image depends on the assumption of a zero water-leaving radiance at 865 and 765 nm, the results are invalid adjacent to the river mouths. The plate shows enhanced biological activity adjacent to the coast, a number of plumes and jets extending into the Pacific and the absence of biological activity in the offshore waters. The image also shows that at the Pacific entrance to the Strait of Juan de Fuca, a plume of a low-chlorophyll water flows into the richer Pacific water. The overall chlorophyll distribution is consistent with field observations, in that coastal upwelling creates productive regions adjacent to the coast, while the offshore regions remain inactive. Because of land contamination from diffuse attenuation, there is little usable data from the interior of Puget Sound.

Figure 6.20 shows the normalized water-leaving radiances $[L_w(\lambda)]_N$ for 670, 555, 510, 490, 443 and 412 nm. Because their water-leaving radiances are zero except at the river mouths, the 865- and 765-nm images are omitted. The radiance scales are to the right of each image; these vary from image to image and are chosen to maximize the contrast. The length of these scales is proportional to the range of radiance values; the shorter the bar, the more sensitive the measurement. Examination of the 670-nm band image shows that the productive regions adjacent to the coast are bright, the river outflows are very bright, and the offshore waters are dark. This suggests that at 670 nm, the bright radiances away from the river mouths are generated by fluorescence; adjacent to the rivers, by reflection from CDOM and sediments. At 555 nm, which is used in the denominator of the R_{rs}-ratios, the image does not have a uniform brightness, rather the river mouths remain bright and away from these regions, the radiances generally follow the Chl-*a* distribution but with less contrast.

At 510 and 490 nm, the river mouths remain bright, while away from these areas, the radiance pattern is reversed from the 670-nm fluorescence image, in that the water-leaving radiance in the high Chl-*a* water adjacent to the coast is now less bright than in the less

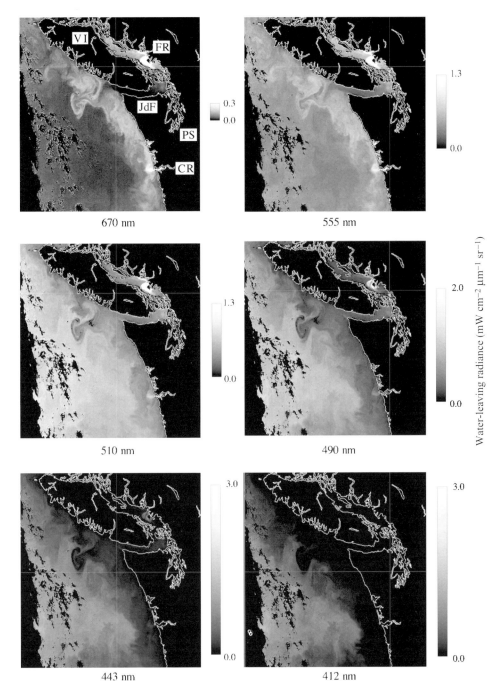

Figure 6.20. The normalized water-leaving radiances for the visible SeaWiFS channels and for the region shown in Plate 5. The gray scale to the right of each image shows the water-leaving radiance. The length of each bar is proportional to the range of the radiance values; the shorter the bar, the more sensitive the measurement. Black corresponds to the land and cloud mask (OrbView-2 Imagery provided by ORBIMAGE, the SeaWiFS Project and NASA/Goddard Spaceflight Center, processing courtesy Brandon Sackmann and Miles Logsdon).

productive, offshore water. At 443 nm, $[L_w(\lambda)]_N$ is zero in the high Chl-*a* waters adjacent to the coast, and there is only a slight hint of a bright region off the Fraser River delta. Finally, at the 412-nm absorption maximum for suspended material and CDOM, the only bright pixels in the image occur in the offshore region of low concentrations of Chl-*a*. This shows that the radiances at the river mouths become smaller at those wavelengths where the enhanced attenuation associated with CDOM and particles predominates. In summary, on a band by band basis, the images are consistent with the algorithms and with the regional oceanography. The above images also illustrate the difficulties that the empirical algorithms have with case 2 sediment-laden waters.

6.7.2 Examples of MODIS retrievals

This section describes the spatial distribution of several parameters derived from 2001 annual-averaged MODIS data. Plate 6 shows four quantities: the empirical and semianalytic distributions of Chl-*a*, the fluorescence line height (FLH) and the CDOM absorption at 400 nm. Plate 7 additionally shows the calcite distribution. For these quantities, the empirical Chl-*a* is from Equation (6.16), the semianalytic distributions of Chl-*a* and a_{CDOM} are from the algorithm described in Section 6.6.4, FLH is derived from the fluorescence triplet algorithm discussed in Section 6.5.2, and the calcite distribution is derived from the coccolithophore algorithm discussed in Section 6.6.5.

   The two Chl-*a* distributions and the FLH image show that along the equator in the Pacific, there is a band of enhanced chlorophyll associated with La Niña-related upwelling. This band is bounded to the north and south by large areas with a dark blue color of the tropical and subtropical ocean. Because the surface solar heating stabilizes the upper ocean and limits the nutrient upwelling, these areas have low biomass. There are also regions of enhanced chlorophyll associated with wind-driven upwelling along the west coast of equatorial South America, along the west coast of southern Africa, near the tip of eastern South America in the vicinity of the Falkland Islands and around Antarctica, and in the marginal seas of the North Pacific. Comparison of the empirical and semianalytic values of Chl-*a* shows that at high latitudes, the semianalytic values are greater than the empirical. Carder *et al.* (1999) state that this difference is due to the NDT-based packaging correction. From their work around Antarctica, they found that the semianalytic values were 40% greater than the empirical and are in better agreement with ship observations. Also, even though the FLH is derived very differently, it shows similar features to the other estimates, with strong fluorescence along the equator and at mid to high latitudes.

   The CDOM absorption image shows that because of river run-off, CDOM has a maximum in the river-rich Northern Hemisphere. In regions of biological productivity, the distribution is also enhanced by the presence of detritus and CDOM. This occurs in the equatorial Pacific and in other regions of biological productivity such as around Antarctica. In contrast, the calcite in Plate 7 has a very different distribution and is large in the North Atlantic, North Pacific and around Antarctica. In contrast to the other images, there is little calcite along the equator. Since *E. huxleyi* tends to occur in high latitudes, this distribution is consistent with surface observations.

6.8 Further reading and acknowledgements

The SeaWiFS mission is described at http://seawifs.gsfc.nasa.gov/SEAWIFS.html; the calibration/validation publications are at http://seawifs.gsfc.nasa.gov/SEAWIFS/CALVAL. publications.html. The prelaunch SeaWiFS Technical Report Series can be ordered and the postlaunch SeaWiFS Technical Report Series can be downloaded at http://seawifs.gsfc.nasa.gov/SEAWIFS/TECH_REPORTS/. The site also contains a list of titles and abstracts of related published papers. The improvements to the algorithms and description of the reprocessing of the global data set are described at http://seawifs.gsfc. nasa.gov/SEAWIFS/RECAL/, which also contains downloadable files of technical reports and papers. There is an excellent collection of SeaWiFS imagery at http://seawifs. gsfc.nasa.gov/SEAWIFS/IMAGES/GALLERY.html. The MODIS Oceans home page is at http://modis-ocean.gsfc.nasa.gov; this site describes the instrument, has links to movies, recent downloadable presentations and provides a publication list. MODIS documents can be found at http://eospso.gsfc.nasa.gov/eos_homepage/for_scientists/index.php, and MODIS presentations can be found at http://modis-ocean.gsfc.nasa.gov/refs.html. The MODIS Algorithm Theoretical Basis Documents (ATBD-MOD) are specifically located at http://modis.gsfc.nasa.gov/data/atbd/ocean_atbd.html. Browse images are in the MODIS Oceans QA Browse 36km Imagery (MQABI) archive at http://jeager.gsfc.nasa.gov/ browse-tool/. Unfortunately, the URLs of these websites tend to change with time; they can be recovered with a search engine. MODIS data are also available on an ftp site at http://podaac.jpl.nasa.gov/, and information about MODIS is available at http://podaac. jpl.nasa.gov/modis/index.html. The Oregon State University website http://picasso.oce. orst.edu contains an excellent summary of all MODIS algorithms. The color instruments of many countries are described at the International Ocean Colour Coordinating Group (IOCCG) website at http://ioccg.org. The spectral properties of the past, present and future ocean color instruments are described at http://www.ioccg.org/sensors/500m.html. In each case, the sensor is described with links to other descriptive material maintained by the relevant space agencies. Recent material on species identification is at http://modis-ocean.gsfc.nasa.gov. The red tide or *K. brevis* is part of a current research effort called Ecology and Oceanography of Harmful Algae Blooms (ECO-HAB) (http://www.redtide.whoi.edu/hab/nationplan/ ECOHAB/ECOHABhtml.html). I thank William Barnes, Jennifer Cannizzaro, Rita Horner, Evelyn Lessard, Mary Jane Perry, Wayne Esaias, Gene Feldman, Charles McClain and Peter Minnett for their help with the material in this chapter.

7

Infrared observations of sea surface temperature (SST)

7.1 Introduction

Satellite SST observations contribute to an understanding of regional variability and global climate change and permit the visualization of a wide variety of oceanic flows. These observations are important for the following reasons. First, because the upper 3 m of the water column has about the same heat capacity and the upper 10 m has about the same mass as an overlying column of atmosphere, the upper ocean moderates the global climate system where SST is proportional to the upper ocean heat storage (Gill, 1982). Second, the spatial and temporal distributions of the atmospheric fluxes of water vapor and heat are functions of surface temperature. Third, the patterns of surface temperature gradients associated with current systems, eddies, jets and upwelling regions make these processes visible in SST imagery.

Beginning in 1981 with the launch of AVHRR/2 on NOAA-7, there now exist two decades of infrared satellite SST observations. These contribute to multiyear global climate studies and to shorter period regional support of fisheries, ship routing, physical oceanography research and weather forecasting (Walton, Pichel and Sapper, 1998). Examples of long-term studies include the changes in SST patterns associated with such interannual climate variations as the La Niña and El Niño cycle in the equatorial Pacific and Atlantic. Also, from examination of surface SST observations over the past five decades, Casey and Cornillon (1999) show that if satellite SSTs were accurate within about ±0.2 K, climate-induced temperature changes would be observable within a two-decade period. Given that the current operational AVHRR accuracy is about ±0.5 K (Walton *et al.*, 1998) and the MODIS night-time SST accuracy is about ±0.3 K (Section 7.5.6), observation of these changes may soon be achievable. Examples of short-term SST applications include delineation of ocean fronts, upwelling regions, equatorial jets and ocean eddies. Identification and tracking of such features require accuracies of at least ±0.5 K and frequent revisits (Walton *et al.*, 1998). The accuracy requirement is presently met, but because the acquisition of cloud-free images has only a 10% success rate, the revisit requirement is not.

The equivalent surface measurements of SST are taken by commercial ships and by arrays of moored and drifting buoys. For example, Figure 7.1 shows for December 1993 the locations of ships reporting seawater intake temperatures to the Japan Meteorological

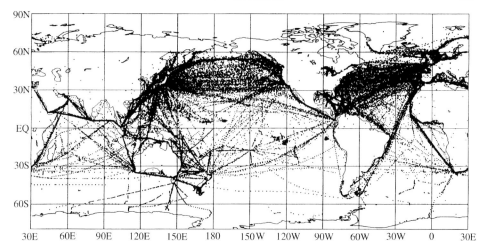

Figure 7.1. Location of ships reporting near-surface ocean temperatures during December 1993 to the Japan Meteorological Agency (JMA) (Figure 21-1 from the Japan Meteorological Agency Monthly Ocean Report, December 1993 (No. 12), courtesy JMA, used with permission).

Agency (JMA) and other national weather services. The intakes are at depths of 3–10 m below the surface. The figure shows that the ship positions are heavily biased toward the Northern Hemisphere and are primarily located on the great circle routes between the US, Europe and Asia. An alternate source of SST observations is provided by arrays of moored and drifting buoys, which as Section 7.5.3 discusses are deployed both globally and in specific regions of climatic importance such as the equatorial Pacific.

This chapter describes the SST retrieval by AVHRR, MODIS and the ESA Along-Track Scanning Radiometer (ATSR). Other instruments that measure SST include atmospheric sounders such as the High Resolution Infrared Radiation Sounder (HIRS) on the NOAA satellites, the infrared imagers and sounders on the GOES satellites, the cloud-independent but lower resolution passive microwave instruments described in Chapter 9, and a variety of infrared satellite instruments with similar observational bands to AVHRR and MODIS. Examples include the Multispectral Visible/IR Scanning Radiometer (MVIRSR) on the Chinese Feng Yun (Wind and Cloud) sun-synchronous satellites, MERIS on ENVISAT, and the Japanese Global Imager (GLI) on ADEOS-2.

Because the MODIS and AVHRR SST retrievals remove the atmospheric radiances generated by tropospheric water vapor but not by aerosols, their SSTs must be continuously calibrated by surface observations. Consequently, their SST product is a hybrid of surface and satellite observations. The AVHRR SST combines satellite-retrieved surface temperatures with near surface temperatures measured by moored and drifting buoys; the MODIS SST combines satellite with buoy and ship-based infrared SST observations. Given that the SSTs depend on atmospheric properties and concurrent near surface measurements, two sets of processes must be considered. The first is the effect of the surface, atmosphere and sun on the received radiances discussed below; the second is the effect of

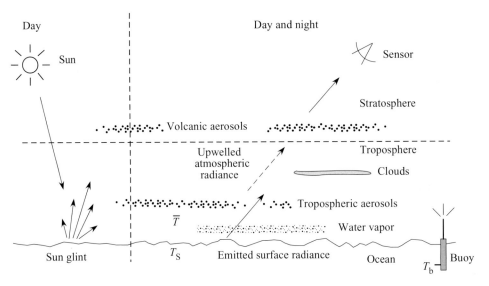

Figure 7.2. Schematic drawing of the radiances and atmospheric and oceanic properties involved in the infrared SST retrieval. See text for further description.

ocean surface and near surface processes on the measured SST discussed in the next section.

Figure 7.2 illustrates the radiances and atmospheric properties involved in the retrieval of the ocean surface radiances. On the figure, $\overline{T}$ is a characteristic temperature of the lower atmosphere, T_b is the buoy or *bulk* temperature measured at depths of 0.3–1 m, and T_S is the surface *skin* temperature, which is the temperature of an infinitesimally thin layer at the very top of the water column. The difference between T_S and the SST is that T_S is the physical skin temperature, while as this chapter discusses, SST is an approximation to T_S that depends on the measurement technique. The processes that govern the retrieval divide into those applicable at all times during day and night, and those applicable only in daytime. During the day and as Section 7.4.2 discusses in detail, a critical concern is the avoidance or masking of the reflection of sunlight into the instrument. Depending on look angle, this reflection dominates the shorter thermal wavelengths, but can be ignored at the longer wavelengths except when the instrument looks directly at the sub-solar spot.

Determination of the transmittance t is a critical part of the retrieval. In the troposphere, t is dominated by the variable water vapor content and by aerosols such as Saharan dust; in the stratosphere, t is altered at sporadic intervals by the stratospheric injection of volcanic aerosols. The received radiance is a combination of the surface-emitted radiance and the atmospheric upwelled radiance, where their relative magnitudes depend on the observing wavelengths and on the atmospheric water vapor and aerosol content. Finally, because clouds can be either opaque and highly attenuating, or thin and slightly attenuating, the day and night identification of clouds is another critical part of the retrieval.

The following sections describe the retrieval of SST from multichannel AVHRR and MODIS observations and from multilook ATSR and AATSR observations. Section 7.2 describes the different definitions of SST and the factors that affect it. Section 7.3 describes the AVHRR and MODIS bands used in the SST retrieval and discusses the various forms of AVHRR data. Section 7.4 discusses the atmospheric emission properties, the emission and reflection properties of the ocean surface, and the problem of sun glint. Section 7.5 describes the operational AVHRR algorithms, the calibration of AVHRR SSTs with moored and drifting buoy temperatures, and the MODIS algorithms. Section 7.6 describes the dual-look ATSR and AATSR retrieval algorithms. Section 7.7 discusses the different day and night cloud discrimination algorithms, and Section 7.8 concludes with examples of AVHRR and MODIS imagery.

7.2 What is SST?

A variety of oceanic surface and near surface processes determine the SST and upper ocean temperature profile. These processes include solar heating and nighttime radiative cooling, evaporative cooling, and wind and wave mixing (Figure 7.3). In general, the upper ocean layer heats up during the day from solar heating and cools at night. Because of these processes, the difference $\Delta T = T_S - T_b$ between the skin and bulk temperature can vary by as much as ± 1 K (Katsaros, 1980).

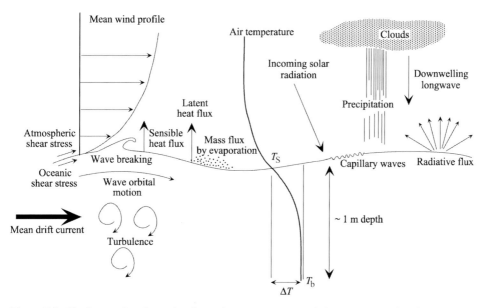

Figure 7.3. The factors that determine the surface temperature and the upper ocean interior temperatures, where ΔT is the temperature difference between the bulk and skin temperatures (Adapted from Figure 1 of Katsaros, 1980).

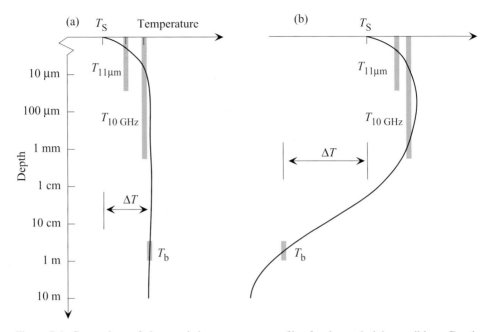

Figure 7.4. Comparison of characteristic temperature profiles for day and night conditions. Depth is on a log scale. (a) Day and night strong wind case, where $U > 6\,\mathrm{ms^{-1}}$. (b) Daytime weak wind and strong solar insolation case that yields a stratified upper ocean. On the figure, T_S is the surface temperature, the bar marked $T_{11\mu m}$ shows the depth range that contributes to the infrared 11-μm SST; the bar marked T_{10GHz} shows the depth range that contributes to the microwave surface temperature, T_b is the buoy or bulk temperature and ΔT is the temperature difference between the buoy and surface temperatures (Adapted from Figure 1, Donlon *et al.*, 2002).

Because different instruments measure SST in different ways, the term SST refers to temperatures measured over several different depth intervals from the surface (Figure 7.4). The first is the desired skin temperature T_S, which contributes to the determination of the ocean/atmosphere heat and moisture fluxes. Because of evaporative and radiative cooling, T_S is generally lower than the interior temperature. The second is the temperature measured by an infrared radiometer operating at 11 μm, shown on the figure as $T_{11\mu m}$, which is the radiative temperature emitted from the top 30 μm of the water column. In general, $T_{11\mu m}$ is slightly higher than T_S, and given current technology, it is the closest measurable approximation to T_S. A similar temperature can be defined at 4 μm. The third is the temperature T_{10GHz}, which is measured at 10 GHz by a microwave radiometer. As Section 9.5 discusses, this is the radiative temperature emitted from the top 1–2 mm of the water column. The final is the bulk temperature T_b that is measured at depths of 0.3–1 m by the moored and drifting buoys.

The terms T_S and $T_{11\mu m}$ respond to changes in the evaporative, conductive and radiative balance within a few seconds; T_{10GHz} is dominated by viscous processes and responds on time scales of order minutes. For weak winds, T_b responds to diurnal heating and cooling on the order of hours; for strong winds, T_b responds at shorter periods (Donlon *et al.*, 2002).

In the AVHRR procedure for SST calibration and removal of aerosols, T_b is a surrogate for T_S. As Section 7.5.4 shows, this limits the accuracy of the AVHRR SST retrieval to about ± 0.5 K. For MODIS and the non-operational Pathfinder AVHRR SST, ship-mounted infrared radiometers such as the Marine-Atmosphere Emitted Radiance Interferometer (M-AERI) are used in the calibration. As Section 7.5.6 describes, the use of M-AERI data reduces the nighttime error to about ± 0.3 K. As Donlon *et al.* (2002) describe, ΔT also depends on wind speed. From radiometer measurements, they show that for wind speeds greater than 6 m s^{-1}, the effect of diurnal heating is minimized so that for both day and night conditions, ΔT is characterized by a cool bias of -0.17 ± 0.07 K (Figure 7.4a). At smaller wind speeds, the daytime thermal stratification of the upper layers means that ΔT is larger (Figure 7.4b). This wind speed dependence further limits the retrieval accuracy.

7.3 Properties of AVHRR and MODIS bands used in SST retrieval

As Chapter 1 and Cracknell (1997) describe, the AVHRR is a whiskbroom scanner, with its scanning rate determined by a mirror rotating at 360 RPM or six scans per second around an axis parallel to the Earth's surface. The AVHRR has a swathwidth of 2700 km, an angular resolution of 1.4 mr and a nadir resolution of 1.1 km. Since the satellite velocity is about 6 km s^{-1}, at this rotation rate the FOVs overlap between scans.

The AVHRR instruments are mounted on the NOAA series of satellites described in Chapter 1. Beginning with the launch of the five channel AVHRR/2 on the NOAA-7 satellite in 1981, replacement satellites have been launched at intervals of 2–3 years, depending on the satellite lifetime. As Chapter 14 describes, AVHRR observations will continue through about 2010, at which time the NPOESS Visible/Infrared Imager/Radiometer Suite (VIIRS) will replace the AVHRR on US satellites. As part of METOP, they may also continue through 2019. The following two subsections discuss the properties of the AVHRR and MODIS bands used in the SST retrieval and the different kinds of AVHRR data.

7.3.1 AVHRR and MODIS thermal bands

The present AVHRR/3 version of AVHRR has six bands, five of which operate at any time. Henceforth, the AVHRR/3 will be referred to as the AVHRR. The bands consist of one in the visible, two in the near infrared and three in the thermal infrared. Table 7.1 lists all of the AVHRR bands and those MODIS bands used in the SST retrieval. Table A.2 in the Appendix lists all the MODIS bands. For each band, the Table 7.1 lists its wavelength range and for the thermal bands, the *NEΔT*. As Figure 4.9 shows, the AVHRR and MODIS thermal bands are located in the infrared windows.

Beginning with AVHRR, the Table 7.1 shows that band 1 is located in the visible at 0.58–0.68 μm, where this wavelength range was chosen to minimize the effect of Rayleigh

Table 7.1. *The properties of the AVHRR/3 bands and the MODIS bands used in SST algorithms*

AVHRR band	Wavelength (μm)	$NE\Delta T$ (K)	MODIS band	Wavelength (μm)	$NE\Delta T$ (K)
1	0.58–0.68				
2	0.725–1.0				
3A	1.58–1.64				
3B	3.55–3.93	0.1	20	3.660–3.840	0.05
			22	3.929–3.989	0.07
			23	4.020–4.080	0.07
4	10.3–11.3	0.1	31	10.78–11.28	0.05
5	11.5–12.5	0.1	32	11.77–12.27	0.05

The $NE\Delta T$ are determined at 300 K (MODIS specifications from http://modis.gsfc.nasa.gov; AVHRR specifications from NOAA KLM Users Guide at http://www2.ncdc.noaa.gov).

scattering (Kevin Engle, private communication, 2002). Band 2 is in the NIR at 0.725–1.0 μm. Because this band is located in a region of negligible oceanic thermal emission and very low water-leaving radiance, it provides good discrimination of the land/ocean boundary and clouds. Band 3A is at 1.58–1.64 μm and operates only during the day when it is used for daytime reflectance discrimination of snow, ice and clouds, and for detection of forest fires. It is not used in the SST retrieval. Band 3B is at 3.55–3.93 μm and operates only at night when it is used for SST retrievals. Finally, bands 4 and 5 at 10.3–11.3 μm and 11.5–12.5 μm are used for day and night SST retrievals. The table also lists the MODIS bands (20, 22, 23, 31, 32) used for SST retrieval; these consist of three bands between 3.6–4.1 μm and two between 10.7–12.3 μm. MODIS band 21, which occupies the same wavelength interval as band 22, is omitted because it is specifically designed for monitoring of forest fires and has a much lower gain than band 22.

AVHRR began as a four and five band instrument, using bands 1, 2, 3B (previously called 3), 4 and sometimes 5 in Table 7.1, where band 5 was omitted on NOAA-6, 8 and 10. As Section 7.4.2 shows, because wavelengths in the vicinity of 4 μm are easily overwhelmed by sun glitter, band 3 is only used for nighttime SST retrievals. Band 3 is also noisier than the other thermal bands. Consequently, beginning with the launch of NOAA-15 in 1998, the AVHRR was upgraded from five bands to the current six-band AVHRR/3.

On each rotation, the AVHRR infrared bands are calibrated by sequentially viewing a constant temperature blackbody and cold space, which is assigned a nominal temperature of 3 K. Bands 1, 2 and 3A are calibrated before launch, but not in space. The units of the calibrated AVHRR data are as follows: bands 3B, 4 and 5 have units of brightness temperature in K. Bands 1, 2 and 3A have units of albedo, which is defined as the ratio of the solar radiance normally incident on the Earth to the radiance received at the instrument.

This means that to calculate the extraterrestrial reflectance in Equation (5.31), the albedo must be modified to include sun angle. As Section 7.7 describes, these three bands are used to identify clouds in daytime. In contrast, and as discussed in the same section, MODIS uses 18 bands for cloud identification.

7.3.2 Forms of AVHRR data

AVHRR data can be obtained in several ways. As the satellite orbits the Earth, the data are broadcast continuously in real time to local ground stations and are also recorded onboard for later broadcast to a US ground station. The simplest way to obtain these data is from the Automatic Picture Transmission (APT) mode, which broadcasts the local visible and infrared imagery in an analog format with a 4-km pixel size to any receiving station. The inexpensive APT receivers require only an omnidirectional antenna and produce fax-like images. The other source of direct broadcast data is the High Resolution Picture Transmission (HRPT) mode, which broadcasts digital data with 1-km pixels. The HRPT station and its tracking antenna is about an order of magnitude more expensive than an APT station.

For regions inaccessible to direct broadcast, such as in the vicinity of Kerguelen Island in the South Pacific which is assumed not to have a ground station, there are two options. First, to obtain 1-km data, the user must request *Local Area Coverage* (LAC) data from NOAA, where if tape recorder space on the satellite permits, the HRPT coverage of the region is recorded for later transmission to a US ground station. LAC data must be requested and paid for by the user ahead of time; the satellite can store about ten minutes of LAC data per orbit. Second, the user can obtain lower resolution *Global Area Coverage* (GAC) data. GAC is a reduced data set that is recorded during only one out of every three mirror rotations. During the data-gathering rotation, the data are averaged into blocks of four adjacent samples, then the fifth sample is skipped, the next four are averaged, and so forth so that the data volume is reduced by an order of magnitude. GAC data are recorded continuously around the globe, have a nominal 4-km pixel size, and are downloaded and archived by NOAA. GAC provides global coverage every 1–2 days; GAC and LAC data can be viewed and retrieved at the NOAA Satellite Active Archive (http://www.saa.noaa.gov).

7.4 Atmosphere and ocean properties in the infrared

In the following, Section 7.4.1 examines the properties of radiances that are emitted or reflected from the ocean surface in the TIR, with particular emphasis on the AVHRR and MODIS bands. The section shows that in the TIR, the radiance is emitted from the top 1–100 μm of the water column, where for θ less than about 40°, the emissivity is approximately constant. Section 7.4.2 discusses solar reflection, with particular emphasis on the ratio of the reflected solar to the thermally emitted radiation. This shows that because of sun glint, thermal bands with $\lambda \le 4$ μm are usable only at night.

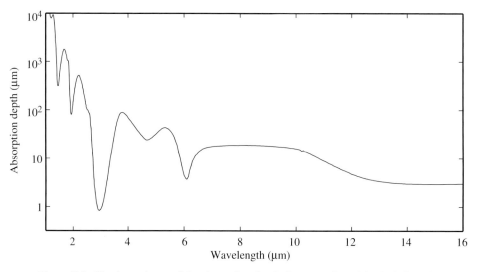

Figure 7.5. The dependence of the absorption depth d_a on wavelength in the infrared.

7.4.1 Thermal emission and reflection

The wavelengths used in the SST retrieval lie between 3 and 13 μm. For this range and from Equation (3.1), Figure 7.5 shows the dependence of the absorption depth d_a on wavelength. For 2–6 μm, d_a is between 10 and 100 μm; for 10–12 μm, it is between 1 and 10 μm. Consequently in the IR the radiance is entirely emitted from the top 1–100 μm of the water column. Since at these wavelengths energy cannot be transmitted through the water column, this also means that $r = 1 - e$, where, following Section 5.2.3, r and e are respectively the unpolarized reflectance and emissivity, with similar relations for the polarized terms.

For $\lambda = 10$ μm, Figure 7.6 shows the dependence on the incidence angle θ of the reflectance and emissivity for V-pol, H-pol and unpolarized radiation. For θ less than about 45°, e is nearly constant at about 0.99, corresponding to $r = 0.01$. Because this θ-dependence approximately holds throughout the TIR, and because $e \cong 0.99$ in this wavelength range, the ocean is approximated as a Lambert surface with the emitted radiance assumed independent of θ. Since for θ less than about 45°, e is nearly constant, this means that even for a rough ocean surface, the emitted radiances are about the same as for a specular surface (Wu and Smith, 1997). The treatment of foam is more complicated. According to Wu and Smith (1997), even though foam is made up of air bubbles surrounded by a thin water film, the detailed physical processes by which foam affects the emissivity are not well enough understood to include its effects. In summary, for moderate look angles and except for large wind speeds, the emissivity is assumed independent of surface roughness.

In the TIR, the emissivity has the following dependence on λ. For normal incidence and using the Segelstein (1981) data from Figure 3.3, the unpolarized values of r and e are derived from Equation (5.6). Figure 7.7 shows the λ-dependence of e and r, where

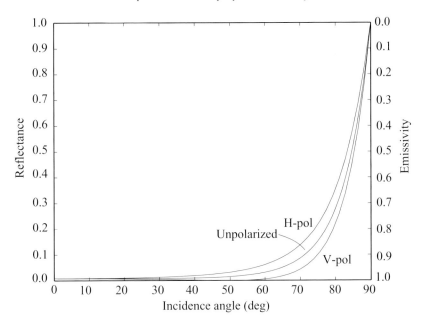

Figure 7.6. The unpolarized and vertically (V) and horizontally (H) polarized reflectance and emissivity versus incidence angle for $\lambda = 10$ μm (Derived from Equation 5.7).

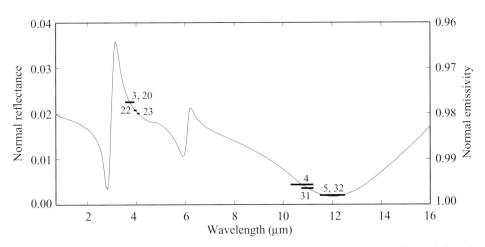

Figure 7.7. The normal reflectance (left-hand scale) and emissivity (right-hand scale) in the infrared. The horizontal bars show the average reflectance and emissivity for the bands on MODIS (20, 22, 23, 31, 32) and on AVHRR (3, 4, 5) listed in Table 7.1. The wavelength ranges of bands 22 and 23 are exaggerated by 50% for clarity. See text for further description.

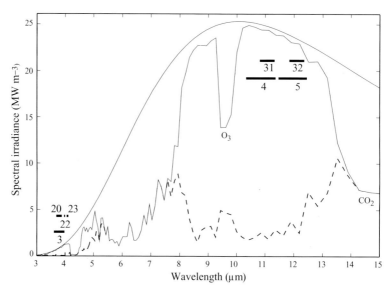

Figure 7.8. The 288 K surface blackbody irradiance (upper solid line), the same blackbody curve at the TOA after a vertical passage through the standard MODTRAN atmosphere (lower solid line), and the atmosphere emission at TOA (dashed line). The numbered bars show the location of the AVHRR and MODIS bands; O_3 and CO_2 identify the regions of ozone and carbon dioxide absorption (Data from MODTRAN, location of absorption regions courtesy Robert Cahalan).

the horizontal lines show their average values at the locations of the MODIS and AVHRR bands. For 3.6–4 μm, $r = 2.2 \times 10^{-2}$ and $e = 0.98$, for 10.8–11.3 μm, $r = 3.5 \times 10^{-3}$ and $e = 0.996$, and for 11.8–12.3 μm, $r = 1.8 \times 10^{-3}$ and $e = 0.998$. This shows that at 4 μm, the magnitude of r is approximately six times its value at 11 μm, and at all IR wavelengths, the emissivities are approximately equal to 1.

As Section 4.8.1 describes, the absorption–emission approximation governs radiative transfer in the IR. Because the temperatures of the atmosphere and surface are both about 300 K and in contrast to the visible bands, the radiation received at the TOA is not only reduced by atmospheric absorption, but is enhanced by atmospheric emission. For observations of an ocean with a 288 K surface temperature corresponding to that of the Standard MODTRAN atmosphere, Figure 7.8 compares the 288 K blackbody irradiance with the ocean-emitted irradiance modified by a vertical atmospheric passage. The upper solid line is the 288 K blackbody irradiance, the lower solid curve shows the same irradiance after passage through an absorbing and emitting atmosphere, the dashed line is the contribution from atmospheric emission at the TOA. When the dashed curve equals the lower solid curve, the surface irradiance is completely attenuated. The symbols on the surface irradiance curve mark the O_3 and CO_2 absorption regions; the numbered black bars show the AVHRR and MODIS thermal bands.

The figure shows that the irradiance observed at the TOA nearly equals the blackbody irradiance in three windows, which are located at approximately 3–4 μm, 8–9 μm and 10–12 μm, respectively called the 4-μm, 8-μm and 11-μm windows. For λ less than 2–3 μm, surface emission is negligible and solar reflection and atmospheric scattering are dominant. For 4 μm, the TOA irradiances are at least an order smaller than at 11 μm, which lies near the peak of the 288 K blackbody curve. Figure 4.9 also shows that the water vapor absorption is much less important at 4 μm than at the other windows. Finally, the reason that the 8-μm window is not generally used for SST retrieval is that the received radiance from the surface is smaller than at 11 μm, but remains sensitive to changes in atmospheric water vapor.

7.4.2 Contribution from reflected solar radiation

For each band, this section discusses the ratio of the reflected solar and thermally emitted radiation. Because in the TIR, the foam reflectance and water-leaving radiances are negligible (Section 5.6 and Chapter 6), the analysis only considers Fresnel surface reflection of the incident solar radiance.

In this calculation, the magnitude of the solar reflectance is calculated by combination of the $r(\lambda)$ shown in Figure 7.7 with the assumption that a given area of wave facets reflects the solar radiance directly into the sensor. The reflecting facets are initially assumed to occupy 0.001% or 10 m² of the 1-km² FOV. Because the relative magnitudes of these emitted and reflected radiances are not affected by an additional atmospheric passage, they are evaluated just above the surface. For this fractional area, Figure 7.9 compares the 288 K blackbody radiance with the reflected solar radiance after a downward passage through the Standard

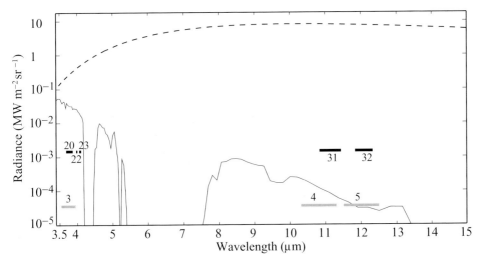

Figure 7.9. Comparison of the 288 K blackbody radiance (dashed line) with a normal solar radiance that passes through the Standard MODTRAN atmosphere, then is reflected by the surface wave facets. The reflected radiance (solid line) is evaluated just above the surface for the case when the reflecting facets occupy 0.001% of the surface. The upper numbered black bars show the location of the MODIS bands; the lower gray bars, the AVHRR bands. See text for further description.

Table 7.2. *The relative contributions to the MODIS and AVHRR*
thermal bands from normal solar reflectance, where the reflecting facets
occupy 0.001% of the FOV

MODIS band	AVHRR band	Relative solar contribution (%)
20	3	12
22		6
23		4
	4	0.002
31		0.001
32	5	0.0004

MODTRAN atmosphere; for the AVHRR and MODIS bands, Table 7.2 lists the ratio of the reflected solar to the emitted blackbody radiances.

Even for this relatively small area of reflecting facets, the table and figure show that for the 4-μm bands, the solar contribution ranges from 12 to 4%. In contrast, for the 11-μm bands, the solar contribution ranges from 0.002–0.0004%. This means that if the reflecting area is increased by an order of magnitude, solar reflectance would dominate the 4-μm bands, but would still make only a very small contribution to the 11-μm bands. Because of this sensitivity to solar reflectance, the 4-μm bands are only used in the nighttime SST retrieval, while except in the immediate vicinity of the sub-solar spot, the 11-μm bands are used at all times.

In summary, the relative advantages of the two groups of bands are as follows. The 11-μm bands can be used at all times and their radiances are an order larger than the 4-μm bands. The width of the 11-μm bands is also greater than that of the 4-μm bands, yielding a greater received power. As Figure 4.15 shows, the 4-μm bands are much less sensitive to water vapor than the 11-μm bands. Also, from the definition of the Planck function, it is easy to show that from calculation of $\frac{1}{L}\frac{dL}{dT}\big|_{\lambda}$ that the 4-μm bands are more sensitive to changes in surface temperature (Stewart, 1985).

7.5 SST operational algorithms

The operational AVHRR SST algorithms have three parts: the theoretical algorithm, the tuning of the algorithm against *in situ* buoy temperatures, and a cloud mask. Because the 4-μm bands cannot be used during daytime, different algorithms are used for day and night, where day is defined as a solar zenith angle $\theta_S < 75°$ and night as $\theta_S > 90°$ (May *et al.*, 1998). Because the SST retrieval contains two unknowns, surface temperature and atmospheric water vapor, for both AVHRR and MODIS, the retrieval requires observations in at least two bands. In the following, Section 7.5.1 gives the theoretical background for the algorithms and Section 7.5.2 describes the specific AVHRR algorithms. Section 7.5.3 discusses the buoys used in the surface matchup data set and Section 7.5.4 discusses the sources of error. Section 7.5.5 describes a specialized data set called the Reynolds SST. Section 7.5.6 discusses the MODIS algorithms. Given that the cloud mask algorithms partially depend on the retrieved SSTs, their discussion is delayed until Section 7.7, after the ATSR discussion in Section 7.6.

7.5.1 Background

Because water vapor attenuation differs from band to band, the MODIS and AVHRR SST algorithms use pairs of bands to remove the effect of the variable columnar water vapor V. This procedure is based on the integrated across-the-atmosphere form of the Schwarzschild equation (Equation 4.45), written as

$$L(\lambda_i, z_H) = L_0(\lambda_i)t_i^{\sec\theta} + f_P(\overline{T}, \lambda_i)(1 - t_i^{\sec\theta}) \tag{7.1}$$

In (7.1), the index i refers to different bands, the left-hand term is the radiance received at the satellite, the first term on the right is the attenuated surface radiance, and the second is the atmospheric path radiance, where the $1 - t$ factor is the atmospheric emissivity. Additionally, the lower troposphere is assumed to be characterized by a mean temperature $\overline{T}$, so that $t = t(\overline{T}, V, \lambda_i)$.

Because at a particular λ the attenuation associated with other atmospheric gases is constant, their contribution to the attenuation is neglected, although it can be easily included in the analysis. The relatively small term generated by reflection of the downwelling atmospheric radiance is also neglected. The analysis further neglects aerosols and assumes that variations in t are caused only by the water vapor concentrated in the lower troposphere (Section 4.2.1). Although this procedure ignores processes such as the injection of volcanic aerosols into the stratosphere, as Section 7.5.3 shows below, the continuous buoy observations correct for them.

Equation (7.1) can be further simplified by rewriting the two radiances in terms of the Planck function. With the assumption that $e = 1$, the radiance L_0 emitted by the surface is $L_0 = f_P(T_S, \lambda_i)$, where f_P is the Planck function and T_S is the skin temperature. Similarly, at the satellite, $L(\lambda_i) = f_P(T_i, \lambda_i)$, where T_i is the blackbody temperature corresponding to the received radiance. Substitution of the Planck function into Equation (7.1) and for convenience omitting the superscript $\sec\theta$ on t yields

$$f_P(T_i, \lambda_i) = f_P(T_S, \lambda_i)t_i + f_P(\overline{T}, \lambda_i)(1 - t_i) \tag{7.2}$$

Equation (7.2) shows that the received radiance L or blackbody temperature T_i is a function of the three unknowns, $\overline{T}$, T_S and t, where t is a function of the columnar water vapor V.

Application of (7.2) to the instruments proceeds as follows. Beginning with AVHRR, its daytime algorithms use bands 4 and 5; its nighttime algorithms use bands 3B, 4 and 5; hereafter band 3B will be referred to as band 3. The following analysis is initially restricted to daytime, where i represents AVHRR bands 4 and 5, so that T_4 and T_5 are the band 4 and

5 radiances received at the TOA. With the addition of band 3, the nighttime algorithm has a similar form.

Following McClain, Pichel and Walton (1985), who derived the initial form of the AVHRR algorithm, the daytime algorithm is derived as follows. As justified below, T_4, T_5, $\overline{T}$ and T_S are of the same order. Consequently, in Equation (7.2), each f_P can be linearized as a first-order Taylor series around the blackbody radiance at T_S, so that for T_4, and with similar equations for T_5 and $\overline{T}$,

$$f_P(T_4) \cong f_P(T_S) + \left.\frac{\partial f_P}{\partial T}\right|_{T_S,\lambda_4} (T_4 - T_S) \tag{7.3}$$

Following McClain *et al.* (1985), the dependence of the transmittance t on V and θ is written as

$$t_i = \exp(-m_i V \sec \theta) \tag{7.4}$$

In (7.4), the subscript i on the constants m_i indicates that the dependence of transmittance on water vapor is a function of wavelength. For later use, ε_i is defined as

$$\varepsilon_i = m_i V \sec \theta \tag{7.5}$$

If $M_i = 1 - t_i$, then substitution of (7.3), (7.4) and M_i into (7.2) yields

$$T_4 - T_S = (\overline{T} - T_S)M_4$$
$$T_5 - T_S = (\overline{T} - T_S)M_5 \tag{7.6}$$

In (7.6), if the upper equation is multiplied by M_5 and the lower by M_4, then subtraction of the lower equation from the upper and reorganization of terms gives

$$T_S = T_4 + \Gamma(T_4 - T_5), \quad \text{where } \Gamma = (1 - t_4)/(t_4 - t_5) \tag{7.7}$$

Equation (7.7) is called the *split window* form of the SST algorithm. The definition of Γ is from Walton *et al.* (1998), where Γ is sometimes written as γ (Barton, 1995). Because band 4 is less affected by water vapor than band 5, the leading term in all of the daytime equations is the 11-μm temperature T_4. If Γ is expanded in terms of small values of ε_i, the leading term in the expansion is

$$\Gamma_0 = m_4/(m_5 - m_4) \tag{7.8}$$

A typical value of Γ_0 is about 2.5 (Barton, 1995). For small ε_i, Equations (7.7) and (7.8) show that T_S is independent of V and θ, so that it is only a function of T_4, T_5 and the water vapor absorption properties of the two bands. The linear relation between temperature and water vapor described by these two equations is the basis for several infrared SST algorithms.

Equation (7.7) contains three temperatures, T_4, T_5 and T_S and the temperature difference $\Delta T_{45} = T_4 - T_5$. For moist tropical atmospheres and under cloud-free conditions, T_4 has a maximum depression from T_S of about 9 K. For the range of MODTRAN atmospheres,

ΔT_{45} ranges from a minimum of about 0.5 K to a maximum of about 4 K, so that T_4, T_5 and T_S are approximately equal (Walton *et al.*, 1998). Similarly, because $\bar{T}$ is representative of the lower troposphere, it is also of order T_S.

7.5.2 AVHRR operational algorithms

The first AVHRR SST algorithm was the Multi-Channel SST (MCSST), which McClain *et al.* (1985) derived from the equivalent of Equation (7.7). In their derivation, Γ was first estimated from a data set consisting of radiosonde profiles of temperature and humidity measured over the ocean. After substitution of this value of Γ into (7.7), the derived T_S were compared with a set of buoy-derived surface temperatures taken at positions and times that closely matched the satellite observations. These surface data are called a *matchup* data set. The comparison showed, however, that the satellite-derived T_S had large biases relative to the buoys.

Motivated by this lack of agreement, they next tried an empirical approach by rewriting Equation (7.7) as

$$SST = C_1 T_4 + C_2 (T_4 - T_5) + C_3 \qquad (7.9)$$

In (7.9), C_1, C_2, C_3 are constants. On the left-hand side of the equation, the replacement of T_S by SST indicates that the retrieved surface temperature now includes contributions from the bulk temperatures and does not equal the skin temperature. The coefficients in (7.9) are determined by least-squares regression of the SSTs derived from satellite observations of T_4 and T_5 against the matchup observations. Equation (7.9) is the simplest form of the two-channel SST retrieval. As an example of the coefficients, for daytime and NOAA-14, Walton *et al.* (1998, Table 2) show that Equation (7.9) becomes

$$SST = 0.95876 T_4 + 2.564 (T_4 - T_5) - 261.68 \qquad (7.10)$$

where T_4 and T_5 are in K and SST is in °C. In (7.10), the first term is T_4 multiplied by a constant nearly equal to unity so that this term approximates the surface temperature; the second removes the effect of water vapor; the third converts K to °C. For $\theta < 30°$, McClain *et al.* (1985) and Brown and Minnett (1999) show that the linear approach described by Equations (7.8) through (7.10) gives results of the desired accuracy.

Because Equation (7.10) is only valid for small values of ε_i, several methods are used to account for larger values of θ and V. Three of these are briefly described: the water vapor SST (WVSST), a revised MCSST and the nonlinear NLSST. First, regarding the WVSST, Emery *et al.* (1994, Appendix 1) find that to second order in ε,

$$\Gamma = [m_4/(m_4 - m_5)][1 + (m_5 V \sec \theta)/2 + \cdots] \qquad (7.11)$$

Substitution of (7.11) into (7.7) yields the WVSST equation with an explicit dependence on V and θ, and allows for direct incorporation into the SST algorithms of values of V derived from radiosonde or passive microwave observations.

Second, beginning in about 1989, to introduce the effect of variable θ into Equation (7.9), the daytime MCSST equation was rewritten as

$$\text{SST} = C_1 T_4 + C_2(T_4 - T_5) + C_3(T_4 - T_5)(\sec \theta - 1) + C_4 \qquad (7.12)$$

where the equivalent nighttime algorithm is discussed below (Walton *et al.*, 1998). In (7.12), the additional $\sec \theta$ term gives the increase in path length with θ, but does not include the effect of large V. The values of the coefficients depend on the specific satellite instrument and are again determined by comparison with the buoy matchup data set. As May *et al.* (1998) describe, the current MCSST algorithm also has the form of Equation (7.12), where operationally it serves as an input to the NLSST described below. Unlike the NLSST, the great advantage of the MCSST is that once its constants are determined, the equation is invariant.

Third, the NLSST algorithm improves on MCSST by additionally and implicitly accounting for V. As Walton *et al.* (1998) describe, a numerical study of the behavior of Γ over a large range of SSTs and marine atmospheric profiles shows that for $0 < \text{SST} < 30\,^{\circ}\text{C}$, Γ increases nearly linearly with SST. The reason for this dependence is that moist atmospheres generally occur over warm oceans, so that atmospheric humidity increases with SST. Because of this dependence, Walton *et al.* (1998) find that in daytime the following NLSST equation gives better agreement with the matchup temperatures.

$$\text{SST} = C_1 T_4 + C_2 T_{\text{sfc}}(T_4 - T_5) + C_3(T_4 - T_5)(\sec \theta - 1) + C_4 \qquad (7.13)$$

In (7.13) the Cs are again arbitrary constants and T_{sfc} is an independent estimate of the surface temperature taken either from a climatological lookup table or from the MCSST (Walton *et al.*, 1998, Table 2). The coefficients in Equation (7.13) are determined by comparison of the retrieved SSTs with matchup data. For NOAA-14 the daytime NLSST has the form

$$\text{SST} = 0.9336 T_4 + 0.079 T_{\text{sfc}}(T_4 - T_5) + 0.77(T_4 - T_5)(\sec \theta - 1) - 253.69 \qquad (7.14)$$

In (7.14) T_4 and T_5 are in K, SST and T_{sfc} are in $^{\circ}$C.

With a suitable choice of coefficients, the MCSST and NLSST algorithms described above are also used at night. Alternatively and to take advantage of band 3, the nighttime forms of the MCSST and NLSST use all three thermal bands in what is called a *triple window* algorithm, where the difference between T_3 and T_5 is used to remove the vapor attenuation. With arguments similar to those used in the daytime derivation and from Walton *et al.* (1988, Table 4), the NOAA-14 NLSST nighttime algorithm is

$$\text{SST} = 0.980064 T_4 + 0.031889 T_{\text{sfc}}(T_3 - T_5) + 1.817861(\sec \theta - 1) - 266.186 \qquad (7.15)$$

In (7.15) and similar to the daytime algorithm, the T_4 term provides the basic SST estimate, with the corrections and conversion to $^{\circ}$C provided by the other terms. Because the third term in (7.15) lacks the product of $(T_3 - T_5)$ with the $\sec \theta$ term, the nighttime equation is simpler than the daytime. The advantage of band 3 over bands 4 and 5 is that it is less sensitive to water vapor, so that over a broad range of atmospheres T_3 is only reduced from T_S by at most 2 K, as opposed to a 9 K reduction for band 4 (Walton *et al.*, 1998).

The analogous MCSST nighttime algorithm has the form

$$SST = C_1 T_4 + C_2(T_3 - T_5) + C_3(\sec\theta - 1) + C_4 \qquad (7.16)$$

where the coefficients are again determined from matchup data (Walton *et al.*, 1988, Table 4).

7.5.3 Surface matchup data set

At daily intervals the operational AVHRR-derived SSTs are compared with temporally and spatially coincident surface temperatures derived from the global distribution of moored and drifting buoys. These buoys use the NOAA satellites to transmit their temperatures and positions to their respective weather services, and to the US NESDIS. As next discussed, this data set contains about 50 moored buoys and 500 to 1000 drifting buoys.

The moored buoys include the equatorial TAO/TRITON array shown in Figure 7.10 (McPhaden *et al.*, 1998). They also include the National Data Buoy Center (NDBC) buoys located off the east and west coasts of the United States, in the Gulf of Mexico and Gulf of Alaska, and around Hawaii shown in Figure 7.11 (Hamilton, 1986; Meindl and Hamilton, 1992). These buoys measure wind direction and magnitude, and air and water temperature. Other moored buoys not shown on the figure include the JMA buoys around Japan and the UK Meteorological Office (UKMO) buoys in the northeast Atlantic. For all of the moored buoys, the water temperature is measured at a depth of about 1 m. The drifting buoys are sponsored by the Global Drifter Program, measure water temperature at depths of about 0.3 m and have lifetimes of 1–2 years. The US buoys are built and deployed by the Drifting

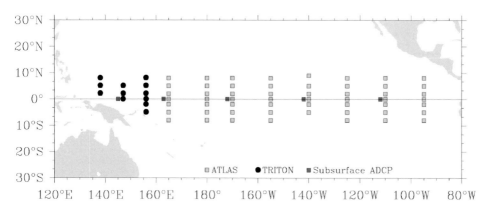

Figure 7.10. The Pacific TAO / TRITON array. The ATLAS and ADCP (Acoustic Doppler Current Profiler) buoys are US; the TRITON buoys are Japanese (Courtesy NOAA/PMEL/TAO Project Office, Dr. Michael J. McPhaden, Director).

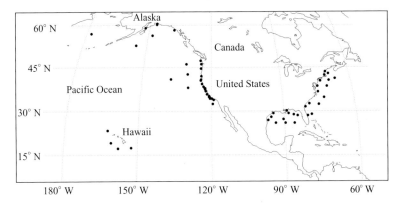

Figure 7.11. The locations of the US NDBC buoys in September 2001 (Buoy locations tabulated at http://www.ndbc.noaa.gov).

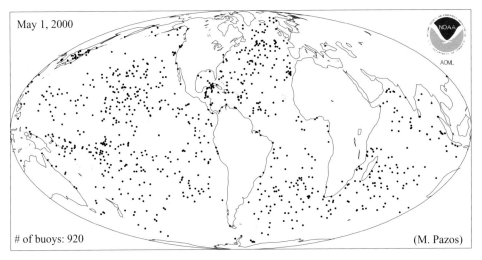

Figure 7.12. The black dots show the surface drifter locations on May 1, 2000 (Figure courtesy Drifting Buoy Data Assembly Center, Atlantic Oceanographic and Meteorological Laboratory, National Oceanic and Atmospheric Administration).

Buoy Data Assembly Center (DAC) at the NOAA Atlantic Oceanographic and Meteorological Laboratory (AOML). Other drifters are deployed by NATO and by the Canadian Maritime Environmental Data Service (MEDS). The locations of all drifters at monthly intervals are given at http://www.aoml.noaa.gov/phod/dac/dacdata.html. For May 1, 2000, Figure 7.12 shows the drifter positions.

The US NESDIS stores the temperatures and positions of all these buoys, then once per day extracts buoy and satellite measurements that are coincident within 4 hours and 25 km, and adds them to the matchup data base. At weekly, monthly and annual intervals,

NESDIS generates statistics such as the standard deviation and bias of the difference between the buoy and satellite temperatures for different geographic areas and for day and night conditions. If necessary, the MCSST and NLSST coefficients are recalculated. As Section 7.7.1 discusses further, cloudiness reduces the annual successful matchups to about 10% of the total possible. As the next section shows, typical standard deviations for the AVHRR SSTs are about ± 0.5 K; the biases range from -0.2 K to $+0.4$ K.

7.5.4 Environmental sources of error

Environmental sources of error in the SST derivation include high thin cirrus clouds, diurnal solar heating, discrepancies between skin and bulk temperatures, and volcanic aerosols. First, the cirrus clouds consist of thin, semitransparent layers of ice crystals that are associated with the penetration of convectively active rain cells into the upper troposphere. At these altitudes the ice crystals spread rapidly and laterally over distances of hundreds of kilometers and persist for hours (Prabhakara *et al.*, 1988). Because cirrus clouds are thin and semitransparent, they are difficult to identify from satellite observations; because they are very cold, they introduce significant errors into the SST retrieval. As Prabhakara *et al.* (1988) show from aircraft observations, the presence of these crystals tends to reduce T_4 and T_5, but with a greater attenuation of T_5 than T_4. The result is that in the presence of cirrus, $\Delta T_{45} = T_4 - T_5$ tends to increase, reducing the retrieved SST below T_S.

Second, short-term solar heating can increase the skin temperature without changing the atmospheric temperature and humidity profiles. Because, for constant atmospheric properties, T_4 has a greater response to an increase in T_S than T_5, ΔT_{45} also increases, which also reduces the retrieved SST. Third, in a related source of error, the daytime solar-driven increase in T_S increases the difference between the buoy temperature and T_S, yielding a larger bias in the matchup data. For these reasons, nighttime SST observations have a greater accuracy than daytime retrievals.

Fourth, episodes of volcanic aerosols create serious problems for the algorithms. Since the beginning of the AVHRR SST time series, two major eruptions have injected large amounts of sulfuric acid droplets into the stratosphere. These were the Mexican El Chichon eruption in April 1982 (Bernstein and Chelton, 1985), and the Philippine Mount Pinatubo eruption in June 1991. The stratospheric aerosols from the Pinatubo eruptions persisted for about two years, first spreading around the globe in the tropical regions, then laterally into the temperate latitudes (Walton *et al.*, 1998). This distribution of aerosols meant that the globally averaged satellite SSTs were 0.5 °C colder than the buoy SSTs, with tropical negative biases exceeding 2 °C. The nighttime algorithm had a similar but smaller negative bias. For both day and night, adjustment of the algorithm coefficients removed this bias. Similar tropospheric aerosol events such as Saharan dust storms that occur at much shorter time scales have a similar effect.

For the 9-year period 1989–1998, Figure 7.13 shows the monthly time series of the global mean and standard deviation of the difference between the satellite SSTs and the

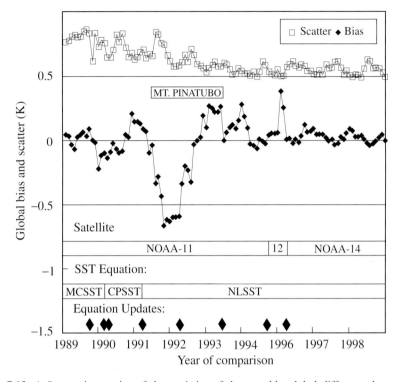

Figure 7.13. A 9-year time series of the statistics of the monthly global difference between the matchup satellite and buoy SSTs. See text for further description (Figure 9 from Walton *et al.*, 1998, published 1998 American Geophysical Union, not subject to US copyright).

buoy temperatures. On the figure, the means are called biases; the standard deviations, scatter. The lower part of the figure lists the satellites used in these measurements, NOAA-11, 12 and 14, and the algorithms used to compute the SST, where CPSST is the briefly used Cross-Product SST (Walton *et al.*, 1998). The diamonds on the bottom line mark the times when either the algorithm or its coefficients were updated. A large increase in cold bias occurs during the period in which the Pinatubo aerosols affected the SST retrieval; following this period, the standard deviation approaches 0.5 K and the bias approaches 0 K. The behavior of the nighttime algorithm, which is omitted, is similar and is described in Walton *et al.* (1998).

7.5.5 *Reynolds SST*

The Reynolds SST is a commonly used product consisting of a weekly and monthly averaged global SST field derived from surface and AVHRR observations that are optimally interpolated to a 1° by 1° latitude and longitude grid (Reynolds and Smith, 1994). This product is used in climate studies, for T_{sfc} in the SST algorithms, and in the passive microwave algorithms described in Chapters 9 and 14. The Reynolds SST is available from 1981 to the present with a one-week time delay; it is available at http://podaac.jpl.nasa.gov/reynolds/reynolds_data.html.

7.5.6 MODIS SST algorithms

As Table 7.1 shows, MODIS employs two sets of thermal bands for SST retrieval; three at 4 μm (bands 20, 22, 23), and two at 11 μm (bands 31 and 32). Brown and Minnett (1999) and Minnett *et al.* (2002) describe the split window SST algorithms used with these bands. First, the 11-μm algorithm has a similar form to the previously undescribed day and night Pathfinder modification of the NLSST algorithm (Kilpatrick, Podestá and Evans, 2001), and is written as

$$SST = C_1 T_{31} + C_2 T_{\text{sfc}}(T_{31} - T_{32}) + C_3(T_{31} - T_{32})(\sec\theta - 1) + C_4 \qquad (7.17)$$

In (7.17), all temperatures have units of °C and T_{sfc} is the Reynolds SST. A difference between the Pathfinder algorithm and the algorithms described in Section 7.5.2 is that Pathfinder uses one set of coefficients for dry atmospheres and another for wet atmospheres. Specifically if $\Delta T = T_{31} - T_{32}$, then for $\Delta T \leq 0.7$ K, the coefficients C_1 through C_4 in (7.17) equal 0.976, 0.126, 1.683, 1.2026; for $\Delta T > 0.7$ K, the Cs equal 0.891, 0.125, 1.109, 2.7478. This equation is used for day and night retrievals; as discussed in Section 7.8.3, Plate 9 shows an example of a MODIS image processed with this algorithm.

Second, at 4 μm, the MODIS SST4 algorithm is written

$$SST4 = C_1 + C_2 T_{22} + C_3(T_{22} - T_{23}) + C_4(\sec\theta - 1) \qquad (7.18)$$

This equation is only used at night and involves only one set of coefficients. For MODIS on TERRA, the Cs equal −0.065, 1.034, 0.723, 0.972. Compared to (7.17), Equation (7.18) is simpler, has one set of coefficients and lacks the T_{sfc} term. The equation shows the advantages of working at 4 μm, where water vapor is less important than at 11 μm (Equation (7.18) and coefficients courtesy Peter Minnett, private communication, 2002).

When compared with buoy temperatures, the daytime 11-μm SST accuracy is ±0.5 K; its nighttime accuracy is ±0.4 K (Brown *et al.*, 2002). Compared with M-AERI, its day and nighttime accuracy is ±0.4 K. The SST4 accuracy compared with buoys is ±0.4 K; compared with M-AERI, its accuracy is ±0.3 K. In summary, the advantages of the 11-μm algorithm are that it is usable at all times and continues the AVHRR SST time series with an improved accuracy. Its disadvantages are that it is affected by water vapor and by volcanic and tropospheric aerosols. The SST4 algorithm is simpler, less sensitive to water vapor and slightly more accurate. However, because of sun glint it is only usable at night, and compared with the 11-μm retrieval, it has a lower signal level, a similar sensitivity to aerosols and lacks continuity with the AVHRR SSTs.

7.6 Along-Track Scanning Radiometer

In contrast to the AVHRR and MODIS split window retrieval, the Along-Track Scanning Radiometer (ATSR) on ERS-1 and ERS-2 and the Advanced ATSR (AATSR) on ENVISAT use a *dual look angle* technique to retrieve SST and remove atmospheric attenuation. This technique works as follows: imagine being able to view the same element of surface area at two different look angles, for example at nadir and at 60°. Assuming that the atmospheric properties are identical along each path and that the observed surface temperature does not vary because of the larger FOV associated with the longer path, then the nadir view corresponds to a passage through one atmosphere; the 60° view, to two atmospheres. The difference in received brightness temperatures between the two paths equals the attenuation and emission associated with a single vertical passage through one atmosphere. Subtraction of this difference from the brightness temperature at nadir yields an accurate measurement of T_S. The advantage of this technique over the split window technique is that it removes *all* of the attenuation along the path, including contributions from tropospheric water vapor and from tropospheric and stratospheric aerosols.

Minnett (1995a, 1995b) describes the ATSR in detail; Birks *et al.* (1999) describe the AATSR. Figure 7.14 shows a schematic drawing of the ATSR operation, where AATSR operates similarly. ATSR operates as a conical scanner that observes the surface in a 500-km wide swath. One side of the cone intersects the surface at nadir; the other at a 55° look angle. Approximately two minutes separate the forward and nadir view of the same area; the nadir FOV has a 1-km diameter. Clouds are identified similarly to the AVHRR procedure described in Section 7.7.

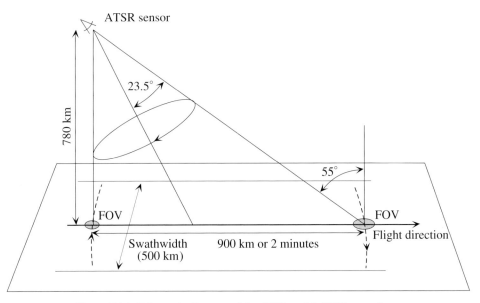

Figure 7.14. Schematic diagram of the ATSR and AATSR operation.

ATSR has four bands that are identical to AVHRR bands 3A, 3B, 4 and 5, where the 3A band at 1.62 μm is used for cloud and land discrimination. AATSR has a total of seven bands, including the four ATSR bands and three additional bands at 550, 670 and 870 nm. As Minnett (1995a) describes, because the ATSR SST algorithm combines a dual window and dual look-angle approach, it does not require correction from a matchup data set. Its rms accuracy is about ±0.3 °C. Unfortunately, in spite of the generally enhanced quality of ATSR data compared with AVHRR, because ATSR data have not been made generally available to the oceanographic community, their usefulness has been limited (Thomas, 2001). The hope is that this distribution problem will be corrected with AATSR.

7.7 Cloud detection algorithms

Identification of clouds depends on whether it is day or night and on whether the ocean surface is ice-free or covered by ice and snow. In general, because of the different reflectance and emissive properties of clouds and open water, cloud discrimination over open ocean is easier than over land or ice. Given the interest in open ocean SST retrieval, the following discussion concentrates on the open ocean day/night case. The day/night AVHRR tests for clouds use all five bands, and MODIS uses a 17-band multispectral approach. Section 7.7.1 describes the AVHRR algorithms; Section 7.7.2 describes the MODIS algorithms.

7.7.1 AVHRR cloud algorithms

McClain *et al.* (1985) and May *et al.* (1998) describe the large variety of operational AVHRR cloud tests and for many of these tests, Saunders and Kriebel (1988) give the theoretical background. Ackerman *et al.* (1997) also describe many of the AVHRR tests and for MODIS cloud tests, describe their physical and observational scientific background. Over open ocean, these tests depend on two factors: the clouds being colder and more reflective than the ocean surface, and for spatial scales of order 100 km, the ocean surface being nearly uniform in temperature and reflectance (Rossow, 1989).

The tests include single band, single pixel *threshold* tests, which eliminate pixels that are more reflective or colder than the ocean surface, and single band, multiple pixel *uniformity* tests. The uniformity tests examine the variance of the temperature or reflectance in a rectangular array of pixels, where, if scattered clouds occur within the array, the variance is greater than for the cloud-free case. Additionally, multiple band tests are used to detect specific cloud types, including high thin cirrus, nighttime uniform low stratus, and fog and sub-pixel broken clouds. Finally, there is a series of tests applied to the retrieved SSTs that include their comparison with climatology, and the comparison of the NLSST-retrieved temperatures with those derived from alternative algorithms such as MCSST.

The cloud tests divide into day and night, where the day tests use bands 1 and 2 and the cloud reflectances, and the night tests use the thermal bands. The *threshold* and *uniformity* tests are the simplest and are designed for clouds that fill a single pixel. For AVHRR, if the band 2 reflectance is greater than the clear sky ocean reflectance plus an uncertainty, the pixel is classified as cloudy. Because the reflectance is a function of sun angle, incidence angle and their difference, the algorithm estimates this threshold from lookup tables that are based on a wide variety of observations (May *et al.*, 1998). Similarly, if the band 4 brightness temperature is less than 270 K, the pixel is also classified as cloudy. The single band uniformity tests are designed to detect broken or scattered clouds, and examine the variance of the brightness temperature or reflectance in an $n \times n$ pixel array, generally measuring 2×2 or 11×11 in extent (May *et al.*, 1998). If the variance exceeds a threshold based on the cloud-free case, the array is classified as cloudy. Near the coasts, because of the presence of strong gradients in SST associated with upwelling and other processes, these uniformity tests must be carefully applied so that a cloud-free ocean surface containing SST gradients is not misclassified as cloud.

An alternative test considers the ratio C_R of the band 2 to the band 1 reflectances, where these reflectances are again adjusted for the sun and satellite incidence angles. Because atmospheric Rayleigh scattering is greater in the visible than in the NIR and because clouds generally occur above the aerosol-laden marine boundary layer, for cloud-free conditions, the band 1 reflectance is nearly twice that at band 2. Consequently, for cloud-free conditions, $C_R < 0.7$, while for clouds, $C_R \cong 1$. Given this response, a pixel is assumed cloudy if $C_R > 0.7$ (Saunders and Kriebel, 1988).

The nighttime IR tests depend on the detailed cloud properties. For clouds consisting of water droplets, the cloud emissivities e_c are functions of λ, the droplet size distribution and the cloud physical and optical thicknesses. For the TIR, Hunt (1973) shows that as the cloud thickness increases, e_c increases and its transmittance decreases, so that, depending on λ and cloud thickness, e_c varies from near zero to about 0.97. For thick clouds, Hunt (1973) also shows that e_c is generally smaller at 4 µm than at 11 µm; specifically, at 3.5 µm, e_c is about 0.80; at 11 µm, e_c is about 0.97. Because of this dependence of e_c on λ and with the definition $\Delta T_{53} = T_5 - T_3$, in the presence of clouds, $\Delta T_{53} > 0$. In contrast, for cloud-free conditions, because for both bands the seawater emissivity is nearly constant at about 0.99 and the water vapor attenuation is greater at band 5, $\Delta T_{53} \leq 0$. This condition provides a nighttime test for uniform low stratus and fog, which can fail uniformity tests. Finally, as Section 7.5.4 shows, the presence of high thin cirrus means that ΔT_{45} is more positive than for cloud-free water vapor attenuation alone. From May *et al.* (1998), the day and night operational criterion used for cirrus clouds is to accept a pixel as cloud-free only if $\Delta T_{45} \leq 3.5$ K.

The above tests rely on the observed reflectances and brightness temperatures. If the pixel or pixel array in question passes these tests, the pixel is subjected to an additional set of tests involving the temperatures derived from the MCSST and NLSST. First, for day and night, the NLSST product is subjected to an unreasonableness test; the SST must lie in the range $-2\,°C < SST < +35\,°C$ to pass the test. Second, during daytime, the MCSST and NLSST retrieved SSTs are compared, and if they differ by more than ± 1.5 K, the pixel is assumed cloudy. Third, at night, the NLSST solution based on bands 3, 4 and 5 is compared with the MCSST solution based on bands 3 and 4, and with an NLSST solution based on bands

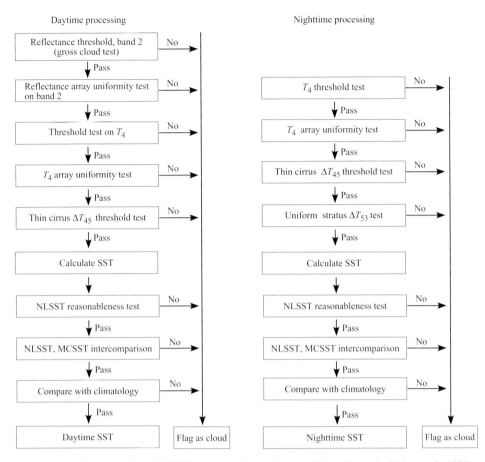

Figure 7.15. Summary of the AVHRR operational tests (Adapted from Figure 3 of May *et al.*, 1998).

4 and 5. For the pixel to be considered cloud-free, the three solutions must agree within 2 K. Finally, for day and night, the retrieved SST must also pass a climatology test, meaning that it must agree with the monthly climatology at its location within 10 K.

Following May *et al.* (1998), Figure 7.15 summarizes the tests in their approximate order of application. If any single test identifies clouds, the pixel or array in question is classified as cloudy. Consequently, these tests are cloud-conservative, and may classify some clear pixels as cloudy. How many cloud-free pixels do the tests produce? At least three investigations address this question. First, McClain *et al.* (1985) discuss the applications of these tests for one day, April 3, 1982. Working with 11×11 nighttime GAC arrays, they begin with 185 000 arrays and end with 18 000 cloud-free arrays, for a 10% survival rate. For daytime, they begin with 1 465 000 arrays and end with 22 000, for a 2% survival rate. Second, for April 14–20, 1985, Saunders and Kriebel (1988) examine seven LAC images around the British Isles and find a 20% survival rate. Third, the AVHRR Pathfinder website (http://www.rsmas.miami.edu/groups/rrsl/pathfinder/index.html) annually compares the number of possible buoy matchups versus the number of successful matchups. For 1995–1998, the success rate is about 13%. Overall, these statistics suggest a cloud-free matchup rate of about 10%.

7.7.2 MODIS cloud algorithms

As Ackerman *et al.* (1997) describe, the MODIS cloud algorithms are designed not only to mask clouds for SST and ocean color retrieval, but also to identify and classify the clouds for radiative balance calculations. Table A.2 in the Appendix lists the MODIS bands, with the 17 bands used for cloud discrimination given in bold face. Bands 1 and 2 have a 250-m resolution; bands 3–7 have a 500-m resolution; all other bands have a 1-km resolution. Bands 1–7 provide reflectances and reflectance ratios for use in threshold and uniformity tests at a finer resolution than the MODIS SST and ocean color bands, which helps in the identification of cloud edges, aircraft contrails and small broken clouds. Bands 18 and 26 occur in strong water vapor absorption bands and, as shown below, are used for daytime reflectance discrimination of thin cirrus. Band 19 is used for detection of cloud shadow, band 27 is used for cloud discrimination in polar regions and band 29 is used in combination with the 11 and 12 μm bands for cloud identification.

From Ackerman *et al.* (1997), the MODIS algorithm applies the following cloud tests to each pixel or pixel array: reflectance thresholds, IR thresholds, spatial and temporal uniformity, reflectance thin cirrus, IR thin cirrus, nighttime low level stratus and fog, and SST reasonableness, climatology and intercomparison. Similar to AVHRR, if any one of these tests yields a high probability of cloudiness, the pixel is considered cloudy. Because many of these tests, specifically the reflectance, threshold, spatial uniformity, IR thin cirrus, IR nighttime low level stratus and the SST comparison tests, are similar to the AVHRR tests, the following primarily discusses those that differ.

General reflectance tests The principal daytime reflective cloud test is based on MODIS bands 1 and 2 at 660 and 870 nm. Threshold and reflectance ratio tests are applied to these two bands. Because the water-leaving radiance at 870 nm is smaller than that at 660 nm, for clear sky, band 1 is brighter than band 2. The reflectance ratio test is similar to that described in Section 7.7.1; namely if C_R is the ratio of the 870 and 660 nm bands, then for cloud-free ocean conditions, $C_R < 0.75$ (Saunders and Kriebel, 1988).

Reflectance tests for thin cirrus MODIS has two bands specifically designed for daytime discrimination of high cirrus clouds. The first is band 26 at 1.375 μm; the second is band 18 at 936 nm. As Figure 4.9 shows, both are located at the centers of strong water vapor absorption bands. For band 26, Gao *et al.* (1993) show that as long as $V > 4$ mm, the surface and near surface radiances are completely attenuated. But because cirrus clouds occur in the upper troposphere and lower stratosphere, they appear bright in contrast to the completely attenuated surface radiance and to clouds in the lower troposphere, whose reflectance is partially attenuated by water vapor. For similar reasons, band 18 is also used for high altitude cloud discrimination.

Spatial and temporal uniformity   The spatial uniformity tests are similar to those used for the AVHRR. In the temporal uniformity test, the brightness temperature or reflectance at a specific pixel is compared to measurements made in the previous seven days. Values which exceed a preset threshold are assumed cloudy and are masked.

7.8 Illustrations and examples

The following discusses three examples of AVHRR and MODIS imagery. Section 7.8.1 examines each band of an AVHRR image, Section 7.8.2 uses AVHRR and ocean color data to examine the transition between El Niño and La Niña conditions in the equatorial Pacific, and Section 7.8.3 describes a global MODIS SST image.

7.8.1 Examination of an AVHRR image

Figure 7.16 shows the five AVHRR bands used in the SST retrieval and the resultant cloud- and land-masked NLSST image. The image is from the same day and same region of the Washington and British Columbia coast as the SeaWiFS image in Plate 5. Although the AVHRR and SeaWiFS images show similar features, the images are derived very differently; the SST is based on thermal emission from the top 10–20 μm of the water column, while the Chl-*a* distributions are derived from water-leaving radiances from the top 10–40 m. On the figure, the upper two panels show the solar reflectance bands 1 and 2. The band 1 image shows the visible reflectance; clouds and land are reflective or white, water is darker. The oceanic pattern of water-leaving radiance shown in Plate 5 is faintly visible in this image, and at the lower left, some sun glint is visible. The band 2 image shows the NIR reflectance, where seawater is black or non-reflective and clouds are gray or white. The sharp contrast between land and water shown in this image illustrates why band 2 is used for discrimination of the land/water boundary.

The next four images show the thermal bands and the NLSST distribution. For all of these images, darker shades of gray are warm, lighter grays are cold. The visible scan lines in the band 3 image are caused by its increased noise relative to the other images. Also, even though the clouds are colder than the seawater, the direct reflection of solar radiation from the clouds and wave facets means that the clouds and sea surface in the region marked C appear warm. As Section 7.4.2 discusses, the cause of the apparent warm clouds and surface is that at these wavelengths solar reflectance overwhelms thermal emission. Examination of band 4 and band 5 shows that in contrast to band 3 thermal emission dominates, so that the clouds are colder and less noisy than at band 3. The last panel shows the NLSST distribution, which is derived using the MCSST for T_{sfc} and a threshold cloud filter based on band 2. All of the thermal images show the oceanic eddy off Vancouver Island, marked by E in the

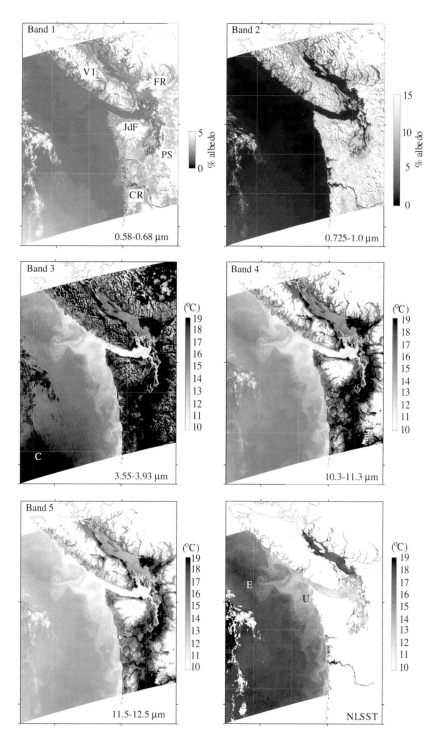

Figure 7.16. Daytime AVHRR image of the Washington and British Columbia coast from NOAA-14, September 1, 1999, 2300 UTC, 1559 Pacific Daylight Time, or nearly coincident with the SeaWiFS

NLSST image, and the cold upwelling adjacent to the coast marked by U, both of which are associated with the biological productivity shown in Plate 5.

7.8.2 Transition from El Niño to La Niña

For January and July 1998, the four panels in Plate 8 compare the monthly Pacific averages of AVHRR SST and SeaWiFS ocean chlorophyll (from Chavez *et al.*, 1999, Figure 1). The SeaWiFS images also show the normalized land difference vegetation index (LDVI), a measure of land chlorophyll described in Chavez *et al.* (1999). The shift from El Niño to La Niña creates the differences between the two sets of panels. January 1998 occurs toward the end of the 1997–98 El Niño, July 1998 shows the return to La Niña conditions.

For La Niña, the July 1998 SST image shows that the easterly trade winds at the equator generate upwelling, which creates an equatorial tongue of cold water extending from the coast of South America across the Pacific. The accompanying Chl-*a* image shows that this cold tongue is accompanied by an equatorial phytoplankton bloom, yielding enhanced values of chlorophyll. El Niño occurs when the equatorial trade winds weaken and allow warm, less biologically productive Pacific water to replace the cold equatorial La Niña upwelling. The January 1998 SST and chlorophyll figures show that for El Niño, the equator is characterized by warm water and low chlorophyll. As Chavez *et al.* (1999) discuss, at some locations the 1998 El Niño equatorial anomalies of SST exceeded 5 K and the equatorial chlorophyll concentrations were the lowest on record. This behavior has important consequences for climate studies, regional fisheries and for understanding the oceanic uptake of carbon. For the same period, Chapter 9 discusses the effect of this El Niño/La Niña transition on atmospheric water vapor and cloud liquid water.

7.8.3 A global MODIS SST image

For May 2001, Plate 9 shows the average global SST distribution derived from MODIS. The image is processed using nighttime data and the 11-μm SST algorithm in Equation (7.17). On the image the broad zonal distributions of SST are illustrated by the dark blue boundaries of the sub-polar fronts, and by the warmer red-to-green boundaries closer to the equator. There are also a number of specific non-zonal features. These include the northward-flowing plume of warm water along the east coast of North America associated with the Gulf Stream (a), and a similar plume adjacent to the Japanese coast generated by the

Caption for Fig.17.16 (cont.) image in Plate 5. The image shows bands 1 through 5, and the SST image processed with the NLSST algorithm initialized with the MCSST. On the band 1 figure, VI is Vancouver Island; FR, Fraser River; JdF, Strait of Juan de Fuca; PS, Puget Sound; CR, Columbia River. The letters C, E and U mark oceanographic features explained in the text; the white cloud and land mask in the NLSST image is a threshold mask based on band 2 (Courtesy Kate Edwards, used with permission).

Kuroshio (b). Along the east coast of South Africa, the southward-flowing Agulhas current extends south of the Cape of Good Hope (c). The image also shows a region of cold water upwelling adjacent to the west coast of South America and the La Niña band of equatorial cold water upwelling in the Pacific (d). A similar cold band extends along the equatorial Atlantic (e). Finally, adjacent to Central America, as discussed further in Chapter 11, strong winds flowing through the mountain gaps generate two localized regions of upwelling (f and g).

7.9 Further reading and acknowledgements

The POES website (http://poes.gsfc.nasa.gov) contains information on the NOAA satellites and the AVHRR, with detailed documentation beginning in 1990. AVHRR SST data are available from http://podaac.jpl.nasa.gov. Section 6.8 lists the websites for MODIS documents, presentations and data. The papers in Njoku and Bernstein (1985) compare different ways of deriving SST from microwave and infrared observations. The NASA archive at http://podaac.jpl.nasa.gov contains some ATSR data; other ATSR data can be ordered through the ATSR project office at http://www.atsr.rl.ac.uk. I thank Peter Minnett for helpful conversations about the MODIS algorithms and Kevin Engle for his help with AVHRR.

Plate 1. Oblique photograph of wave breaking and foam generation on the Japan/East Sea taken through the front window of a Twin Otter meteorological flight on February 28, 2000. The ambient air temperature was about −8 °C, the aircraft altitude was about 38 m, the flight direction was 330° and the wind speed was 17 m s^{-1} from 340°, so that the camera is looking into the wind and toward Russia. (Meteorological and flight data courtesy Djamal Khelif; photograph courtesy Jon Stairs, used with permission).

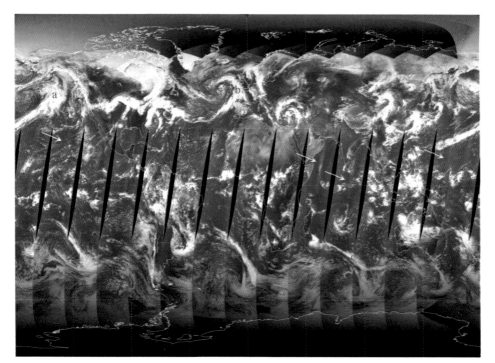

Plate 2. True color composite image of the Earth taken by the Moderate-Resolution Imaging Spectroradiometer (MODIS) on April 19, 2000. The image consists of one day of sun-synchronous orbital passes from the sunlit side of Earth. For several swaths, the arrows mark the direct reflection of the sun from sun glint; the letters a, b, c mark storms that are also shown on Plate 10. See text for additional description (Courtesy Mark Gray, MODIS Atmosphere Team, NASA GSFC, and the Earth Observatory website at http://earthobservatory.nasa.gov).

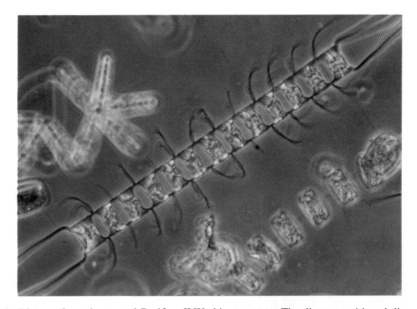

Plate 3. Diatoms from the coastal Pacific off Washington state. The diatom positioned diagonally across the picture center is a member of the *chaetoceros* species; it consists of a chain of silica-shelled single cells with spines protruding from each cell and from both ends of the chain. For this species, the width of each cell is 20–25 μm. Inside each cell, the chloroplast contains the photosynthetic pigments, a mixture of green chlorophyll and the brownish yellow carotenoids. Other diatom species are adjacent, including a chain of cells below the *chaetoceros* (Courtesy Rita Horner; used with permission).

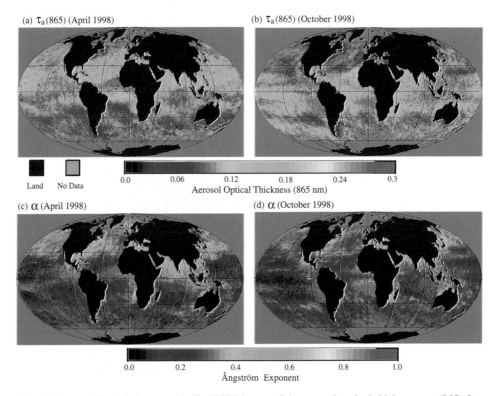

Plate 4. A monthly global composite SeaWiFS image of the aerosol optical thickness τ_A (865) for April (a) and October (b) 1998, and the Ångström exponent α for the same periods (c and d). The color bars show the scales; land is black and regions with no data are gray. See text for further description (Figure 1 from Wang *et al.*, 2000, © 2000 American Geophysical Union, reproduced/modified by permission of AGU, courtesy Menghua Wang; OrbView-2 Imagery provided by ORBIMAGE, the SeaWiFS Project and NASA/Goddard Spaceflight Center).

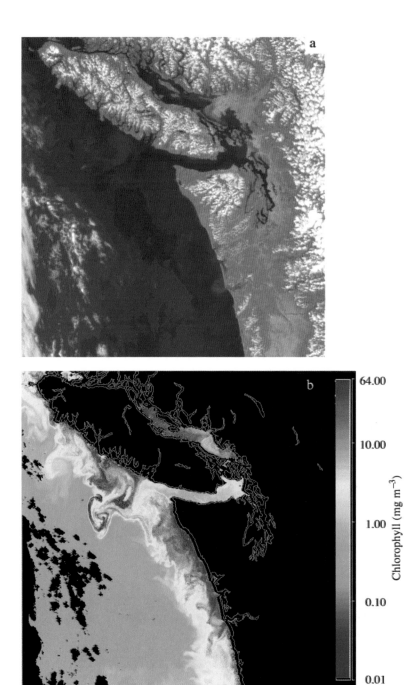

Plate 5. Composite SeaWiFS image of the northeast Pacific adjacent to British Columbia, Canada and Washington and Oregon, United States on September 1, 1999. (a) A true color image, mixed from SeaWiFS bands at 410, 555 and 670 nm, with Rayleigh scattering removed. (b) The Chl-a distribution for the same region in mg m^{-3}. Black corresponds to land and to cloud mask (OrbView-2 Imagery provided by ORBIMAGE, the SeaWiFS Project and NASA/Goddard Spaceflight Center; processing courtesy of Brandon Sackmann and Miles Logsdon).

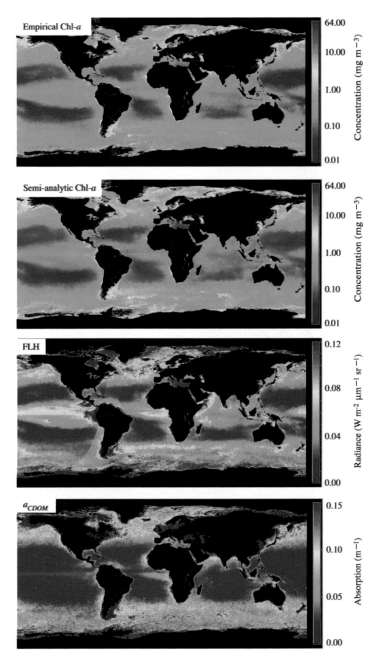

Plate 6. The average 2001 MODIS empirical and semianalytic Chl-*a* concentrations, fluorescence line height (FLH) and semi-analytic CDOM absorption at 400 nm. See text for further description (Images courtesy MODIS Ocean Team, NASA Goddard Space Flight Center, processing courtesy Brandon Sackmann).

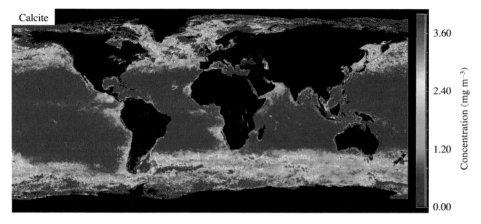

Plate 7. The average 2001 MODIS calcite concentration. See text for further description (Courtesy MODIS Ocean Team, NASA Goddard Space Flight Center, processing courtesy Brandon Sackmann).

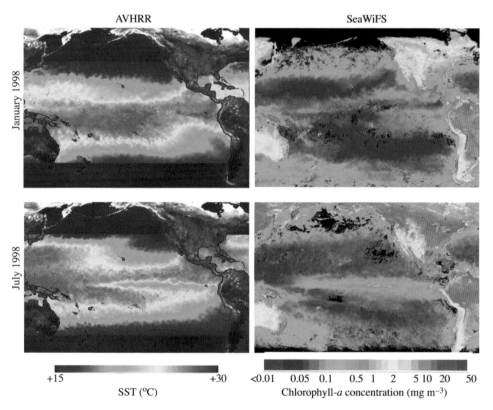

Plate 8. Comparison of AVHRR SST with OrbView SeaWiFS ocean chlorophyll for January 1998 and July 1998. See text for further description (Courtesy Francisco Chavez, reprinted with permission from Chavez *et al.*, 1999, Figure 1; ©1999 AAAS; OrbView-2 Imagery provided by ORBIMAGE, the SeaWiFS Project and NASA/Goddard Spaceflight Center).

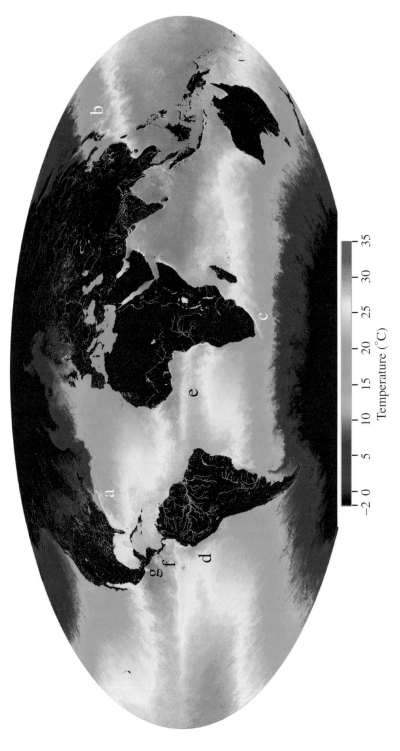

Plate 9. One month composite of the MODIS derived SST for May 2001. Black is land, the colors correspond to the temperature scale. The letters identify physical features discussed in the text (Courtesy MODIS Ocean Group, NASA GSFC and the University of Miami).

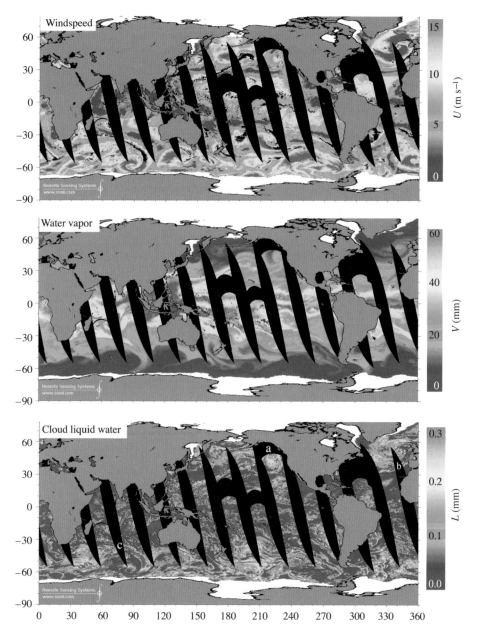

Plate 10. Composite image of the distribution of SSM/I wind magnitude, water vapor, and cloud liquid water for April 19, 2000. The swaths are the ascending, evening passes at 1800 local time. The color bars to the right give the scale for the distribution of each variable; gray is land, white is sea ice, black is missing data or the masked rain rate (SSM/I data and images are produced by Remote Sensing Systems and sponsored by the NASA Pathfinder Program for early Earth Observing System (EOS) products. Used with permission).

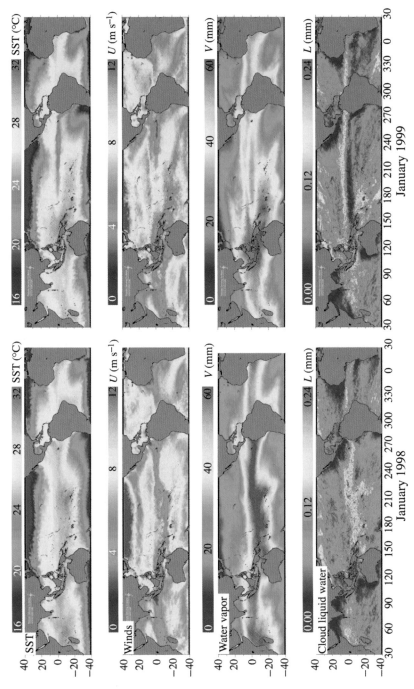

Plate 11. Composite monthly averages of TMI sea surface temperatures, winds, water vapor and cloud liquid water for January 1998 and January 1999 (TMI data and images are produced by Remote Sensing Systems and sponsored by NASA's Earth Science Information Partnerships (ESIP): a federation of information sites for Earth Science; and by NASA's TRMM Science Team. Used with permission).

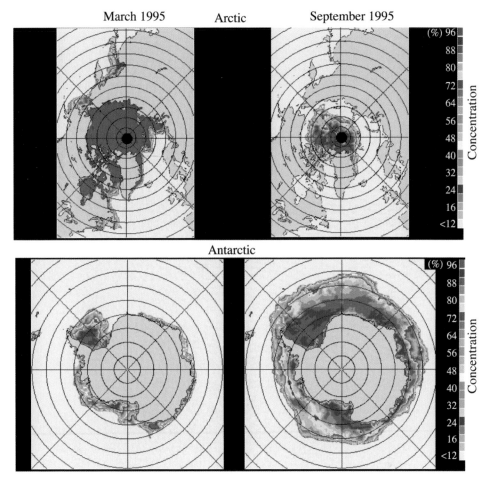

Plate 12. The 1995 monthly averaged maximum and minimum SSM/I ice extent for the Arctic and Antarctic, where gray is land or glacial ice, and light blue is open water, and the black circle at the North Pole is missing data. The images are generated with the NASA Team algorithm (Images from Cavalieri *et al.* (1999, updated 2002). Sea ice concentrations from DMSP SSM/I passive microwave data, courtesy National Snow and Ice Data Center (NSIDC), Boulder, CO, USA. Used with permission).

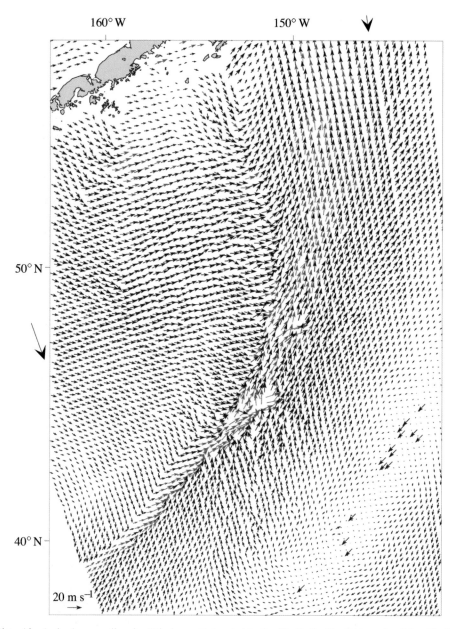

Plate 13. A single ascending SeaWinds swath located in the North Pacific, just south of the Alaska Peninsula and acquired on September 2, 1999 at 1530 UTC. The wind vectors are given at intervals of 25 km; the vectors are color coded so that black vectors are rain-free, red vectors are rain-contaminated. The inset arrow shows the wind scale (Courtesy Jérôme Patoux and Robert Brown, used with permission).

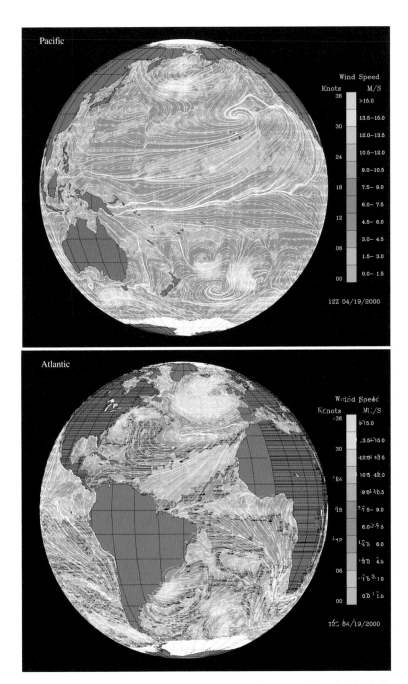

Plate 14. The QuikSCAT ocean wind field for April 19, 2000 for the Pacific and Atlantic Oceans. The lines and arrows show the wind direction; the colors show the wind speed (The images are obtained from the NASA/NOAA sponsored data system Seaflux at JPL through the courtesy of W. Timothy Liu and Wenqing Tang, used with permission).

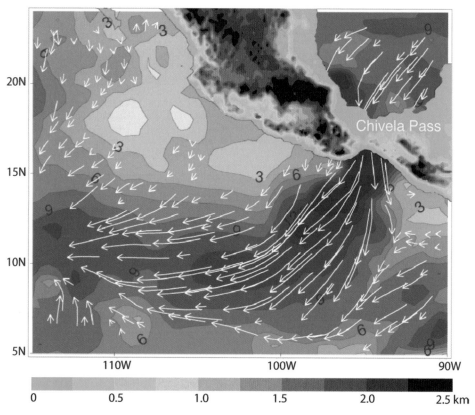

Plate 15. A QuikSCAT image of a Tehuanos event at Chivela Pass in southern Mexico taken at 00 UTC on December 1, 1999. The Atlantic is on the upper right, the Pacific is on the left. The color bar gives the scale of the land topography; the contoured shades of blue and arrows show the wind speed and direction, where darker shades of blue and longer arrows indicate greater speeds. The wind speed contours are at intervals of 1.5 m s^{-1} (Courtesy Mark Bourassa and Josh Grant, used with permission).

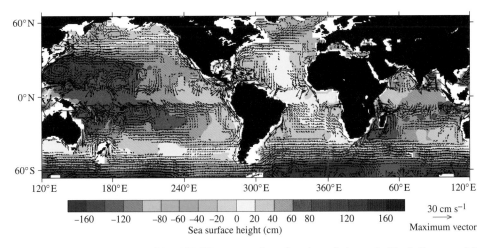

Plate 16. The 4-year average of the TOPEX ocean surface elevation relative to the Earth Geopotential Model 96 (EGM96) geoid. The arrows show the geostrophic velocities. Near equatorial values are omitted because of the breakdown of the geostrophic relation; small velocities are omitted for clarity. Because they are dominated by geoid error, all flows with length scales less than 500 km are omitted (Courtesy Detlef Stammer, Figure 6a from Wunsch and Stammer, 1998, with permission, from the *Annual Review of Earth and Planetary Sciences*, Volume 26, ©1998, by Annual Reviews).

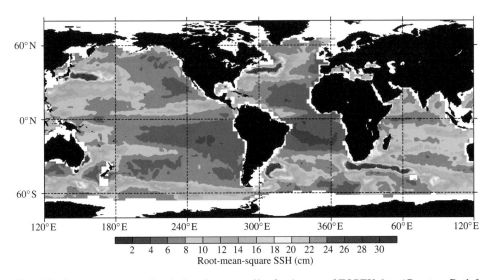

Plate 17. Root-mean-square (rms) elevation anomalies for 4 years of TOPEX data (Courtesy Detlef Stammer, Figure 8a from Wunsch and Stammer, 1998, with permission, from the *Annual Review of Earth and Planetary Sciences*, Volume 26, ©1998, by Annual Reviews).

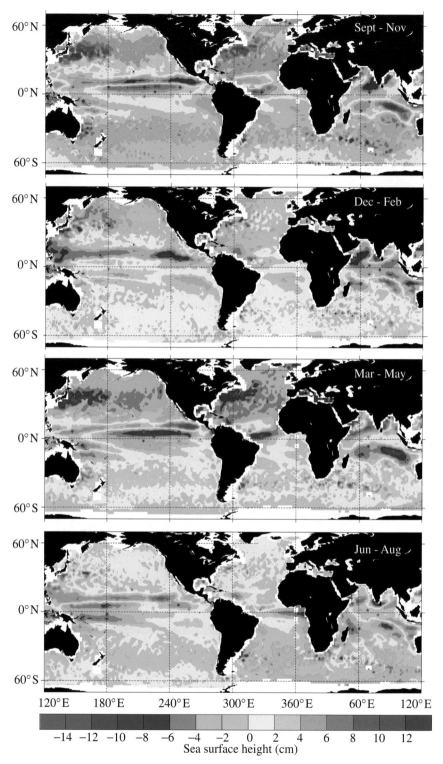

Plate 18. Seasonal mean anomalies of the TOPEX sea surface heights relative to the 9-year mean field. Top image is September–November 1992–2000; second is December 1992–2000 through February 1993–2001, third is March–May 1993–2001; fourth is June–August 1993–2001. Contour interval is 2 cm (Courtesy Detlef Stammer, used with permission).

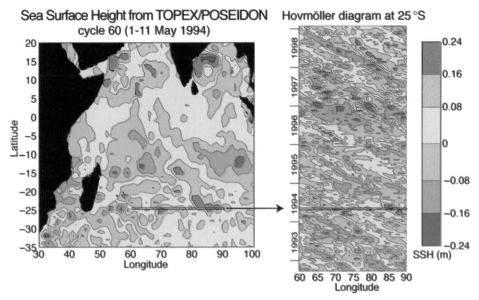

Plate 19. Rossby waves in the Indian Ocean. The left-hand figure shows the geographic distribution of the anomaly in sea surface height in the Indian Ocean for TOPEX cycle 60 corresponding to the 10-day period May 1–11, 1994. The rectangular strip outlined in black at 25° S is also outlined in the right-hand Hovmöller diagram. On this diagram, the horizontal axis corresponds to the central portion of the Indian Ocean; the vertical axis is the TOPEX cycle number. In each case, the colors correspond to SSH defined in the right-hand scale. The characteristic upper left to lower right tilt within the Hovmöller diagram illustrates the westward propagation of Rossby waves (Courtesy Paolo Cipollini; Figure 5 from Killworth, 2001, © 2001, with permission from Elsevier Science).

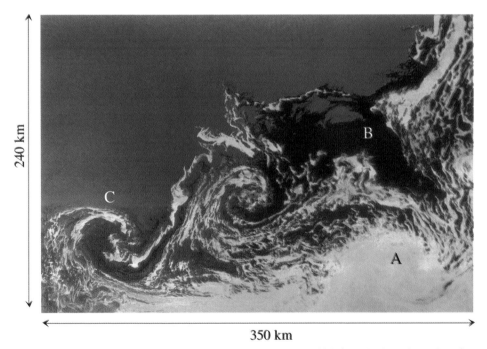

Plate 20. SAR image of the Antarctic sea ice taken on October 5, 1994, from the Spaceborne Imaging Radar C/X-Band Synthetic Aperture Radar (SIR-C/X SAR) onboard the Space Shuttle Endeavour. The image is oriented approximately east–west, with a center latitude and longitude of about 56.6° S and 6.5° W; its dimensions are 240 km by 350 km (Courtesy of NASA/JPL/Caltech, used with permission).

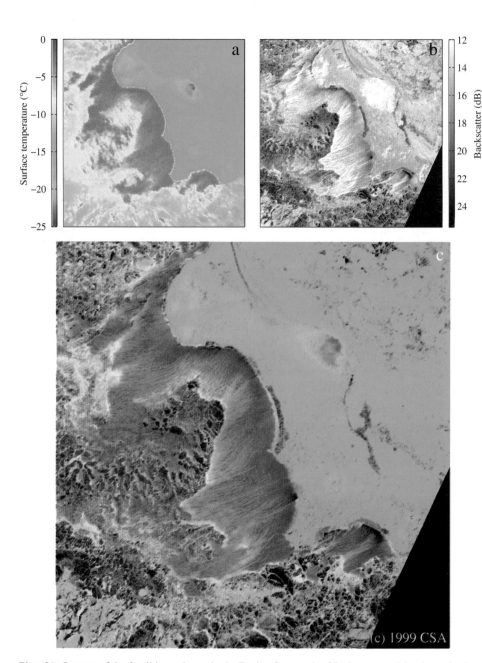

Plate 21. Images of the frazil ice polynya in the Bering Sea south of St. Lawrence Island acquired on January 9, 1999. Upper left, AVHRR image processed for ice surface temperature and acquired at 0431 UTC; upper right, RADARSAT ScanSAR image acquired at 0504 UTC, so that the two images are 33 minutes apart. The long axis of the island measures about 200 km. See text for further description (RADARSAT data © Canadian Space Agency/Agence Spatiale Canadienne 1999. Processed and distributed by RADARSAT International; image processing by Robert Drucker and the author).

Composite infrared image of the Gulf of Alaska, from NOAA-10 AVHRR imagery acquired on March 18 and 19, 1990. Red is warm, blue is cold, and a swath of cold clouds is visible to the lower left. The warm Alaska coastal current and its turbulent eddies are clearly visible (Courtesy Kevin Engle, used with permission).

8

Introduction to microwave imagers

8.1 Introduction

Passive microwave radiometers provide a powerful, nearly all-weather technique for retrieval of a wide variety of ocean, sea ice and atmospheric geophysical variables. In contrast to the lenses and mirrors used in the VIR, antennas are used with passive microwave radiometers to receive the Earth-emitted radiances, and with radars to transmit and receive energy pulses. The present chapter describes the radiometer antennas and their terminology, discusses how radiometers work, and summarizes the properties of the past, present and near future operational passive microwave imagers. The next chapter describes the atmospheric transmissivity in the microwave, discusses the dependence of the surface emissivity on a variety of geophysical variables, gives examples of the retrieval algorithms and concludes with examples.

Microwaves occupy that part of the electromagnetic spectrum between 1 and 500 GHz in frequency, or 0.3 m–1 mm in wavelength. The microwave band is bounded at low frequencies by the television broadcast bands, the presence of which makes Earth observations difficult, and at high frequencies by the far infrared wavelengths where alternative methods exist for detection. Because of the frequency dependence of the atmospheric transmissivity and interference from other users, the frequencies used in retrieval of oceanic variables are restricted to specific windows in the range 1–90 GHz. The importance of microwave observations to ocean remote sensing is that the ocean surface emissivities and atmospheric transmissivities depend both on frequency and on such variables as atmospheric water vapor and liquid water, rain rate, sea surface temperature and salinity, wind speed, and sea ice type and extent. As Chapter 9 discusses in detail, to retrieve these variables, the instruments observe the ocean at several different frequencies.

Compared to the VIR, the microwave has advantages and disadvantages. First and as Chapter 9 shows, because the atmosphere is much more transparent than in the VIR, especially in the range $1 < f < 10$ GHz, microwave instruments can view the surface through clouds and gather data under almost all weather conditions except heavy rain. Second, with the exception of the sun, the brightness temperatures of the surface and of the atmospheric and extraterrestrial sources of radiation are less than or order of 300 K. At these temperatures and in the microwave, the Rayleigh–Jeans approximation to Planck's law applies, so that

197

the radiative transfer equation can be written in terms of brightness temperatures instead of radiances. Third, the disadvantages are due to the long microwave wavelengths relative to the VIR. From the Rayleigh criterion in Section 3.5.1, these long wavelengths mean that the antenna size or aperture must be much larger to obtain the same spatial resolution. Also, because the Earth approximately radiates as a 300 K blackbody with its maximum radiance at a wavelength of about 10 μm, from Planck's equation, the emitted microwave radiances are much smaller than in the VIR. Thus for an antenna to receive the same radiant flux as an optical instrument also requires either a larger aperture or larger FOV. At present, because the size constraints imposed by launch vehicles restrict the diameters of passive microwave antennas to 1–4 m, the microwave resolution is in the range 5–100 km. Finally, because the ocean surface microwave emissivity is strongly look-angle dependent, a whiskbroom scanner cannot be used. Instead, most microwave imagers employ a conical scanner that observes the surface at a constant angle.

In the following, Section 8.2 discusses the properties of antennas. Although this discussion primarily applies to passive imagers, some of it also applies to the active radars discussed in Chapter 10. Section 8.3 describes how passive microwave antennas retrieve the surface radiances. Section 8.4 describes the design of the conical scanner and the dependence of the surface emissivity on incidence angle. Section 8.5 describes the process of antenna pattern correction. Section 8.6 describes the design and properties of the four past and present microwave imagers used for retrieval of oceanic properties: the Scanning Multichannel Microwave Radiometer (SMMR), the Special Sensor Microwave/Imager (SSM/I), the TRMM (Tropical Rainfall Measuring Mission) Microwave Imager (TMI) and the Advanced Microwave Scanning Radiometer (AMSR). The chapter concludes with a brief summary of future instruments.

8.2 General antenna properties

Ulaby, Moore and Fung (1981, p. 93) define an antenna as "a region of transition from electromagnetic radiation propagating in free space, to a guided wave propagating in a transmission line." Figure 8.1 shows three common antennas, including a horn, a front-feed Cassegrain dish antenna and a front-feed paraboloid. For a transmitting antenna, the energy from the waveguide is radiated into space with a non-uniform directional distribution; for receiving, the antennas collect the incident radiation and focus it into a waveguide. The Reciprocity Theorem (Balanis, 1982, Sections 3.8.0, 3.8.1) states that the directional distribution of the energy transmitted and received by the same antenna is identical. Because of this reciprocity between transmission and reception, the antenna discussion begins with the derivation of their directional properties during transmission, then continues with the application of these properties to the receiving case.

Figure 8.2 shows an idealized radiating two-dimensional antenna of aperture width D, where $D > \lambda$. The antenna is illuminated by an electric field E, written as

$$E(x, t) = f(x)e^{i\omega t} \tag{8.1}$$

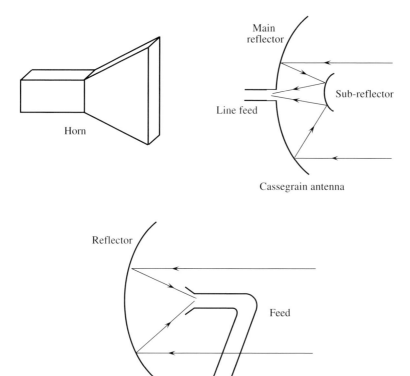

Figure 8.1. Different kinds of antennas used in satellite applications (Adapted from Figure 3.1, Ulaby *et al.*, 1981).

In (8.1), $f(x)$ is the antenna *illumination pattern*. On Figure 8.2, r is range, θ is elevation angle, ϕ is azimuth angle and the exitance $M(r, \theta, \phi)$ is the power density. At any given r, the *boresight* direction is that direction, generally at $\theta = 0$ and $\phi = 0$, along which M is a maximum.

For transmission and in the far field where $r > 2D^2/\lambda$, M decreases as r^{-2}, so that it can be written as

$$M(r, \theta, \phi) = I(\theta, \phi)/r^2 \tag{8.2}$$

where the intensity $I(\theta, \phi)$ is independent of r (Balanis, 1982). If Φ_T is the total radiant flux transmitted by the antenna, then

$$\Phi_T = \int_{4\pi} \int I(\theta, \phi) \mathrm{d}\Omega \tag{8.3}$$

where $\mathrm{d}\Omega = \sin\theta \, \mathrm{d}\theta \mathrm{d}\phi$ and the integration is over the entire sphere. From these definitions, and as the next sections show, the antenna properties are characterized by the *power pattern*,

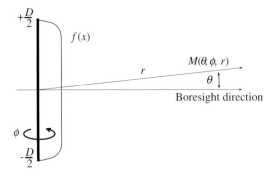

Figure 8.2. Diagram of an idealized antenna.

the *pattern solid angle*, the *main beam* and *sidelobe solid angles*, the *main beam efficiency* and the *gain*.

8.2.1 Power pattern

One of the differences between antennas and the lenses used in the VIR is that antennas have sidelobes, which means that they transmit and receive energy at angles well away from the boresight direction. Following Ulaby *et al.* (1981), the sidelobe properties are defined in terms of the *normalized power* or *radiation pattern* $F_n(\theta, \phi)$, given by

$$F_n(\theta, \phi) = I(\theta, \phi)/I_0 \tag{8.4}$$

where I_0 is the maximum intensity, which is in the boresight direction. For a particular antenna, F_n is either calculated numerically or analytically, or is determined experimentally in an antenna test facility. From the Reciprocity Theorem described above, the power pattern F_n for receiving and transmission is identical, so that the F_n of a receiving antenna can be determined from its transmission properties.

For the 85-GHz H-pol channel of the SSM/I microwave imager, Figure 8.3 shows the power pattern, which consists of the dominant *main lobe*, smaller *sidelobes* and although the figure does not show these, much smaller *back lobes*. As the figure shows, the width of the main lobe is given by the angular distance between the minima closest to the boresight direction; the sidelobe widths are similarly defined by their respective minima. The magnitude of F_n is given in terms of decibels or dB as defined in Equation (4.7), and the *half-power points* are those angles at which the radiated power is reduced by a factor of $1/2$ from its peak value, or where $F_n = -3$ dB. From Ulaby *et al.* (1981, Section 3.11), the *half-power beamwidth* $\Delta\theta_{1/2}$ is the angle between the two half-power points and is approximately given by

$$\Delta\theta_{1/2} \sim \lambda/D \tag{8.5}$$

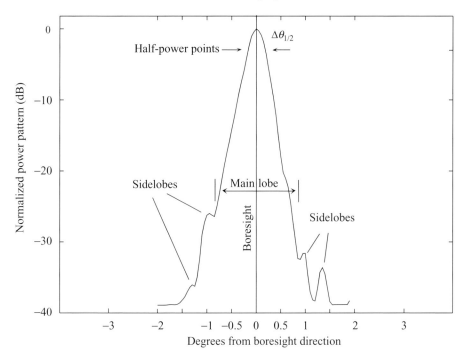

Figure 8.3. Power pattern for the SSM/I 85-GHz, H-pol antenna, plotted around the boresight direction from − 2 to + 2°, and showing the main and sidelobes. The half-power beamwidth is 0.35° (Data courtesy Gene Poe).

In Figure 8.3, the power pattern consists of a sharp peak between −2 and +2° and a half-power beamwidth of 0.35°. The narrow beamwidth and small sidelobes are typical of satellite antennas. The shape of the power pattern means that the antenna not only receives power within $\Delta\theta_{1/2}$ but also from solid angles outside this beamwidth. For example, if a very bright object such as the sun fills one of the sidelobes, its radiance can overwhelm the contribution from the main lobe.

For a microwave antenna, $\Delta\theta_{1/2}$ defines the size of the surface *half-power field-of-view*, or equivalently, the 3-dB FOV, which is also called the *surface footprint*. Equation (8.5) shows that the smaller the wavelength or the greater the frequency, the better the resolution. However, because the surface radiance is very small, it is frequently necessary to average the received radiances over a period of time to enhance the signal-to-noise ratio. As Sections 1.6 and 8.6.1 describe in greater detail, because of this averaging, the FOV discussion divides into the instantaneous FOV (IFOV) or the FOV determined at any instant, and the effective FOV (EFOV), which is its time-averaged value.

The power pattern has two limiting cases. The first corresponds to an optical telescope or to what is called a *pencil-beam* antenna, which lacks sidelobes and gathers radiance only from within a specified solid angle around the boresight direction. For this case, the power pattern becomes

$$F_n(\theta, \phi) = 1, \qquad \theta \leq \Delta\theta/2, \qquad 0 \leq \phi \leq 2\pi$$

$$F_n(\theta, \phi) = 0, \qquad \text{otherwise} \tag{8.6}$$

Figure 8.3 is an example of an antenna pattern that is nearly a pencil beam.

The second case is the ideal isotropic antenna, which transmits or receives uniformly at all angles around the sphere, so that its power pattern is

$$F_n(\theta, \phi) = 1, \qquad \text{all angles} \tag{8.7}$$

For this isotropic case, the average normalized intensity is

$$I_{ave} = \Phi_T/4\pi \tag{8.8}$$

Even though it is impossible to build an isotropic antenna, it is a useful limiting concept.

8.2.2 Solid angles associated with the power pattern

The antenna properties are also described by a number of solid angles, including the pattern solid angle, the main beam solid angle and the sidelobe and backlobe solid angles. The pattern solid angle Ω_P is a measure of the width of the antenna pattern and is defined as the integral of the power pattern over all solid angles,

$$\Omega_P = \int_{4\pi} \int F_n(\theta, \phi) d\Omega \tag{8.9}$$

For an isotropic antenna, $\Omega_P = 4\pi$, and for the pencil-beam antenna described in (8.6),

$$\Omega_P = 2\pi \, \Delta\theta^2/8 \tag{8.10}$$

Similarly, the main beam solid angle Ω_M is defined as the integral of F_n over the main lobe, as in

$$\Omega_M = \int \int_{\text{main lobe}} F_n(\theta, \phi) d\Omega \tag{8.11}$$

The sidelobe and backlobe solid angles Ω_S and Ω_B have similar definitions. From (8.9) and (8.11), the main beam efficiency η_M is defined as

$$\eta_M = \Omega_M/\Omega_P \tag{8.12}$$

Generally speaking, the closer η_M is to 1, the smaller the sidelobes and the larger the contribution from the half-power beamwidth. As the following sections show for the different channels on the microwave imagers, in most cases, $\eta_M > 0.9$. For the antenna pattern shown in Figure 8.3, $\eta_M = 0.92$.

8.2.3 Gain

The gain $G(\theta, \phi)$ describes the antenna *directionality*, and is defined as the ratio of the actual intensity received from a given direction $I(\theta, \phi)$ to the average intensity I_{ave} in (8.8):

$$G(\theta, \phi) = I(\theta, \phi)/I_{ave} \tag{8.13}$$

Division of the top and bottom of this equation by I_{max} and use of Equations (8.3), (8.4), (8.8) and (8.9) transforms (8.13) to

$$G(\theta, \phi) = 4\pi F_n(\theta, \phi)/\Omega_P \tag{8.14}$$

The maximum gain, called G_0, occurs for $F_n = 1$, where

$$G_0 = 4\pi/\Omega_P \tag{8.15}$$

so that G_0 is the ratio of the solid angle occupied by a sphere to the pattern solid angle. Since a large gain implies a small Ω_P, a pencil-beam antenna has high gain.

8.3 Measurement of the surface radiance with an antenna

This and the next two sections describe the method used by scanning radiometers to determine the surface brightness temperature T_B within the half-power beamwidth. First, the present section discusses how a microwave antenna retrieves T_B, where this retrieval includes contributions from both the main beam FOV and the various sidelobes. Second, Section 8.4 describes why microwave imagers operate as conical scanners. Third, Section 8.5 discusses the process of *antenna pattern correction* (APC), which involves the use of the values of T_B taken from the adjacent FOVs produced by the scanner to remove the sidelobe contributions and to improve the accuracy of the retrieved T_B. In the following, the V-pol and H-pol brightness temperatures are written as T_{BV} and T_{BH}, or as $T_{B(V, H)}$ where the subscript can be either V or H.

Consider the relation between the solid angle distribution of radiances incident on an antenna and the received radiant flux. The derivation assumes that there are no losses within the antenna, and that the antenna is sensitive to radiation at a center frequency f_0 with a bandwidth $\Delta f \ll f_0$. From Ulaby *et al.* (1981, Section 4.2), if $L(\theta, \phi, f)$ is the angular distribution of radiance observed by the antenna and $F_n(\theta, \phi)$ is the power pattern, then the received radiant flux $\Phi_{(V, H)}$ is approximately given by the product of L and the power pattern F_n integrated over the entire sphere:

$$\Phi_{(V,H)} = \frac{1}{2} A_e \Delta f \int\int_{4\pi} L(\theta, \phi, f_0) F_n(\theta, \phi) \, d\Omega \tag{8.16}$$

In (8.16), $\Phi_{(V,H)}$ is the polarization-dependent radiant flux received by the antenna. Because microwave antennas operate at only one polarization, then depending on the instrument design, the subscript on Φ is either V or H. This restriction means that the antennas receive only half the incident power, which explains the factor of $1/2$ in front of the integral (Ulaby

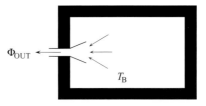

Figure 8.4. An antenna extending into a black box held at a constant temperature.

et al., 1981). Also in (8.16), A_e is the effective aperture area, which depends on the nature of the antenna and the properties of the incident radiation. For example, an antenna made up of a wire grid can have an effective area nearly equal to that of a solid antenna. Finally, Δf is assumed sufficiently small that the integral over f can be linearized.

For an antenna located within a black box, the received radiant flux is calculated from (8.16) for two cases: the inner surface of the box at a uniform temperature, and the surface at spatially non-uniform temperature. For the first case, Figure 8.4 shows the antenna penetrating through the wall of a black box with its inner surface held at a constant temperature T_B. The Rayleigh–Jeans law is assumed valid, so that

$$L(f_0) = 2k_B T_B f_0^2 / c^2 = 2k_B T_B / \lambda_0^2 \tag{8.17}$$

where $f_0 \lambda_0 = c$ and λ_0 is the center wavelength. After substitution of (8.17) into (8.16) and dropping the polarization subscript to simplify the resultant equation, the radiant flux Φ_{OUT} out of the antenna becomes

$$\Phi_{OUT} = \frac{1}{2} A_e \Delta f \int\!\!\int_{4\pi} \left(2k_B T_B / \lambda_0^2\right) F_n(\theta, \phi)\, d\Omega \tag{8.18}$$

Integration of (8.18) and substitution of the definition of Ω_P from (8.9) gives

$$\Phi_{OUT} = \left(A_e \Delta f k_B T_B / \lambda_0^2\right) \int\!\!\int_{4\pi} F_n(\theta, \phi)\, d\Omega = A_e \Omega_P \Delta f k_B T_B / \lambda_0^2 \tag{8.19}$$

From (8.19) and assuming that the antenna properties A_e, Δf, λ_0 and Ω_P are known, measurement of Φ_{OUT} permits solution for the unknown T_B.

Assume next that the inside of the box has a non-uniform surface or scene temperature $T_{sc}(\theta, \phi)$. For this case,

$$\Phi_{OUT} = \left(A_e \Delta f k_B / \lambda_0^2\right) \int\!\!\int_{4\pi} T_{sc}(\theta, \phi) F_n(\theta, \phi)\, d\Omega = A_e \Omega_P \Delta f k_B T_A / \lambda_0^2 \tag{8.20}$$

For the non-uniform temperature distribution, Equation (8.20) shows that the box interior appears to have a uniform temperature T_A, where

$$T_A = \frac{1}{\Omega_P} \int\!\!\int_{4\pi} T_{sc}(\theta, \phi)\, F_n(\theta, \phi)\, d\Omega \tag{8.21}$$

The temperature T_A is called the *antenna radiometric temperature* or simply the *antenna temperature*, and is the weighted integral of $T_{sc}(\theta, \phi)$, with either a V or H polarization.

Despite its name, it is not the physical antenna temperature; in fact, because antennas are designed to be highly reflective with correspondingly low emissivities, their brightness temperatures are small.

For a specific antenna, there are great advantages to be able to correct the measured T_A such that it corresponds only to the brightness temperature that is emitted from within the half-power FOV. Because the APC procedure depends on the use of scanning radiometers, Section 8.4 next describes why conical scanners are used in the microwave.

8.4 Conical scanners and microwave surface emissivity

Because the microwave emissivities of the atmosphere and ocean strongly depend on inci-dence angle, the current operational microwave imagers are conical scanners. As Figure 8.5 shows, these instruments view the surface at a fixed incidence angle θ and rotate at a con-stant rate about their nadir axis so that their FOVs lie along successive arcs. For the reasons discussed in Section 9.4.2 and for all conical scanners, the incidence angles are in the range $50°$–$55°$.

For the representative observational frequencies of 6, 18, 37 and 85 GHz, and for a specular fresh water surface, Figure 8.6 shows the θ-dependence of the V- and H-pol values of the reflectivity and emissivity. The curves are derived by substitution of the fresh water index of refraction into the Fresnel equations (Equations 5.7). As Section 9.5 shows, the reason for using fresh water is that for $f > 5$ GHz, the emissivity is

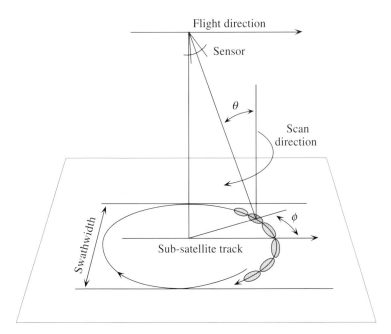

Figure 8.5. The geometry of the conical scanner and its surface scanning pattern. The figure shows the incidence angle θ, azimuth angle ϕ, the swathwidth and a few representative half-power FOVs.

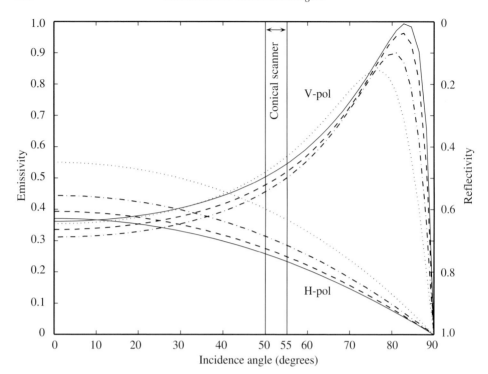

Figure 8.6. The dependence on incidence angle of the microwave emissivity and reflectivity of a specular fresh water surface. The figure shows the emissivity (left-hand scale) and reflectivity (right-hand scale) for 6 GHz (solid line), 18 GHz (dashed line), 37 GHz (dot-dash line), and 85 GHz (dotted line) and for a V- and H-polarization. The vertical lines show the 50–55° operating range of the conical scanners. See text for additional description.

independent of surface salinity. The figure also shows the 50°–55° range of the scanner incidence angles. Examination of the figure shows that the emissivity and reflectivity have a very different and stronger θ-dependence than the infrared properties shown in Figure 7.6. For $\theta = 50°–55°$, the reflectivities are on the order of the emissivities. These large reflectivities mean that unlike the infrared, the microwave radiative transfer model must include the reflected atmospheric and extraterrestrial radiances. Finally, even though the conical scanner avoids the problems associated with θ-dependent emissivities, small spacecraft oscillations can alter θ and e, creating uncertainties in the received brightness temperatures.

8.5 Antenna pattern correction (APC)

This section describes how the combination of the antenna scanning pattern and the received antenna temperatures permits retrieval of the surface or atmospheric brightness temperature

T_B from within the half-power FOV, while minimizing unwanted radiances. As Wentz (1992) and Colton and Poe (1999) show, these unwanted radiances include the following:

1. surface radiances originating from outside the half-power FOV, called sidelobe contamination;
2. reflected extraterrestrial sources of radiance;
3. radiances generated by cross-polarization coupling within the instrument so that a radiance received at one polarization has contributions from the opposite polarization.

Depending on the antenna design and on the magnitude of the brightness temperatures within the surrounding FOVs, these additional radiances can cause the antenna temperatures to differ significantly from the desired T_B.

The APC procedure uses the antenna properties in combination with the FOVs generated by the antenna scan to reduce the contributions of the unwanted radiances, and to reformat the retrieved data into regularly spaced Earth-located grid cells. Njoku, Christensen and Cofield (1980) review the general APC procedure and apply it to the SMMR; Wentz (1992) briefly describes the application of APC to the SSM/I; Colton and Poe (1999) describe this application in more detail. This analysis follows the treatment of Colton and Poe, which is valid for antennas with main beam efficiencies $\eta_M \geq 0.9$. For this case, the concentration of most of the received power within one or two beamwidths of the boresight direction simplifies the APC.

Figure 8.7 shows a series of circles representing three successive radiometer scans, where each circle represents a half-power FOV. The figure illustrates three cases: (1) an ocean FOV with its surrounding FOVs completely inside the swath; (2) an ocean FOV at the swath edge; (3) an ocean FOV adjacent to land. For the first case, the T_A measured at the surrounding FOVs are used to estimate the sidelobe radiances. Following Colton and Poe (1999) and for the nth FOV shown in Figure 8.7, it is assumed that $T_{B(V,H)}(n)$ is the desired surface brightness temperature and $T_{A(V,H)}(n)$ is the measured antenna temperature. For the H-pol radiances, the solution for T_{BH} can be written as

$$T_{BH}(n) = c_0 T_{AH}(n) + c_1 T_{AV}(n) + c_2 T_{AH}(n-1) + c_3 T_{AH}(n+1) \qquad (8.22)$$

with a similar solution for T_{BV}. On the right-hand side of (8.22), the first term is the corrected measurement from within the FOV; the second is the cross-talk term, which is the contribution from the radiance with the opposite polarization; the third and fourth terms are sidelobe contributions from the surrounding FOVs. For the sidelobe contributions, the derivation implicitly assumes that the radiances at the $(n-1)$ and $(n+1)$ FOVs are representative of all the surrounding FOVs. For the 85-GHz H-pol antenna pattern in Figure 8.3, typical values of the coefficients in (8.22) are $c_0 = 1.03$, $c_1 = 0.013$ and $c_2 = c_3 = 0.0015$, where these coefficients were determined in an antenna measurement facility (Colton and Poe, 1999, Table II). The reason c_0 is greater than 1 is to compensate for the falloff in the power pattern around the boresight direction.

Estimation of the magnitude of the corrections generated by Equation (8.22) proceeds as follows. For a surface temperature of 300 K, a non-interfering atmosphere and based on the emissivities in Figure 8.6, T_{AH} is approximated as 120 K and T_{AV} as 160 K. Substitution of

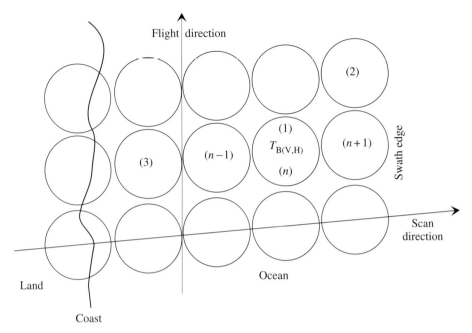

Figure 8.7. The surface pattern of the main beam FOV for a conically scanned microwave antenna. See text for further description.

these values and the 85H coefficients into (8.22) shows that the first term on the right-hand side yields a temperature increase of 3.6 K, the second term yields an increase of 2.1 K, and the third and fourth terms yield increases of 0.2 K. Summation of these terms yields $T_{BH} = 125.9$ K, which is 5.9 K larger than T_{AH}. The size of this correction verifies that the APC procedure must be applied to obtain the desired accuracy of 0.5– 1 K.

Because under most conditions the surface T_B distribution varies negligibly over the scale of the FOV diameter, Equation (8.22) is further simplified by combining c_2 and c_3 into c_0, in which case the APC is only applied to the FOV of interest. This simplification means that even for FOVs adjacent to the swath edge, the APC is still applicable. Alternatively for the swath edge case, when η_M is less than 0.9 as occurred with the SMMR 6.6-GHz channel, the sidelobe radiances from beyond the swath edge are accounted for with a *classification map*. This consists of a mean climatology or lookup table of ocean surface brightness temperatures that are used to estimate the sidelobe contributions, where the temperatures are functions of location and season. The third case that is not amenable to correction occurs when the FOV is adjacent to land or to an ice edge, ocean storm, or to any region with a step or strong spatial gradient in surface brightness temperatures. Because the APC smooths out such steps, the correction is not applied. Instead, the FOVs adjacent to these gradient regions are discarded or masked, so that the imager data are used only at distances of at least one FOV away from a land or ice edge boundary. Similar to the discussion in Section 4.9.1, such data are described as *contaminated*.

8.6 Passive microwave imagers

The series of research and operational passive microwave satellite imagers began with the US Electrically Scanned Microwave Radiometer (ESMR), which was a single channel 19-GHz cross-track scanner that operated from 1973 to 1976. Since ESMR, all of the imagers have been conically scanned. The first of the conically scanned instruments was the US Scanning Multichannel Microwave Radiometer (SMMR) that operated on SEASAT for 3 months in 1978, and on the NASA NIMBUS-7 spacecraft from 1978 to 1987, and briefly on SEASAT in 1979; the second is the US Special Sensor Microwave/Imager (SSM/I) that with a number of replacements has operated on the US Department of Defense DMSP satellite since June 1987. The third is the US/Japan TRMM Microwave Imager (TMI) that was launched on the Tropical Rainfall Measuring Mission (TRMM) in November 1997 and is still operating in 2004. The fourth is the Japanese Advanced Microwave Scanning Radiometer-EOS (AMSR-E) launched on AQUA in May 2002, which is a modified version of the AMSR launched on the Japanese short-lived ADEOS-2 in December 2002.

For these instruments, Table 8.1 compares their operating frequencies and incidence angles, and shows that they employ similar, but not identical channels. As the next chapter describes, the choice of frequencies is determined both by the atmospheric transmissivity and by the frequency sensitivity of the desired atmosphere and ocean variables. The largest change in channel selections occurs between SMMR and SSM/I, where the SSM/I dropped the 6.6 and 10.7 GHz channels, shifted the frequencies of the 18 and 21 GHz channels to 19.3 and 22.2 GHz, and added high frequency channels at 85 GHz. As the next chapter shows, the lack of the 6.6 and 10.7 GHz channels means that SSM/I cannot retrieve SST. These low frequency channels remained absent until the 1997 launch of TMI with its 10.7-GHz channels. The next sections briefly survey these instruments.

8.6.1 Scanning Multichannel Microwave Radiometer (SMMR)

SMMR was launched on the NIMBUS-7 satellite and provided data for the period 1978–1987. NIMBUS-7 was in a noon–midnight sun-synchronous orbit at an altitude of 955 km. Preceding its launch, another SMMR on the SEASAT satellite operated for a 99-day period

Table 8.1. *Comparison of the frequencies, polarizations and incidence angle θ for the SMMR, SSM/I, TMI and AMSR-E*

Instrument	Frequencies and polarization (GHz, V, H)						θ (deg)
SMMR	6.6 V, H	10.7 V, H	18.0 V, H	21.0 V, H	37.0 V, H	—	51
SSM/I	—	—	19.3 V, H	22.2 V	37.0 V, H	85.5 V, H	53
TMI	—	10.7 V, H	19.3 V, H	21.3 V	37.0 V, H	85.5 V, H	53
AMSR-E	6.9 V, H	10.7 V, H	18.7 V, H	23.8 V, H	36.5 V, H	89.0 V, H	55

Adapted from Wentz and Meissner, 1999

Table 8.2. *SMMR properties*

Frequency (GHz)	6.6 V, H	10.7 V, H	18 V, H	21 V, H	37 V, H
3-dB beamwidth (deg)	4.5	2.9	1.8	1.5	0.9
Bandwidth Δf (MHz)	250	250	250	250	250
$NE\Delta T$ at 300 K (K)	0.9	0.9	1.2	1.5	1.5
Integration time τ_I (ms)	126	62	62	62	30
Main beam efficiency η_M	0.82	0.85	0.87	0.85	0.89
3-dB EFOV (km × km)	148 × 95	90 × 60	45 × 30	40 × 25	20 × 15
(along-scan × along-track)					

From Stewart, 1985; Gloersen and Barath, 1977; Njoku, Stacey and Barath, 1980

in 1978. Gloersen and Barath (1977), Massom (1991) and Gloersen *et al.* (1992) describe the instrument; Table 8.2 lists its properties. The instrument consisted of an oscillating 1.1 m × 0.8 m elliptical antenna that reflected the Earth radiances into fixed microwave feed horns. The antenna was the only part of SMMR that rotated relative to the spacecraft. On NIMBUS-7 the antenna scanned in the forward direction across a swathwidth of 780 km. The scanning was sinusoidal, in that the reflector swung to one side of the flight path, paused, swung back to the other side and paused again, over a 4.096-s period.

For each frequency and polarization, Table 8.2 lists the SMMR 3-dB beamwidth, the bandwidth Δf, the $NE\Delta T$, the integration time τ_I, the main beam efficiency η_M and the EFOV. As described earlier, the EFOV consists of the IFOV averaged over the integration time, where the purpose of this integration is to reduce instrument noise. From Stewart (1985, Section 9.3), the noise reduction is calculated as follows: for a bandwidth Δf, the correlation time τ_c of the received radiance is given by $\tau_c \sim \Delta f^{-1}$. For $\Delta f = 250$ MHz, $\tau_c \sim 4 \times 10^{-6}$ ms, so that an integration over $\tau_I = 126$ ms is equivalent to averaging over $N = 6 \times 10^7$ independent observations. Since the uncertainty is proportional to $\sqrt{N}$, the integration reduces the noise by a factor of 10^4. As the table shows, this integration also makes the along-scan axis of the EFOV ellipse larger than the along-track. Since the swathwidth is only 780 km, at 6.6 GHz, the swath contains only 5 EFOV. The table also shows that for all frequencies, $\eta_M < 0.9$. Because $\eta_M = 0.82$ at 6.6 GHz, the sidelobe contributions meant that SST could not be retrieved within 1–4 EFOV or about 600 km of the ocean/land boundary, which greatly restricted its value (Njoku, *et al.* 1980a).

There were several problems with the SMMR design. First, the combination of the fixed feedhorns and rotating antenna generated cross-talk between different polarizations at the same frequency. Second, because NIMBUS-7 was in a sun-synchronous orbit with a local noon and midnight equator crossing time, daytime heating and nighttime cooling generated noise within the instrument. Also, as the satellite passed over the South Pole, SMMR experienced severe transients generated by the sun shining directly into the feedhorns. Third, because SMMR operated on alternate days to conserve power, a near global coverage was achieved only at six-day intervals. Fourth, when the instrument was turned on, there was a 1-hour transient during which the data had to be discarded; fifth, the SMMR was inadequately calibrated. In spite of these problems, the SMMR served as a valuable test bed for future instruments and contributed greatly to the understanding of the polar sea ice cover.

8.6.2 Special Sensor Microwave/Imager (SSM/I)

The SSM/I corrected many of the problems associated with SMMR, and provided the basis for the TMI and AMSR design. Hollinger, Peirce and Poe (1990) and Massom (1991) describe the SSM/I; on June 19, 1987, the first SSM/I was launched on the US Air Force DMSP satellite and with occasional replacements, remains in orbit. DMSP is in a dawn–dusk sun-synchronous orbit at an altitude of 860 km and a period of 102 minutes. This orbit provides complete Earth coverage except for two circular areas of 2.4° centered at the poles. SSM/I is supported by the US Department of Defense through 2003. In late 2003, it was replaced by the Special Sensor Microwave Imager/Sounder (SSMI/S) that combines the SSM/I imager and the temperature and humidity sounder into a single instrument that uses the present SSM/I antenna. Then about 2009 and as part of NPOESS, the SSMI/S will be replaced by the Conical Microwave Imager/Sounder (CMIS) discussed in Chapter 14.

Figure 8.8 shows a photograph of the SSM/I; it consists of an offset parabolic reflector of dimensions 61 cm by 66 cm that focuses microwave radiation into a seven-port antenna feedhorn. In a design feature common to TMI and AMSR, SSM/I is mounted on top of the DMSP satellite. Relative to the spacecraft, the reflector and feedhorns rotate with a uniform period of 1.90 s, where the data pass through a set of slip-rings into the spacecraft body. SSM/I has two non-rotating calibration sources, a cold space reflector and a hot reference load held at a temperature of about 300 K. These sources are fixed to the spacecraft, so that to calibrate the instrument, once per scan, their radiances are sequentially reflected into the feedhorns. The hot load is independently measured with precision thermometry and the cold space temperature is assumed constant at the 2.7 K background temperature of the universe (Colton and Poe, 1999).

Figure 8.9 shows the scan geometry; surface observations are taken during a 102.4° arc when the SSM/I is looking aft. The arc is centered on the spacecraft track and corresponds to a swathwidth of 1394 km. During the 1.90-s antenna rotation period, the spacecraft advances 12.5 km along the surface. The ellipses on the surface show the IFOVs; they become progressively smaller with increasing frequency. The scans are divided into A- and B-scans that alternate in time; the A-scan includes all channels, the B-scan includes only 85 GHz. For both scans, the 85-GHz channels are sampled 128 times over the arc, where each sample is integrated over 3.89 ms, during which time the antenna boresight moves about 12 km on the surface in the along-scan direction. Because of their larger resolution, the three lower frequency channels are only sampled during the A-scan, where they are averaged into 64 EFOVs along the arc. Table 8.3 lists the instrument properties and the 3-dB EFOVs for

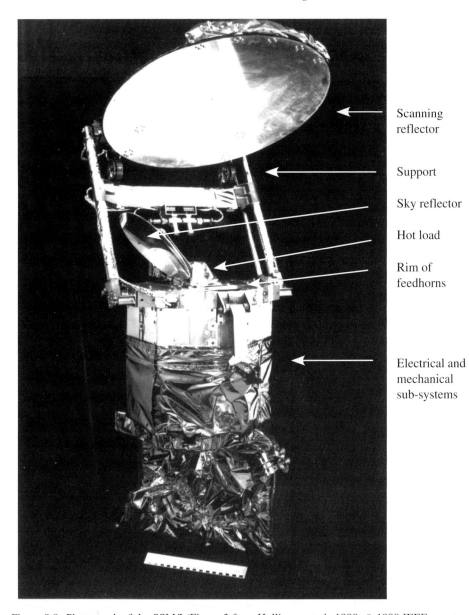

Scanning
reflector

Support

Sky reflector

Hot load

Rim of
feedhorns

Electrical and
mechanical
sub-systems

Figure 8.8. Photograph of the SSM/I (Figure 3 from Hollinger *et al.*, 1990, © 1990 IEEE, courtesy Gene Poe).

Table 8.3. *Properties of the SSM/I channels*

Frequency (GHz)	19.35 V, H	22.235 V	37.0 V, H	85.5 V, H
3-dB beamwidth (deg)	1.9	1.6	1.0	0.42
Bandwidth Δf (MHz)	100	100	200	600
$NE\Delta T$ (K)	0.8	0.8	0.6	1.1
Integration time τ_1 (ms)	7.95	7.95	7.95	3.89
Main beam efficiency η_M	0.96	0.95	0.93	0.92
3-dB EFOV (km × km)	45 × 70	40 × 50	30 × 37	13 × 15
(along-scan × along-track)				

Hollinger *et al.*, 1990; Wentz, 1992

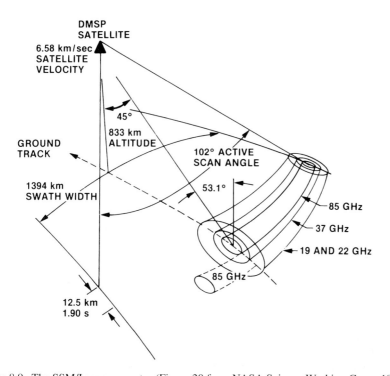

Figure 8.9. The SSM/I scan geometry (Figure 20 from NASA Science Working Group 1984).

the individual channels. Because the SSM/I integration times are much shorter than for SMMR, the SSM/I EFOVs are ellipses with their long axis in the along-track direction and their short axis in the along-scan direction, which is the reverse of the SMMR case.

The SSM/I design corrected many of the problems that occurred with SMMR. First, the SSM/I dawn–dusk orbit minimized the SMMR diurnal heating and cooling problems. Second, because the SSM/I feedhorns and antenna rotate together, the position of the feedhorns relative to the reflector is fixed, which eliminates the rotation-dependent cross-talk between polarizations that occurred with SMMR. Third, because SSM/I operates continuously, there are no start-up transients. Fourth, because the SSM/I swathwidth is twice as large as the SMMR, and since the SSM/I operates continuously instead of on alternate days, it has much greater coverage and produces about four times as much data. Fifth and as the next chapter discusses, the use of a dawn–dusk orbit greatly reduces sun glint. Its major disadvantage occurs because SSM/I lacks channels at frequencies lower than 19 GHz, it cannot retrieve SST.

8.6.3 TRMM Microwave Imager (TMI)

TMI is a nine-channel radiometer designed to investigate tropical regions of heavy precip-itation. TMI is mounted on the TRMM satellite, where TRMM is a joint mission between NASA and the National Space Development Agency of Japan (NASDA), which in October 2003 was incorporated into the Japan Aerospace Exploration Agency (JAXA). TRMM is in a low-inclination circular orbit at an altitude of 350 km and an inclination angle of 35°, which covers an area slightly greater than half the globe. The orbit is not sun-synchronous, rather it was chosen so that over a month, it samples the tropics at uniform intervals through-out the day. This permits determination of the rainfall dependence on local time-of-day. Its lower inclination orbit also means that its surface sampling rate is roughly twice that of a polar orbiter.

As Table 8.1 shows, the locations of TMI 19, 21, 37 and 85 GHz channels are almost identical to the SSM/I (Kummerow *et al.*, 1998). Differences between the two instruments include the addition of TMI channels at 10.7V and 10.7H, and a shift of the 22.235-GHz channel to 21.3 GHz. The purpose of this shift was to move the channel onto the shoulder of the 22.235-GHz water vapor absorption line described in Section 9.2.1, so that the channel would not saturate in the tropical atmosphere. Similar to SSM/I, the rotating part of the TMI includes the antenna and feedhorns. These rotate uniformly about the nadir axis with a period of 1.9 s, during which time the satellite advances 13.9 km along the surface. TMI antenna is an offset 61-cm diameter parabolic reflector that takes surface observations within a 130° arc, yielding a swathwidth of 786 km. Because the entire TRMM spacecraft is occasionally rotated by 180° about its nadir axis to maintain thermal stability, the instrument may point either forward or backward relative to the flight direction. To calibrate the TMI, once per rotation the feedhorns are moved to point sequentially at a hot load and a cold space reflector that are fixed to the spacecraft. Unlike SSM/I, TMI does not divide the scans into A and B scans, but accepts the gaps in the 85-GHz EFOVs. Table 8.4 lists the TMI characteristics and shows that because of the lower TRMM altitude and at any specific frequency, the TMI EFOVs have about half the area of the SSM/I EFOVs.

8.6.4 Advanced Microwave Scanning Radiometer-EOS (AMSR-E)

AMSR-E is a NASDA instrument launched on the AQUA satellite in May 2002; the com-panion AMSR instrument was launched in December 2002 on the Japanese Advanced Earth

Table 8.4. *TMI characteristics*

Frequency (GHz)	10.65 V, H	19.35 V, H	21.3 V	37.0 V, H	85.5 V, H
3-dB beamwidth (deg)	3.7	1.9	1.7	1.0	0.42
Bandwidth Δf (MHz)	100	500	200	2000	3000
$NE\Delta T$ (K)	0.6	0.5	0.7	0.3	0.7
Integration time τ_I (ms)	6.6	6.6	6.6	6.6	3.3
Main beam efficiency η_M	0.93	0.96	0.98	0.91	0.83
3-dB EFOV (km × km)	37 × 63	18 × 30	18 × 23	9 × 16	5 × 7
(along-scan × along-track)					

Adapted from Kummerow *et al.*, 1998

Observing Satellite-2 (ADEOS-2), which failed prematurely in October 2003. The AMSR-E design is slightly modified from AMSR; the difference between the two instruments is that AMSR has two additional V-pol channels at 50.3 and 52.8 GHz designed for atmospheric sounding. AQUA is in a 1330 ascending sun-synchronous orbit at an altitude of 705 km; ADEOS-2 is in a 1030 descending sun-synchronous orbit at 800 km.

Figure 8.10 shows the configuration of the AMSR-E instrument on AQUA; the instrument rotates continuously around its nadir axis with a 1.5-s period, during which time the spacecraft travels 10 km along its surface track. AMSR-E measures the upwelling surface radiances over an angular sector of ±61° about the sub-satellite track, yielding a swathwidth of 1445 km. AMSR-E is a twelve-channel, six-frequency conically scanned radiometer similar to SSM/I; the major differences are that AMSR-E has more channels, a larger 1.6-m diameter parabolic reflector, and a slightly different choice of frequencies (Table 8.5). The AMSR-E parabolic reflector focuses the surface radiances into an array of six feedhorns that are amplified by twelve separate receivers. The 18.7 and 23.8 GHz receivers share a feedhorn. To avoid having A- and B-scans, two offset feedhorns are used for the 85-GHz channels, which produce two 85-GHz FOVs that are separated in the along-track direction by 5 km. This gives the 85-GHz FOVs an along-track separation of 5 km; at the other channels, the FOVs are separated by 10 km.

The AMSR-E calibration is provided by two non-rotating external sources, a hot reference load maintained at a physical temperature of about 300 K, and a mirror that reflects the cold space brightness temperature into the instrument. The mirror and reference load are fixed to the spacecraft so that once per rotation, they pass in sequence between the feedhorn array and the parabolic reflector and provide a calibration. The zenith angle of the parabolic reflector is fixed at 47.4°, which results in an incidence angle of 55° ± 0.3°. The small variation in θ is due to the slight eccentricity of the orbit and the oblateness of the Earth.

8.6.5 *Future passive microwave radiometers*

There are two additional passive microwave radiometers; the WindSat radiometer on the Coriolis satellite launched in January 2003, and the CMIS radiometer scheduled for launch

Table 8.5. *AMSR-E properties*

f (GHz)	6.9 V, H	10.7 V, H	18.7 V, H	23.8 V, H	36.5 V, H	89.0 V, H
3-dB beamwidth (deg)	2.2	1.4	0.89	0.9	0.4	0.18
Bandwidth Δf (MHz)	350	100	200	400	1000	3000
$NE\Delta T$ (K)	0.3	0.6	0.6	0.6	0.6	1.1
Integration time τ_I (ms)	2.6	2.6	2.6	2.6	2.6	1.3
Main beam efficiency η_M	0.95	0.95	0.96	0.96	0.95	0.96
EFOV (km $\times$ km) (along-scan $\times$ along-track)	43×75	27×48	16×27	18×31	8×14	4×6

From http://www.ghcc.msfc.nasa.gov/AMSR

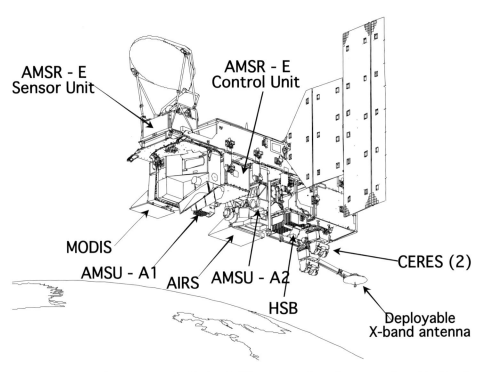

Figure 8.10. The AQUA spacecraft with the AMSR mounted above the spacecraft, and looking in the flight direction (Figure courtesy Claire Parkinson and the AQUA project, © Northrop Grumman Corporation, 2000, all rights reserved. Republished by kind permission of Northrop Grumman).

on the NPOESS Preparatory Project (NPP) in about 2005. Both instruments are part of NPOESS and are designed to produce vector wind speeds from passive microwave measurements. As Chapter 14 discusses, WindSat and CMIS have channels similar to those listed in Table 8.1; their major difference is that at several frequencies, instead of simply measuring

V- and H-pol, they measure all four Stokes parameters (Section 3.2.4). Such instruments are called *polarimetric* radiometers. The next chapter discusses how these measurements can be used to retrieve vector wind speeds.

8.7 Further reading and acknowledgements

Ulaby *et al.* (1981,1982) and Balanis (1982) give a detailed description of antenna properties. The AMSR-E program is described at http://www.ghcc.msfc.nasa.gov/AMSR/, where GHCC is the Global Hydrology and Climate Center, located at the Marshall Space Flight Center in Alabama. The site contains a number of Algorithm Theoretical Basis Documents (ATBD) for the AMSR-E. These documents are also at the ATBD home page at http://eospso.gsfc.nasa.gov/eos_homepage/for_scientists/publications.php. The Japanese AMSR is described at http://sharaku.eorc.nasda.go.jp/ADEOS2/index.html. The TRMM spacecraft and mission are described in both English and Japanese at http://www.eorc.nasda.go.jp/TRMM/. I thank Maria Colton, Son Nghiem and Gene Poe for their help in the preparation of this chapter.

9

Passive microwave observations of the atmosphere and ocean surface

9.1 Background

The importance of the multifrequency passive microwave imagers is that, independent of cloud cover, they can retrieve a large variety of surface and atmospheric variables. Figure 9.1 shows the atmospheric and surface variables that affect the transmissivities and emissivities, and are retrievable by passive microwave. In the atmosphere, these include the rain rate and the columnar water vapor and cloud liquid water. At the surface, they include sea ice properties, sea surface temperature (T_S) and salinity (S_S) and the scalar and vector wind speeds, which are derived from the changes in wave properties and foam extent with increasing wind speed.

As this chapter shows, because the emissivity or transmissivity associated with each constituent has a different frequency dependence, the variables are retrieved from a set of multifrequency, multivariable simultaneous equations. The disadvantage of this formulation is that in most cases the solutions are not separable, so that if, for example, the only variable of interest is T_S, by necessity many of the other variables must also be retrieved. The advantage is that except in regions with heavy rain the retrievals are independent of cloud cover. In the following, Section 9.2 discusses the frequency dependence of the atmospheric absorption and transmission, and shows that for frequencies less than 10 GHz, the effects of clouds and water vapor are negligible. Section 9.3 discusses the microwave form of the radiative transfer equation and the problem of sun glint. Section 9.4 describes the effect on the emissivity of ocean waves, surface roughness and foam, and shows how their azimuthal distribution of roughness relative to the wind direction permits retrieval of the vector wind speed. Section 9.5 describes the effect on the emissivity of sea surface temperature and salinity. Section 9.6 describes the multichannel algorithms for the simultaneous retrieval of the different oceanic and atmospheric variables. Section 9.7 discusses the sea ice algorithms.

9.2 Atmospheric absorption and transmissivity in the microwave

The next three sub-sections describe the contributions of oxygen, water vapor and liquid water droplets to the atmospheric microwave attenuation and transmissivity. The

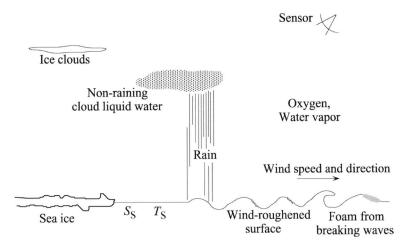

Figure 9.1. The properties of the sea surface and the atmosphere that affect the atmospheric attenuation and emission and the surface emissivity in the microwave.

results show that the best conditions for viewing the surface occur at the lower frequencies.

9.2.1 Atmospheric absorption by oxygen and water vapor

The oxygen absorption κ_{oxy} and water vapor absorption κ_{vap} are functions of the air temperature T_a and pressure p; κ_{vap} additionally depends on the vapor density ρ_v. For oxygen, and following Ulaby *et al.* (1981, Section 5.5), Figure 9.2 plots κ_{oxy} versus f for $p = 1013$ mb and $T_a = 270$ and 300 K.

On the figure, κ_{oxy} is given in units of Np (*neper*) km^{-1}, where nepers are a dimensionless quantity frequently used in the microwave literature to describe atmosphere absorption. The word "neper" is derived from "Naperian"; its name comes from the fact that the transmissivity t is proportional to $\exp(-\kappa \Delta z)$, where Δz is a length scale, so that to recover κ from t involves taking the natural or Naperian log of t. Neper is dimensionless, its only purpose is to designate that the quantity in question is atmospheric microwave absorption. Absorption is also described in units of dB km^{-1}, where Np km^{-1} and dB km^{-1} are linearly related (Ulaby *et al.*, 1981, Section 5.6).

Examination of Figure 9.2 shows that κ_{oxy} has two absorption peaks; the first is the *60-GHz oxygen complex*, which consists of the large number of absorption lines in the range 50–70 GHz; the second is a single line at 119 GHz. The figure also shows that κ_{oxy} increases with T_a and suggests that the best observing windows occur at 1–40 GHz and 80–105 GHz.

For water vapor, following Ulaby *et al.* (1981, Section 5.4), Figure 9.3 shows a plot of κ_{vap} versus f at $T_a = 300$ K and $p = 1013$ mb and for several values of ρ_v. As Section 4.2.1 describes, at sea level ρ_{v0} ranges from near zero in the polar regions to a maximum of about

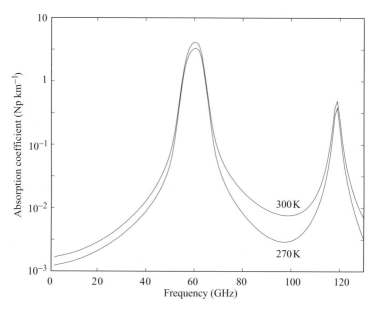

Figure 9.2. The dependence of the oxygen absorption κ_{oxy} on f for $p = 1013$ mb and for the two different temperatures on the curves, from formula in Ulaby *et al.* (1981, Section 5.5). See text for additional description.

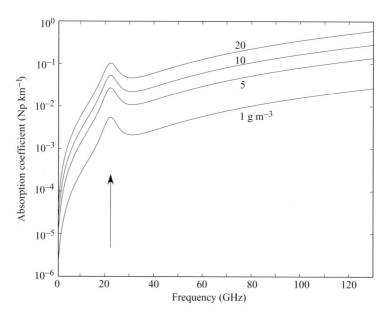

Figure 9.3. The dependence of the water vapor absorption κ_{vap} on f for $p = 1013$ mb, $T_a = 300$ K, and on several values of ρ_v from formula in Ulaby *et al.* (1981, Section 5.4). The numbers on each curve give ρ_{v0} in units of g m^{-3}; the arrow marks the absorption peak at 22.235 GHz. See text for further description.

30 g m^{-3} in the tropics; the values of ρ_{v0} on the curves lie within this range. For $0 < f <$ 130 GHz, the figure shows that the frequency dependence of κ_{vap} has two parts, a strong absorption peak at 22.235 GHz and a general increase of κ_{vap} with frequency generated by water vapor lines at 183.31 GHz and higher frequencies. Because the 22-GHz absorption peak occurs for water vapor but not for liquid water, the presence or absence of this peak makes it possible to distinguish water vapor from the liquid. The figure also shows that κ_{vap} has a strong dependence on ρ_v and that the most transparent part of the atmosphere occurs for f less than about 10 GHz.

9.2.2 Atmospheric transmissivity of oxygen and water vapor

Integration of κ_{oxy} and κ_{vap} vertically across the atmosphere yields the respective contributions of oxygen and water vapor to the atmospheric transmissivity. Figure 9.4 shows the resultant oxygen and water vapor transmissivities, and their total. For three MODTRAN atmospheres, Sub-arctic winter, Standard and Tropical, Figure 9.4a shows the dependence of the oxygen transmissivity t_{oxy} on frequency, where t_{oxy} is derived by integration of κ_{oxy} across the first 25 km of the atmosphere. Because κ_{oxy} increases with temperature, the tropical case is slightly less transmissive.

Figures 9.4b and 9.4c display an additional Arctic winter case derived from running the MODTRAN Sub-arctic case with $V = 0.5$ mm. Figure 9.4b shows the dependence of water vapor transmissivity t_{vap} on the columnar vapor density V; Figure 9.4c shows the sum of t_{oxy} and t_{vap}. Because water vapor is primarily concentrated in the lower troposphere, t_{vap} is derived by integration of κ_{vap} across the first 10 km. As expected, the transmissivity decreases with increasing V; this is most apparent at high frequencies and in the vicinity of the 22-GHz absorption peak. In Figure 9.4c and for all cases, the most transmissive frequencies occur for $f < 10$ GHz, where t is nearly independent of water vapor, and for the arctic dry atmosphere, where for $f < 40$ GHz, $t > 0.9$.

The figure also shows the location of several frequencies in use or proposed for use in oceanographic microwave radiometry, including 1.4, 6, 10, 18, 21, 37 and 85 GHz. With the exception of 18 and 21 GHz, these frequencies occur in the minima of the absorption curves. The 21-GHz channel is very close to the 22-GHz absorption peak, and 18 GHz lies on its shoulder, so that 21 GHz is used for water vapor retrieval at small vapor concentrations and 18 GHz is used at large concentrations. In summary, the figures show that while regions of strong oxygen absorption are avoidable by a proper choice of frequencies, for $f > 10$ GHz, the effect of atmospheric water vapor must generally be included in the modeling of atmosphere attenuation.

9.2.3 Transmissivity of water droplets

The liquid water droplets in clouds and rain affect the transmissivity by scattering the incident radiation. Ulaby *et al.* (1981) show for non-raining clouds that the droplet radius

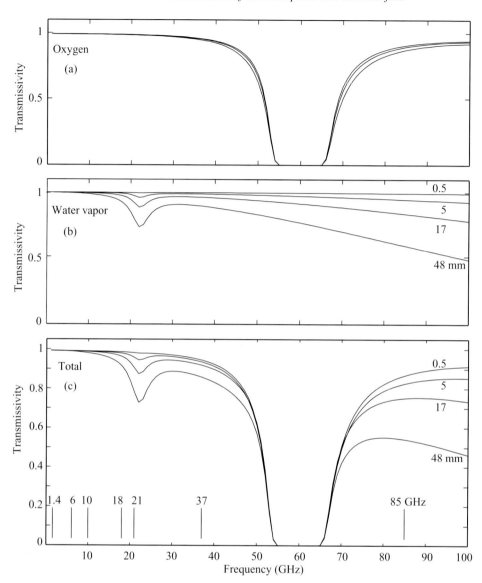

Figure 9.4. The atmospheric transmissivity for (a) oxygen, (b) water vapor, and (c) the sum of the oxygen and vapor transmissivity. Several typical observing frequencies are shown at the bottom of the figure; the numbers on the curves in (b) and (c) give the columnar water vapor V in mm. See text for further description.

is generally less than 0.1 mm, while for rain the droplet radius can be as large as 3 mm. Because of this difference in droplet size, liquid water attenuation occurs from two kinds of scatter; cloud droplets are Rayleigh scatterers, while especially at the higher observational frequencies the larger rain droplets are Mie scatterers. Given this difference, clouds and

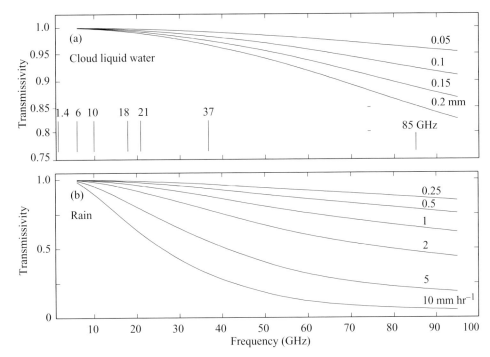

Figure 9.5. The effect of (a) non-raining cloud liquid water L and (b) rain rate R_R on the transmissivity. In (a), the curves are labeled in mm of columnar liquid water; in (b), in mm hr^{-1} of rain rate. To separate the liquid water curves from each other, the vertical scale of the upper figure extends from 0.75 to 1.05 and is four times the scale of the lower figure. See text for additional description (Curves derived from formula in Wentz and Meissner, 1999).

rain are separately considered. As Section 4.2.2 discusses, cloud liquid water is described in terms of the columnar liquid water L measured in mm. The range of L is much smaller than that of V and varies from 0 to 0.25 mm. Rain is described in terms of the rain rate R_R with units of mm hr^{-1}. Over the ocean, only 3% of the SSM/I observations of R_R exceed 2 mm h^{-1} (Wentz and Spencer, 1998).

Figure 9.5a shows the frequency dependence of transmissivity on L, and Figure 9.5b shows its dependence on R_R, where the curves are derived from a formula based on the AMSR range of frequencies in Wentz and Meissner (1999). The curves for $R_R >$ 2 mm h^{-1} lie outside the range of their solution and are only qualitatively correct. Since Rayleigh scattering is proportional to λ^{-4}, or equivalently to f^4, in Figure 9.5a, the transmissivity associated with cloud liquid water decreases as L and f increase. In contrast, for f < 10 GHz or long wavelengths, the effect of L on the transmissivity is negligible. For the larger rain droplets, Mie scattering becomes increasingly important at high frequencies and at large droplet concentrations. Consistent with this behavior, Figure 9.5b shows that, for f < 10 GHz and R_R < 1 mm h^{-1}, rain attenuation can be neglected, while for f > 10 GHz, the transmissivity decreases dramatically with both f and R_R. At high frequencies, the increase in attenuation with increasing R_R shows why the rain rate must be either determined or masked. Finally, both parts of Figure 9.5 show that for frequencies less than 10–15 GHz, the transmissivity is unaffected by light rain and clouds.

Because cirrus clouds are composed of ice crystals instead of liquid water, they have a negligible effect on the transmissivity (Ulaby *et al.*, 1981, Chapter 5.11). Since these crystals have characteristic radii $r < 0.2$ mm, Rayleigh scattering applies. In the microwave, ice crystals also have a smaller index of refraction than water droplets, so that for constant f, the attenuation associated with the crystals is an order smaller than for droplets. Given that columnar concentrations of these crystals are also about an order less than cloud liquid water, the attenuation associated with cirrus crystals is generally neglected. This is very different from the VIR, where, as Sections 7.5.4 and 7.7 describe, complicated procedures are required to identify and mask thin cirrus clouds.

In summary, comparison of Figures 9.4b and 9.5 shows that the transmissivities of water vapor, cloud liquid water, and rain rate each have a different frequency dependence. The water vapor absorption peak at 22 GHz does not occur for rain and cloud liquid water. Also, because of their respective derivations from Mie and Rayleigh scattering, the transmissivities of rain and cloud liquid water differ from the vapor and each other. As the following sections show, these differences make possible their retrieval from microwave observations.

9.3 Radiative transfer in the microwave

For small rain rates and for frequencies less than about 25 GHz, Figure 9.5 shows that scattering is negligible in the microwave. For these conditions, the radiative transfer equation (RTE) can be approximated as an absorption–emission balance similar to that used in the infrared (Section 4.8.1). The difference between radiative transfer in the infrared and microwave windows is that, as Figure 8.6 shows for the microwave, a large range of incidence angles exists for which the surface reflectivity cannot be neglected. Consequently, the radiance received at the satellite has significant contributions from the reflection of the downwelled atmospheric and extraterrestrial radiances.

The derivation of the microwave RTE assumes that for θ constant, the ocean has an emissivity e and a reflectivity $1 - e$, where e depends on the sea surface salinity (SSS), SST and roughness. The atmosphere is also assumed to be plane parallel, so that from Equation (4.31), the transmissivity is given by $t(\theta) = t^{\sec \theta}$. Given that θ is constant for the conical scanner, the superscript on t is dropped in the following. Finally, because the Rayleigh–Jeans approximation applies to the microwave, brightness temperatures can replace the radiances in the RTE.

Following Stewart (1985, Section 9.4) and as Figure 9.6 illustrates, the absorption–emission form of the RTE can be written as

$$T_A = etT_S + (1 - t)\overline{T} + (1 - t)(1 - e)t\overline{T} + (1 - e)t^2(T_{ext} + T_{sol}) \tag{9.1}$$

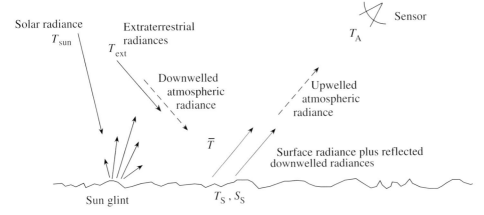

Figure 9.6. The solar, extraterrestrial and atmospheric radiances, their reflection from the rough ocean surface and the emission and attenuation of the surface brightness temperature.

In (9.1), T_A is the brightness temperature observed at the satellite, T_S is the surface temperature, $\overline{T}$ is the vertical average of the tropospheric temperature profile, T_{ext} is the extraterrestrial brightness temperature exclusive of the sun and as described below, T_{sol} corresponds to the solar brightness temperature T_{sun} modified by the antenna pattern angle. Except for heavy rain and at the higher frequencies where the equation breaks down, Equation (9.1) shows that T_A can be written as the sum of the following terms: the upwelled surface radiance, the upwelled atmospheric radiance, and the reflection of the downwelled atmospheric, extraterrestrial and solar radiances. The upper boundary on the RTE is cold space; the lower is the rough ocean surface. As shown below, the surface emissivity and reflectivity are determined by T_S, S_S and the wind-induced surface wave field.

The upwelled and downwelled atmospheric radiances and the extraterrestrial terms in Equation (9.1) merit further discussion. First, although Equation (9.1) does not include this effect, because the atmospheric temperature profile decreases with height, the $\overline{T}$ associated with the downwelled atmospheric radiance is generally 1–2 K warmer than that associated with the upwelled radiance. This occurs because the warmer, lower atmospheric layers make a greater contribution to the downwelled radiance, while the colder upper layers similarly contribute to the upwelled radiance (Wentz, 1992). Second, the extraterrestrial radiance T_{ext} consists of two terms:

$$T_{ext} = T_{univ} + T_{gal} \tag{9.2}$$

where T_{univ} is the nearly isotropic 2.7 K background temperature of the universe and T_{gal} is the temperature of the Milky Way galaxy. T_{gal} has a strong directional dependence and is strongest in the plane of the zodiac. Ulaby *et al.* (1981, Section 5.6.2) show that the galactic radiance decreases approximately as f^{-3}, and for $f > 5$ GHz, T_{gal} can be neglected relative to the atmospheric downwelling radiance. For $f \sim 1$ GHz, however,

Table 9.1. *Solar brightness temperatures
as a function of frequency*

f (GHz)	T_{sun} (K)
1.4	$\sim 2 \times 10^5$
6.6	$\sim 2.2 \times 10^4$
10.7	$\sim 1.5 \times 10^4$
18	$\sim 1.1 \times 10^4$
21	$\sim 1.0 \times 10^4$
37	$\sim 7 \times 10^3$

Adapted from Wentz, 1978, 1981

T_{gal} is relatively large with a strong angular dependence relative to the galactic center (Le Vine and Abraham, 2001). For this reason and because of the additional interference from terrestrial sources and the ionosphere, remote sensing observations are rarely made for $f < 1$ GHz.

Third, the contribution of T_{sun} to T_{sol} depends on four factors: the number of reflecting surface facets, the solid angle Ω_S subtended by the solar disk, the antenna pattern solid angle Ω_p and the frequency. If the antenna points directly at the sun, then from Equation (8.21),

$$T_{sol} = T_{sun}\Omega_S/\Omega_p \tag{9.3}$$

In (9.3), T_{sun} is assumed independent of position on the solar disk. If, away from the sun, the blackbody sky temperatures can be neglected, then the ratio Ω_S/Ω_p determines the relative importance of T_{sol} (Ulaby *et al.*, 1981). Although for $f > 37$ GHz, T_{sun} is approximately given by 5900 K; for $f \leq 37$ GHz, the solar brightness temperature is no longer constant. Instead, as Table 9.1 shows, T_{sun} increases dramatically as f decreases. Because of this increase, the solar contribution becomes increasingly important at lower frequencies. At 1.4 GHz, which as Section 9.5 shows is used for retrieval of SSS, the large solar brightness temperature means that care must be taken to avoid reflection of the sun into the instrument.

Ulaby *et al.* (1986) review the microwave sun glint algorithms, Wentz (1978, 1981) presents a sun glint model based on observational data, and Wentz, Cardone and Fedor (1982) describe the sun glint mask used with SMMR observations. Because of its noon–midnight orbit, SMMR was particularly vulnerable to sun glint. Figure 9.7 shows the coordinate system from Wentz *et al.* (1982); they define the relative solar angle $\Delta\theta_S$ as the angle between the conjugate solar zenith angle θ_0 and the instrument look angle θ.

For $\Delta\theta_S < 15°$ and $U < 15$ m s^{-1}, they find that there is more sun glint than at larger velocities. The result is consistent with Figure 5.7, which shows that although the angular

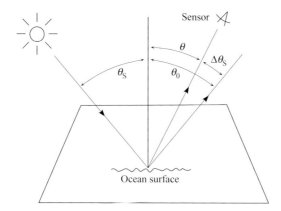

Figure 9.7. Coordinate system used in the discussion of the microwave sun glint mask.

extent of the reflected radiances increases with the wind speed, there is less reflection into any given direction. Consequently, their sun glint filter has two parts. If $\Delta\theta_S < 10°$, a pixel is masked for all values of U; if $\Delta\theta_S < 15°$, it is masked for $U < 15$ m s^{-1}. They also find a residual solar contribution for angles as large as $\Delta\theta_S = 25°$, where this residual corresponds to a 0.1 m s^{-1} error in their retrieval of a 10 m s^{-1} wind speed. Their work is based on SMMR; different calculations apply to the newer instruments. For instruments in dawn–dusk polar orbits such as the SSM/I where the sun angle θ_S is close to 90° or near grazing, problems with solar reflectivity are minimized. Because the TMI on TRMM is not in a sun-synchronous orbit, it employs a sun glint filter that is currently unpublished (Wentz, private communication, 2001).

Over the ocean, the atmospheric and surface variables retrievable from the V- and H-pol passive microwave observations include the mean atmospheric temperature $\overline{T}$, the columnar water vapor V, the columnar non-raining cloud liquid water L, the rain rate R_R, the sea surface temperature and salinity T_S and S_S, and the 10-m scalar wind speed U. Because the distribution of wave slopes varies with azimuth angle relative to the wind direction, the emissivity also varies with azimuth angle, polarization and the Stokes parameters. As the next section shows, this means that from a single look at the sea surface, a radiometer operating at V- and H-pol can retrieve both the wind magnitude U and the component of wind speed in the radiometer azimuthal look direction, called the *line-of-sight* wind speed U_{LOS}. It also means that if a V- and H-pol radiometer takes two looks at the same area from different directions, it can retrieve the vector wind speed. Furthermore, for polarimetric radiometers that measure all four Stokes parameters, under certain conditions, the vector wind speed can be determined from a single look.

Because none of the above atmospheric and oceanic variables occur in isolation, solution for any of them involves either solving for all of them, or providing masks or climatological estimates for those variables that cannot be retrieved, then solving for the rest. Given that the lower boundary common to all these retrievals is the ocean surface, the next section describes the dependence of the emissivity on surface waves, roughness and foam, and on the azimuthal look angle relative to the wind direction.

9.4 Dependence of the emissivity on surface waves and foam

At the ocean surface, the wind generates waves and foam, both of which affect the emissivity. Additionally, because the wave slopes and amplitudes are azimuthally distributed around the wind direction, the emissivity dependence on U divides into two parts: an isotropic term that is independent of azimuth and depends only on U, and an anisotropic term that depends on both U and the azimuthal angle.

In order of increasing complexity, the next five sub-sections describe the effect of waves and foam on the emissivity. Section 9.4.1 discusses the terminology used in the description of their contributions to the emissivity and describes the two-scale approximation and its dependence on the observing wavelength. Section 9.4.2 describes the azimuthally averaged emissivity of a foam-free, two-scale surface and justifies the choice of the 50° conical scanner look angle. Section 9.4.3 discusses the contribution of foam to the azimuthally averaged emissivity. Section 9.4.4 discusses the azimuthal dependence of the V and H emissivity components on U, and Section 9.4.5 describes the azimuthal dependence of all four Stokes components on U, where both sections discuss the use of this dependence in retrieval of the vector wind speed. As in Chapter 8, V-pol and H-pol brightness temperatures are written as T_{BV} and T_{BH} and the 18-GHz V-pol instrument channel is abbreviated as 18V, with similar notation for the other channels.

9.4.1 Contributors to the wave-induced emissivity

If e_0 is the temperature- and salinity-dependent specular emissivity and Δe_{WF} is the emissivity contribution of waves and foam, then the total surface emissivity e can be written as

$$e = e_0 + \Delta e_{WF} \tag{9.4}$$

In Equation (9.4) and following Wentz (1983), if F_F is the fractional areal coverage of foam within the instrument FOV, Δe_{WF} can be written as

$$\Delta e_{WF} = (1 - F_F)\Delta e_W + F_F \Delta e_F \tag{9.5}$$

where Δe_W is the wave-induced emissivity and Δe_F is the foam emissivity.

For the case of no foam, Wentz (1975, 1997) theoretically models the scattering and emission from the ocean surface using the *two-scale scattering* approximation, which divides the surface into a large scale surface Σ_L and a small scale surface Σ_S, where the water wavelength separating the two scales is called the *cutoff* wavelength (Figure 9.8).

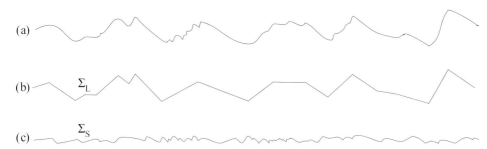

Figure 9.8. The two-scale division of (a) a wave-covered surface into (b) a faceted surface Σ_L and (c) roughness Σ_S, where the upper surface is the sum of the lower two. The vertical amplitude is exaggerated; see text for further description (notation from Wentz, 1975).

Physically, Σ_L consists of long gravity waves with surfaces that can be approximated as specularly reflecting facets; Σ_S consists of short capillary-gravity waves that generate the roughness associated with scattering. On Σ_L, the radius-of-curvature of the surface must be large compared with λ, so that Equation (5.9) is satisfied. Additionally, if σ^2 is the mean-square slope of Σ_L and σ_η^2 is the mean-square amplitude of Σ_S, σ must be much less than unity and σ_η much less than λ. With these assumptions, the emitted and scattered radiation can be described by perturbation theory.

Wentz (1975, 1992) shows that the factors contributing to Δe_W are the large scale surface tilt associated with Σ_L and the small scale surface roughness associated with Σ_S, where the emissivity contribution of both factors increases with U. The tilted facets of Σ_L act as independent specular surfaces, mix the horizontal and vertical polarization and change the local incidence angles of the emitted and reflected downwelled radiances (Ulaby *et al.*, 1982, Chapters 11 and 12), while the roughness associated with Σ_S diffracts and scatters the emitted and reflected radiation.

Wentz (1975) also shows that the cutoff wavelength increases with roughness and wind speed and is larger than, but of order λ. Waves that are short compared with the cutoff wavelength do not contribute to σ^2. Because λ decreases with increasing f, σ^2 increases with f, reaching its maximum limiting value, called the *optical limit*, at about 37 GHz (Wentz, 1997). Because the lower observational frequencies exclude from σ^2 the short 1–10 cm ocean wavelengths that are most responsive to changes in wind speed and direction, for successful wind retrievals, f must be greater than about 10 GHz (Wilheit, 1978).

In addition to their f-dependence, σ^2 and σ_η^2 are functions of the azimuthal angle relative to the wind direction, where both quantities are larger in the upwind and downwind directions than in the crosswind. Also, because the parasitic capillary waves form on the downwind face of the wave crests, there is an upwind/downwind anisotropy in their distribution and in their resultant contributions to the emissivity.

9.4.2 The azimuthally averaged emissivity of a wind-roughened foam-free surface

For a downward-looking radiometer above a foam-free wind-roughened ocean surface, Wentz (1975) derives the dependence of T_{BV} and T_{BH} on f, U and θ. His derivation uses the two-scale approximation, averages over all azimuth angles to remove the directional effects of wind, and includes the scattering and reflection of the downwelled atmospheric radiance. From his solution, Figure 9.9 shows the dependence of T_B on polarization and θ for $U = 13.5$ m s^{-1} and for the $U = 0$ specular surface case corresponding to Figure 8.6. The figure also shows the values of T_B derived from field measurements at $U = 0.5$ and 13.5 m s^{-1}, where the 13.5 m s^{-1} observations were filtered to remove foam. At $\theta = 0°$, small scale roughness accounts for the elevation of the 13.5 m s^{-1} curves above the specular.

An important result shown in this figure is that at both frequencies and for $\theta = 50-55°$, the values of T_{BV} at $U = 0$ and 13.5 m s^{-1} are equal, so that given an incidence angle of about 50° and for foam-free conditions, T_{BV} is independent of wind speed. Even though this result neglects foam, the advantages associated with the decoupling of T_{BV} from U at $\theta \cong 50°$ means that, as the previous chapter shows, all of the conically scanned microwave imagers operate at incidence angles close to 50°. In contrast, at the same

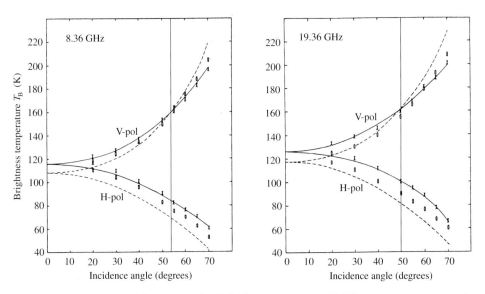

Figure 9.9. Comparison of the computed and field measurements of brightness temperatures made from a foam-free, wind-driven sea for V- and H-pol at 8.36 GHz and 19.3 GHz and averaged over all azimuth angles. On each panel, the dashed line corresponds to the $U = 0$ specular surface case; the solid line to $U = 13.5$ m s^{-1}. The vertical lines mark the angles at which the V-pol curves are independent of U. The sea surface temperature is 291 K; the surface salinity, 35 psu, where psu is the abbreviation for precision salinity units. The ellipses show observations at 0.5 m s^{-1}; the Xs, at 13.5 m s^{-1}, where foam from breaking waves is excluded. See text for further description (Figures 3 and 4 from Wentz 1975, © 1975 American Geophysical Union, reproduced/modified by permission of AGU).

Table 9.2. *The emissivity increase* Δe_F
observed from passive microwave aircraft
observations of foam, where in these
measurements, the foam patches are
larger than the beam footprint

Channel	Δe_F
19V	0.15
37V	0.15
37H	0.28

From Table III, Smith, 1988

incidence angle, T_{BH} strongly depends on U. For example, at 19 GHz, as U increases from 0 to 13.5 m s^{-1}, T_{BH} increases by about 20 K. Given a radiometer accuracy of about ±0.5 K, this suggests that a wind speed retrieval algorithm should have an accuracy of about ±0.5 m s^{-1}.

9.4.3 Contribution of foam

Smith (1988) discusses aircraft passive microwave observations of oceanic foam at 19V, 37V and 37H. For the case when foam fills the antenna footprint, Table 9.2 summarizes his observations and shows at 37H, the value of Δe_F is nearly twice its value at 37V and 19V. The combination of this foam-induced emissivity change with the increase in foam extent with U means that once foam starts to appear, Δe_{WF} in Equation (9.5) increases more rapidly with U than for waves alone. The result is that at low wind speeds the emissivity is dominated by the mean-square slopes and roughness; at greater wind speeds, by the additional contribution of foam and its increase in fractional area.

To illustrate the contributions of waves and foam, Figure 9.10 shows the U-dependence of the V-pol and H-pol components of Δe_{WF} derived from co-located SMMR and SEASAT scatterometer measurements (Wentz *et al.*, 1986). For four frequencies and two polarizations, the figure shows that the dependence of Δe_{WF} on U divides into two linear curves, separated by a slope break located between 7 and 12 m s^{-1}. The cause of the slope break is the onset of significant breaking and foam. The low wind speed linear regime for $U < 7$ m s^{-1} corresponds to the foam-free case in Figure 9.9; the high wind speed linear regime for $U > 12$ m s^{-1}, to wind speeds where the area of wave breaking and foam formation increase rapidly with U. Because of the 51° SMMR look angle, for the low wind speed regime and consistent with Figure 9.9, the V-pol values of Δe_{WF} are almost independent of U, while the H-pol values have a much stronger U-dependence. Wentz (1997) models the U-dependence of the emissivity as a linear increase for $U < 7$ m s^{-1}, a quadratic increase for $7 < U < 12$ m s^{-1}, and a steeper linear increase for $U > 12$ m s^{-1}. For all frequencies,

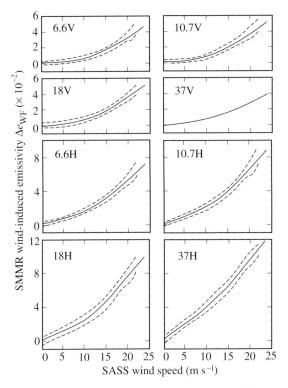

Figure 9.10. Dependence of the wave and foam contribution to the emissivity Δe_{WF} on U at $\theta = 51°$ for V- (upper four figures) and H-polarization (lower four figures). The curves are derived from co-located SMMR and SEASAT open ocean scatterometer wind measurements; the figures are plotted to the same vertical scale. Solid lines show the mean values; dashed curves, the one standard deviation envelopes. The 37V standard deviations are missing on the original. See text for further description (Redrafted from Figures 9 and 10, Wentz, Mattox and Peteherych, 1986, © 1986, American Geophysical Union; reproduced/modified by permission of AGU).

the figure shows that the contributions to the emissivity of both roughness and foam are less for V than for H, so that the H-pol component is more sensitive to changes in U. Also, for f less than about 10 GHz and especially for H-pol, the sensitivity of the emissivity to changes in U increases with frequency. As Section 9.6.1 shows below, this response is the basis for the algorithms that retrieve the wind magnitude.

9.4.4 Azimuthal dependence of the V- and H-pol emissivities

For a radiometer azimuthal look angle ϕ and a wind direction ϕ_W, the wind direction relative to ϕ is

$$\phi_R = \phi_W - \phi \tag{9.6}$$

so that $\phi_R = 0$ in the upwind direction. Because conically scanned radiometers operate at a fixed incidence angle but a large range of ϕ, these instruments unavoidably retrieve the surface and atmospheric variables at different ϕ_R. The present section discusses the observed dependence of the emissivity on polarization, U and ϕ_R, then shows how this dependence permits the retrieval of two quantities. The first is the wind component U_{LOS} in the radiometer look direction that is derived from SSM/I, TMI and AMSR observations; the second is the vector wind, which should be derived from WindSat and the forthcoming CMIS observations. Both these retrievals are important because of the obvious advantages of better wind retrievals, and because the accuracies of the other retrieved variables improve with more accurate winds.

The emissivity dependence on polarization, U and ϕ_R was derived from Russian and US aircraft experiments (Irisov *et al.*, 1991; Yueh *et al.*, 1999) and from comparison of satellite and buoy winds (Wentz, 1992; Meissner and Wentz, 2002a). For the satellite case, Wentz's (1992) derivation is based on about 3000 spatially and temporally coincident pairs of SSM/I and NDBC winds, where SSM/I data are also used to remove the effects of atmospheric liquid water and water vapor. Later comparison with aircraft experiments showed that at small wind velocities, his results were too large (Meissner and Wentz, 2002a). This difference occurs because the retrieved atmospheric variables and surface emissivities both depend on the same distribution of surface roughness, so that at small wind velocities, the use of a single radiometer to retrieve all these variables generates a systematic error. To eliminate this error, Meissner and Wentz (2002a) redo the analysis using one satellite radiometer for atmospheric variable retrieval and another for directional wind retrieval. Specifically, they use about 8×10^5 coincident pairs of SSM/I and QuikSCAT winds and 8000 pairs of SSM/I and buoy winds for the directional wind retrieval, both with TMI for atmospheric correction, and 10^6 pairs of TMI and QuikSCAT winds with SSM/I for atmospheric correction.

For 37 GHz and both polarizations, Figure 9.11 shows their resultant dependence of ΔT_{BV} and ΔT_{BH} on ϕ_R. On the figure, ΔT_{BV} and ΔT_{BH} are grouped into 20° bins and into three wind speed ranges: 0–6 m s^{-1}, 6–10 m s^{-1}, and 10–14 m s^{-1}; the dotted, dashed and dot-dash lines show the Meissner–Wentz results, the solid lines show Wentz's (1992) results. For $U < 6$ m s^{-1}, the Meissner–Wentz results are much less than Wentz's; for $U > 10$ m s^{-1}, the two approximately agree. The Meissner–Wentz observations show that ΔT_{BV} and ΔT_{BH} increase with U and are symmetric around the wind direction. Specifically, ΔT_{BV} has a dominant cos ϕ_R dependence with its maximum and minimum respectively in the upwind and downwind directions, while at the larger velocities, ΔT_{BH} has a cos $2\phi_R$ dependence with its minima and maxima respectively in the upwind, downwind and crosswind directions.

For 10–14 m s^{-1}, the upwind/downwind range of ΔT_{BV} is about 3 K; the cross-wind/downwind range of ΔT_{BH} is about 4 K. Although Figure 9.10 shows that the azimuthally averaged values of ΔT_{BH} are more sensitive to changes in U than ΔT_{BH}, Figure 9.11 shows that when their azimuthal variability is considered, both polarizations have about the same sensitivity. From Figure 9.10, the ΔT_B associated with a 0 to 12 m s^{-1} increase in the azimuthally averaged wind magnitude is 15 K for 37H and 3 K for 37V. This means

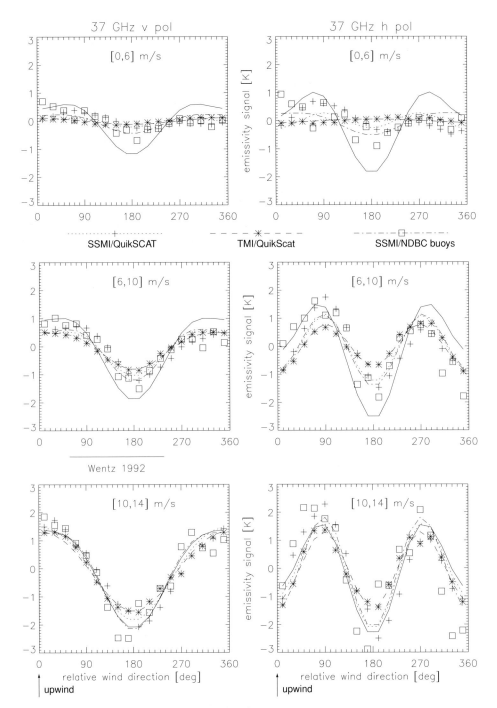

Figure 9.11. Azimuthal dependence of the brightness temperature difference ΔT_{B} for 37 GHz, both polarizations and three different wind speeds, derived from a comparison between passive microwave radiometer and NDBC winds. The upwind direction corresponds to $\phi_{\mathrm{R}} = 0$. The solid line shows the best fit to Wentz's (1992) observations; the other symbols are defined on the figure. See text for additional description (Figure 1 from Meissner and Wentz, 2002a, © 2002 IEEE, courtesy Thomas Meissner).

that for 37 GHz and a velocity increase of 12 m s^{-1}, the azimuthal dependence creates a small but non-negligible perturbation in the azimuthally averaged ΔT_{BH}, while for ΔT_{BV}, the azimuthal perturbation is the same size as the azimuthally averaged increase. Using TMI data, Meissner and Wentz also examine the azimuthal response at 10.6 and 19 GHz, and find that the amplitude of the return decreases approximately linearly with frequency, such that at 10.6 and 19 GHz, the amplitudes are respectively 40 and 70% of the 37-GHz amplitude.

Even though Wentz's (1992) results are overstated for $U < 6$ m s^{-1}, Wentz (1997) found that if the wind retrieval algorithm described below in Section 9.6 did not consider the azimuthal variability, the errors in U depended on ϕ_R. In the upwind direction, the SSM/I estimate of U is 2.5 m s^{-1} less than the buoy estimate; in the downwind direction, the SSM/I estimate is 1.2 m s^{-1} greater. Over all directions, the rms difference between the buoy and the SSM/I wind magnitudes is about 1.6 m s^{-1}. Because the accuracies of all of the retrieved variables depend on each other, unless the algorithm takes this azimuthal dependence into account, the accuracies of the other retrievals are reduced. To improve the accuracy, Wentz (1997) introduced the line-of-sight wind magnitude, U_{LOS}, which is the wind magnitude in the radiometer azimuthal look direction, $U_{LOS} = U \cos \phi_R$. Even though ϕ_R cannot be retrieved separately from U_{LOS}, Wentz shows that the inclusion of U_{LOS} in the SSM/I algorithm reduces the error of all of the retrieved variables, and in particular reduces the error in U by a factor of 3.

From the azimuthal dependence of ΔT_{BV} and ΔT_{BH} given in Figure 9.11, Wentz (1992) also shows that the wind direction could be retrieved through use of two satellite radiometers. In this retrieval, the first radiometer looks forward and the second looks backward, so that the same ocean area is observed at two different times and two values of ϕ_R separated by about 180°. His analysis shows that such an instrument could retrieve the vector wind speeds and generate more accurate measurements of the other variables. As Chapter 14 describes, the WindSat instrument on the Coriolis satellite launched in January 2003 will provide a test of this concept. Because WindSat also measures the third and fourth Stokes parameters as well as V-pol and H-pol components of the emissivity (Section 3.2.4), the next section discusses the ϕ_R-dependence of the emissivity at all four Stokes parameters.

9.4.5 Azimuthal dependence of the four Stokes parameters

Based on a series of aircraft experiments, Yueh (1997) and Yueh *et al.* (1999) describe the dependence on ϕ_R and U of the brightness temperatures measured at the four Stokes parameters and Yueh (1997) theoretically models this dependence. The observational data were taken from an aircraft-mounted Jet Propulsion Laboratory (JPL) 19- and 37-GHz polarimetric radiometer that at 19 GHz measured all four Stokes parameters, and at 37 GHz measured the first three parameters. In the experiments, the aircraft flew in circles around an NDBC anemometer-equipped buoy off the California coast, where the aircraft was oriented such that the radiometer viewed the surface at a constant incidence angle.

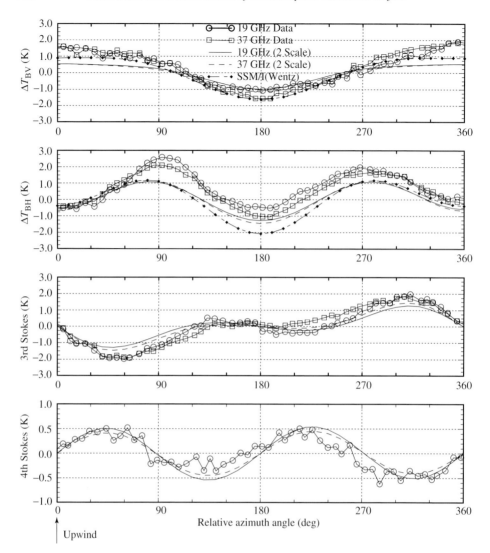

Figure 9.12. Comparison of theoretical and observed values of the Stokes parameters for JPL WIN-DRAD'94 aircraft radiometer data taken at a 55° look angle and at a 10-m wind speed of about 10 m s^{-1}. The vertical arrow marks the upwind direction at 0°. The figure shows the 19 and 37 GHz data and model results, and for V- and H-pol, Wentz's (1992) results. See text for further description (Figure 4 from Yueh, 1997, © 1997 IEEE, courtesy Simon Yueh).

The observations acquired data at $\theta = 45°$, 55° and 65°; for comparison with the satellite radiometer observations, the following discusses the observations at 55°.

For 19 and 37 GHz, Figure 9.12 compares the model results with observational data and with Wentz's (1992) V- and H-pol results, where the azimuthally averaged response is removed from the data. Examination of the figure shows that the V-pol and H-pol curves are similar in appearance to those in Figure 9.11; they respectively show a $\cos \phi_R$ and $\cos 2\phi_R$ response. Consistent with theory, the curves for the third and fourth Stokes parameters are antisymmetric about the wind direction. Specifically, the third Stokes parameter has a mixed $\sin \phi_R$ and $\sin 2\phi_R$ dependence and the fourth has a dominant $\sin 2\phi_R$ dependence, although its magnitude is only about 25% of the other three. Given that for each Stokes parameter the observations differ in their azimuthal response, then depending on wind speed, a single look with a polarimetric radiometer at any azimuthal angle provides a unique solution for the vector wind speed.

Meissner and Wentz (2002a) investigate the range of U for which a polarimetric radiometer can retrieve vector wind speeds. Their results show that the best retrievals would occur for $U > 10$ m s^{-1}, where all four Stokes parameters have a strong directional response. From Figure 2.1, this includes about 20% of the measured global winds. For $U < 5$ m s^{-1}, they show that there is no azimuthal wind signal at V-pol, H-pol or the fourth Stokes parameter, and that the variability of the third parameter is very small. This strongly suggests that a vector wind retrieval is impossible for $U < 5$ m s^{-1}, which excludes about 30% of the winds. Finally, for 5 m s$^{-1} < U < 10$ m s^{-1}, the range of application of the wind speed retrieval is uncertain, although modeling results of Meissner and Wentz (2002b) suggest that for $U > 7$ m s^{-1}, the directional error of the retrieval will be less than 20°. As Chapter 14 discusses further, the WindSat polarimetric observations will test Wentz's concept of a two-look retrieval of the vector wind using V-pol and H-pol, and should resolve the range and accuracy of polarimetric retrieval.

9.5 Temperature and salinity

The Klein–Swift formulation (Klein and Swift, 1977; Swift and McIntosh, 1983), as updated by Wentz and Meissner (1999), describes for a specular surface the dependence of the emissivity on sea surface temperature T_S and salinity S_S. Following Wilheit (1978), the *sensitivity* of T_B to changes in T_S and S_S is given by the partial derivatives of T_B with respect to these variables. For $\theta = 53°$, $T_S = 20\,°C$ and $S_S = 30$ psu, Figure 9.13 shows the frequency dependence of the V- and H-pol components of these partial derivatives. For T_S, the T_{BV} sensitivity has a peak at 5.6 GHz, with zero crossings at 1.25 and 32.2 GHz. From these curves, the T_{BH} sensitivity is smaller than the T_{BV} and has zero crossings at 1.5 and 24 GHz. The optimum frequency and polarization for retrieving T_S, while avoiding attenuation by atmospheric water, occurs for V-pol at about 6 GHz, although 10 GHz also provides a satisfactory T_S estimate. The two lower curves in Figure 9.13 give the dependence of the sensitivity of T_B on S_S, and show that for f increasing from 1 to about 5 GHz, the sensitivity decreases rapidly. This suggests that a frequency of about 1 GHz is best suited for salinity retrieval.

As Section 7.2 describes, the retrieved values of T_B are approximations to the desired skin temperatures. From Equation (3.1), the absorption depth d_a has the following dependence

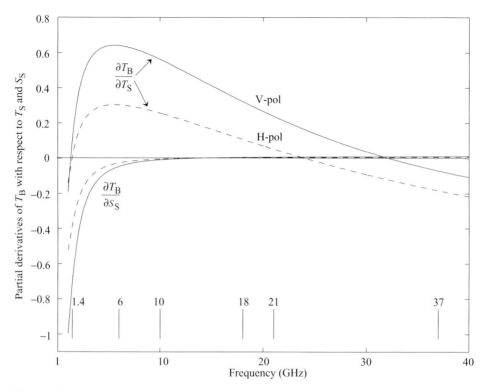

Figure 9.13. The partial derivatives of T_B with respect to T_S and S_S for a specular surface, and for $\theta = 53°$, as derived from the Klein–Swift relations cited in the text. Solid lines are V-pol; dashed lines are H-pol (Computer code courtesy Gary Lagerloef).

on frequency: at 6.9 GHz, $d_a = 2.5$ mm and at 10.7 GHz, $d_a = 1.2$ mm, where d_a continues to decrease at higher frequencies. For contrast, in the infrared at 10–12 μm, d_a is about 10 μm. Because of this dependence of the absorption depth on frequency, the infrared SST is dominated by conductive and diffusive processes, the microwave SST is dominated by viscous processes (Section 7.2, Donlon *et al.*, 2002). Consequently, the microwave and infrared SSTs differ not only from the bulk temperature but also from each other. As future SST data sets consisting of blended infrared and microwave observations are developed, an understanding of these differences will become necessary.

For salinity retrieval and for $f > 1$ GHz, the protected 1.41-GHz radio astronomy band is the lowest available observational frequency. At this frequency, Figure 9.14 shows the T_B dependence on T_S and S_S. On the figure, the shaded region shows the range of the open ocean surface temperatures and salinities. Across this region, the maximum change in brightness temperature is about 3 K. The figure also shows that for 15–30 °C and 22–37 psu, T_B depends primarily on S_S with a weak dependence on T_S. Even though Figure 9.14 is derived for normal incidence, the effect of changing the incidence angle is to move the pattern of curves vertically without altering their spacing (Lagerloef, private

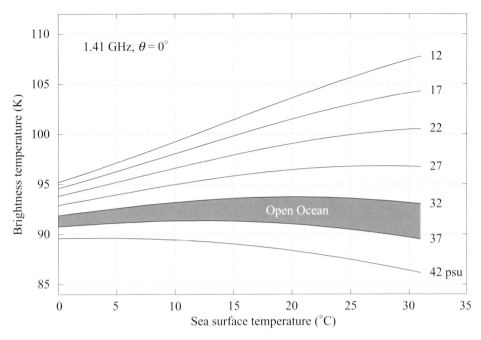

Figure 9.14. The dependence of brightness temperature on surface salinity and temperature at 1.41 GHz and normal incidence, where psu is the abbreviation for precision salinity units (Figure 2 from Lagerloef, 2000, © 2000, with permission from Elsevier Science, courtesy Gary Lagerloef).

communication, 2002). Given the oceanographic requirement that surface salinity be determined to within 0.2 psu, for retrieval purposes, T_B must be determined to within about 0.1 K.

In the SSS retrieval at 1.41 GHz, because the emissivity also depends on surface roughness and SST, these variables must be measured. Another problem occurs because the low observing frequency creates a large surface FOV. For example, the resolution of the SSM/I antenna operating at 1.4 GHz would be of order 500 km. To achieve a spatial resolution of 50 km or less, the antenna diameter must be greater than 4 m. Chapter 14 describes two satellites planned for the next decade that use different means to achieve this resolution and to determine surface roughness and SST, the European Soil Moisture and Ocean Salinity (SMOS) satellite and the NASA Aquarius satellite.

9.6 Open ocean algorithms

The above discussion shows that SSS is most sensitive to observations at 1.41 GHz, SST to the V-pol observations at 6–10 GHz, and from Figure 9.10, U to H-pol observations at $f \geq 10$ GHz. Given the relative transparency of the atmosphere for $f \leq 10$ GHz, a hypothetical instrument for measurement of S_S, T_S and U could be constructed using 1.41V (S_S), 6.6V (T_S) and 10H (U). At these frequencies where water vapor and cloud liquid water are unimportant, the combined measurements yield three equations for three unknowns: U, SST and SSS.

The further determination of the water vapor V, cloud liquid water L and rain rate R_R requires additional channels at 19, 21 and 37 GHz. Specifically, the retrieval of small water vapor concentrations depends on a V-pol measurement in the immediate vicinity of the 22-GHz absorption peak, while for larger concentrations, the retrieval depends on channels located on the shoulders of the peak at 18, 19 or 24 GHz. For retrieval of L, Figure 9.5a shows that because the variability of t with L increases with f, an appropriate frequency for its retrieval is 37 GHz. Finally, by masking pixels with large attenuations at either 37 or 85 GHz, regions of heavy rain can be excluded. This qualitative discussion gives a general description of an algorithm for retrieval of all of the above variables. With this as background, the next sections describe the SSM/I, TMI and AMSR-E algorithms.

9.6.1 SSM/I algorithm

Wentz (1997) describes his NASA-selected algorithm that is used to generate the open ocean SSM/I Pathfinder data sets. For each pixel, his retrieval uses the 19V, 22V, 37V and 37H channels, yielding four equations for L, V, U and U_{LOS}. Because, as Table 8.3 shows, the 37- and 22-GHz channels have a greater resolution than the 19-GHz, the higher frequency channels are spatially averaged so that all channels have the same pixel size. For these channels, Wentz (1997) lists the coefficients that describe the frequency dependence of their associated emissivities and transmissivities.

For the non-retrieved variables of SST and rain, his algorithm uses a climatological SST lookup table for T_S and masks rain. The reason that an SST lookup table must be used is that, as Figure 9.13 shows for 19–37 GHz, the dependence of T_B on T_S is too weak to allow retrieval of T_S, but too strong to ignore (Wentz, 1997). Typically for this frequency range,

$$\partial T_B / \partial T_S \sim 0.3 \tag{9.7}$$

so that if T_S is ignored it produces significant errors in the other retrieved variables. Since T_S cannot be determined from the SSM/I frequencies, an estimate must be supplied from climatology.

Second, a rain mask is applied whenever L exceeds a threshold. The columnar water vapor L ranges from 0 to about 0.25 mm, and as Wentz (1997) shows at 37 GHz, a 0.1-mm change in L yields at H-pol a $\Delta T_B \approx 9$ K. Wentz assumes that when $L \geq 0.18$ mm it is no longer a reliable measure of liquid cloud water but is rather an indicator of rain. Therefore, when $L \geq 0.18$ mm, a rain flag is set and the pixel is masked. Thus, the use of a climatological SST field and a rain mask allows solution for the four variables. In the algorithm development, the resulting values of U and U_{LOS} were tuned against buoy vector winds; V and L were tuned against a global array of radiosonde measurements made from island stations. Wentz's results show that the rms error in U is about 0.9 m s^{-1}, in V about 1.2 mm, and in L about 0.025 mm, with additional but smaller systematic errors of respectively 0.3 m s^{-1}, 0.6 mm and 0.005 mm.

As an example of these retrievals, for April 19, 2000, Plate 10 shows a composite image of U, V and L for the ascending SSM/I swaths taken at 1800 local time. These swaths, which cover more than half the ocean, were chosen because they are closest in time to the MODIS image in Plate 2 and show similar phenomena. The top image gives the wind magnitude U and similar to Plate 2, shows storms around Antarctica, a strong cyclonic storm in the North Atlantic and another storm approaching the Pacific coast of North America. The middle image gives the columnar water vapor V and shows that the vapor is concentrated in the tropics, especially in the vicinity of Indonesia and New Guinea, and in the storm-associated bands of vapor extending from the equatorial region into the temperate and subpolar latitudes. The bottom image displays the cloud liquid water L and shows that the regions of storm-associated strong winds are regions of enhanced L. In this image, the white letters mark cloud patterns and storms common both to this image and to Plate 2.

9.6.2 TMI and AMSR-E algorithms

Because TMI has channels at 10.7 GHz and AMSR has channels at 6.9 and 10.7 GHz, these instruments can retrieve a cloud-independent SST. Although the SMMR channels at 6 and 10 GHz demonstrated the feasibility of retrieving SST and generated some scientific results (Liu, 1988), because of the SMMR problems with low resolution, sidelobe contamination and instrument noise described in Section 8.6.1, its SST retrieval had large errors. Consequently, the TMI 10.7-GHz channels provided the first real opportunity to retrieve SST.

For the frequency range 19–85 GHz, Table 8.1 shows that the TMI and SSM/I frequencies are nearly identical. Given this agreement, the TMI algorithm used for SST retrieval is an extension of the SSM/I four parameter algorithm, with an additional equation for T_S based on 10.7V and H. The algorithm also retrieves the wind magnitude U at both 10.7H and 37H. Within the operational model, the retrieved T_S were calibrated against coincident moored buoy temperatures for the period December 1997–January 1999, so that similar to the AVHRR SST algorithms, there is an uncertainty associated with the difference between the satellite-retrieved temperatures and the buoy temperatures. Further, unlike the AVHRR algorithms and because of the greatly reduced atmospheric interference in the microwave, the SST portion of the algorithm was tuned only once against buoy measurements and is not continually updated.

For January 1998 and 1999, where January 1998 is toward the end of the 1997–98 El Niño and January 1999 is in a La Niña, Plate 11 shows a composite image of the monthly averaged TMI-retrieved values of SST, U, V and L. Because of the TRMM near equatorial orbit, the images run from 40° N to 40° S. Comparison of the two sets of images shows dramatic differences between El Niño and La Niña conditions. The El Niño SST image shows a warm equatorial Pacific with no upwelling off South America. In contrast, the corresponding La Niña image shows cold upwelling along the South American Pacific coast with an additional cold upwelling equatorial tongue extending from the coast into the central Pacific. Between these images, the decrease in equatorial temperatures is about 10 K.

The El Niño wind image shows weaker equatorial winds and stronger North Pacific winds than the La Niña image, where the enhanced La Niña winds drive the cold water equatorial upwelling. The El Niño water vapor image shows a nearly uniform distribution of large values of V along the equator and south along the South American coast. The large equatorial values of V are correlated with the warmer equatorial SSTs, which permit more evaporation. For La Niña, the cold equatorial SSTs means that the evaporation and the corresponding values of V are smaller. Also, and in part because of the strong easterly equatorial winds, the maximum values of V for La Niña are concentrated in the western Pacific and Indian Ocean. For cloud liquid water in the Pacific, the El Niño image shows that L has an irregular equatorial distribution. In contrast, the La Niña image shows that L is concentrated in the doldrum region at $4°$ N of the equator, while the equator remains cloud-free.

Because of the shifts in the AMSR-E frequencies shown in Table 8.1, the coefficients developed for the TMI and SSM/I algorithms cannot be used with AMSR-E. For the AMSR-E frequencies, Wentz and Meissner (1999) describe in detail the frequency dependence of the atmosphere and ocean emissivities and transmissivities. They show that the AMSR-E algorithms are similar to the SSM/I and TMI algorithms, and are designed to retrieve T_S, U, V and L, and to apply a rain mask. The AMSR-E algorithms retrieve two estimates of T_S at 6.9V and 10.7V, and two estimates of U at 10.7H and 18.7H. For AMSR-E, the final form of the algorithm coefficients will be determined in a way similar to SSM/I and TMI; the atmospheric parameters by regression against radiosonde data, the ocean surface parameters by regression against buoy observations. Because there is presently no information on the wind directional effects at frequencies lower than 19 GHz, the algorithm for U_{LOS} is a postlaunch product currently under development. The rain flag is set for $L \geq 0.18$ mm. Table 9.3 lists the AMSR-E standard and special products with their accuracies, and shows that AMSR-E produces a global temperate ocean SST product from the 6.9V GHz channel at a 58-km resolution. AMSR-E should yield a near global SST data set at alternate day intervals.

9.7 Sea ice algorithms

The determination of sea ice extent and ice type is one of the great successes of the passive microwave imagers. This section discusses the form of the RTE used in the polar atmosphere, describes the algorithms used for the retrieval of ice properties and gives examples. Because at this writing the AMSR algorithms remain under development, the discussion is based on the SMMR and SSM/I algorithms.

Table 9.3. *The AMSR-E standard and special products and their accuracies and resolutions*

Variable	Accuracy	Resolution
T_S	0.5 K	58 km (standard product)
T_S	0.5 K	38 km (special product)
U	1.0 m s^{-1}	38 km (standard product)
U	1.0 m s^{-1}	24 km (special product)
U_{LOS}		Under development
V	1.0 mm	24 km
L	2×10^{-2} mm	13 km

Wentz and Meissner, 1999

In the polar regions, the RTE in Equation (9.1) can be further simplified. First, because over winter pack ice the polar atmosphere is very dry, from Figure 9.4, $t_{vap} \cong 1$. Second, the extraterrestrial brightness temperature is generally neglected (Cavalieri, Gloersen and Campbell, 1984). These simplifications mean that over winter polar pack ice, Equation (9.1) reduces to

$$T_B \cong eT_S \tag{9.8}$$

In (9.8), e represents the emissivity of open water and the different ice types, where T_S is the water and ice surface temperature. Although the assumptions underlying Equation (9.8) break down at the ice edge where liquid water and water vapor become important, this simple formulation permits the retrieval of many ice properties. These include time series of the areal sea ice extent in the Northern and Southern Hemisphere, and in the Northern Hemisphere, the relative concentrations of open water, first-year ice and multiyear ice. The SMMR frequencies used in this retrieval are 18 and 37 GHz; the SSM/I frequencies are 19, 37 and sometimes 85 GHz.

As Comiso *et al.* (1997) describe in detail, two of the algorithms used in this retrieval are called the NASA Team and Bootstrap algorithms. Both these algorithms use the different instrument channels to take advantage of the frequency-dependent emissivity differences between open water and the pack ice categories. As Chapter 2 discusses, for the Arctic, these categories include first-year and multiyear ice, where first-year ice is less than one year old and multiyear ice has survived one summer. For the Antarctic, the ice categories are called type A and type B ice, where at this time the kinds of physical ice corresponding to types A and B are not known. The reason the Arctic ice types have different emissivities is that the upper surface of first-year ice is saline, while the surface of multiyear ice is nearly fresh and contains many air bubbles.

For both hemispheres, the large emissivity differences between open water and sea ice simplify the retrieval. For SMMR and SSM/I and beginning with the Arctic, Figure 9.15 shows the dependence of the emissivities of open water, first-year (FY) and multiyear

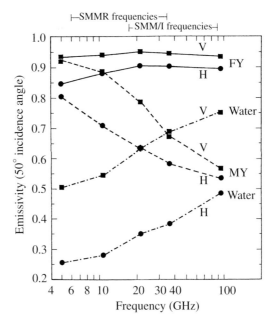

Figure 9.15. Dependence of the emissivity of the northern hemisphere sea ice on frequency for the SMMR and SSM/I frequencies. In the figure V and H are polarization, Water refers to open water, FY to first-year ice, MY to multiyear ice (Figure 1 from Comiso *et al.* 1997, © 1997 with permission from Elsevier Science).

(MY) ice on frequency and polarization. Examination of the figure shows that the difference between the V-pol and H-pol emissivity of open water is much larger than for first-year or multiyear ice, and that the open water emissivities increase with frequency. The ice emissivities are very different. For first-year ice, the V- and H-pol emissivities are large, almost equal to each other and nearly independent of f. For multiyear ice, the V-pol emissivity is greater than the H-pol where both emissivities decrease with increasing frequency. Comparison of the first-year and open water emissivities shows if the surface temperatures of open water and first-year ice are at the seawater freezing point, the open water brightness temperature is smaller than that of first-year ice. For multiyear ice with the same surface temperature as open water, at low V-pol frequencies, ice has the greater brightness temperature, at high V-pol frequencies, open water is brighter. These differences in the responses of the emissivities to frequency and polarization form the basis for the algorithms.

For 19V, 19H and 37V, Table 9.4 lists some characteristic brightness temperatures for the three Arctic categories. The table also lists the values of two variables that are functions of the brightness temperatures, the *polarization ratio* P_R and the *gradient ratio* G_R, which are used in the NASA algorithms. P_R and G_R are defined as follows:

$$P_R = (T_{BV19} - T_{BH19})/(T_{BV19} + T_{BH19})$$

$$G_R = (T_{BV37} - T_{BV19})/(T_{BV37} + T_{BV19})$$

(9.9)

Table 9.4. *Northern Hemisphere brightness temperatures for open water, first-year (FY) ice and multiyear (MY) ice used in the SSM/I algorithm*

f (GHz)	Open water (K)	FY ice (K)	MY ice (K)
19V	177.1	258.2	203.9
19H	100.8	242.8	203.9
37V	201.7	252.8	186.3
P_R ($\times 10^3$)	275	31	45
G_R ($\times 10^3$)	65	-11	-90

Adapted from Table 1, Comiso *et al.*, 1997

The advantages of using P_R and G_R are that to first order they are independent of the ice surface temperature; further, the use of the V-pol terms in G_R minimizes its wind speed dependence.

The success of these algorithms is due to the large temperature differences between the different ice and water categories. For example, the table shows that for T_{B19V}, the brightness temperature difference ΔT_{B19V} between open water and first-year ice is 80 K, and between open water and multiyear ice is 30 K. Similar large differences occur at the other frequencies and for P_R and G_R. For comparison, the oceanic range of SST is about 30 K, which from Figure 8.6 corresponds to a ΔT_{B19V} of about 15 K. Since the ΔT_{B19V} between open water and sea ice is four times this value, retrieval of the areal ice extent is relatively simple.

Within a pixel, the algorithms retrieve ice concentrations in the following way. Consider the simplified case of only two ice categories, open water and first-year ice. For this case, the algorithm can be written in terms of the relative concentrations C_W of open water and C_I of ice, where $C_I = 1 - C_W$. If T_{BW} is the open water brightness temperature, and T_{BI} is the sea ice brightness temperature, T_B becomes

$$T_B = T_{BW}C_W + T_{BI}C_I \qquad (9.10)$$

Thus if within each pixel, the brightness temperatures of open water and ice are known, their respective concentrations can be calculated.

In a similar manner and depending on hemisphere, the NASA Team algorithm uses P_R and G_R to retrieve the concentrations of open water and two classes of ice. For the Team algorithm and in both hemispheres, Figure 9.16 shows plots of P_R and G_R. On the figure, the small numbers represent the base-10 logarithm of the number of observations, while the curved lines make up the triangles used to define the ice concentrations. The triangle vertices are the algorithm tie-points and represent 100% concentrations of the indicated ice type or water. For the Northern Hemisphere, the tie-points consist of open water, first-year and multiyear ice; for the Southern Hemisphere, the tie-points consist of open water and type A and B ice. In the upper portion of both figures, the large number of points to the left of the open water tie-points are a weather effect that is associated with the atmospheric water

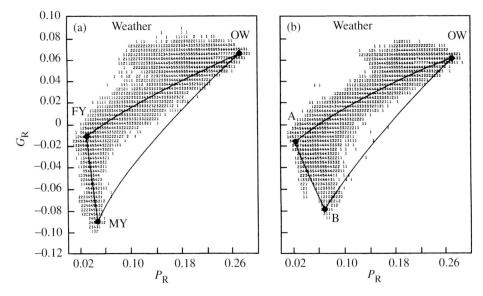

Figure 9.16. The P_R, G_R plots of data from (a) the Northern Hemisphere ice cover and (b) the Southern Hemisphere. The solid lines represent the curves used to define the ice concentrations in the NASA Team algorithm. The small numbers on the plots are the logarithms of the number of observations at each point. See text for further description (Figure 2 from Comiso *et al.*, 1997, © 1997, with permission from Elsevier Science).

vapor and liquid water at the ice margins. This effect is often filtered out by ignoring values above a fixed G_R threshold (Gloersen and Cavalieri, 1986). When, for any pixel, the values of P_R, G_R lie off the triangle, its location relative to the triangle permits solution of the relative contribution of each type.

Problems with the algorithms are as follows. First, as the air temperature rises above freezing during late spring and early summer, the ice surface becomes covered with melt ponds that contain nearly fresh water and, above the ice, the amount of atmospheric water vapor increases. The melt ponds and the associated water vapor cause the algorithm to generate apparent open water in the ice interior. Second, because the ice edge is a mixture of ice, ocean and atmospheric water vapor, the algorithm also breaks down in this region. In the Team algorithm, the ice edge problem is dealt with by choosing the edge as the 30% open water contour.

For an example of images processed with the SSM/I Team algorithm, Plate 12 shows the March and September 1995 monthly mean ice extent for the Arctic and Antarctic. In both hemispheres, the images show the respective maximum and minimum ice extent. In winter, the Arctic images show that marginal seas such as the Bering, Okhotsk, Hudson Bay and the peripheral seas of the Siberian coast are ice-covered, while in summer, they are ice-free. The Antarctic images show that much less ice survives the summer with the largest amount occurring in the Weddell Sea.

For examples of time series of ice extent for both hemispheres, Figure 9.17 shows a passive microwave time series derived from SMMR and SSM/I imagery similar to that shown in Plate 12. The figure shows mean monthly and annual time series of ice extent, plus a least-squares linear fit to the annual series. The small figures to the right of 9.17a and 9.17c show the monthly averaged annual cycle. On average, the Arctic ice extent has a March maximum of 15×10^6 km^2 and a September minimum of 6×10^6 km^2. The Antarctic has a September maximum of 19×10^6 km^2 and a February minimum of 4×10^6 km^2. The two pack ice regions differ because the Arctic Ocean is surrounded by land and has a strong oceanic surface stratification, while the Antarctic Ocean surrounding Antarctica has a weak stratification and an open ocean boundary so that the ice edge is subject to divergence from wind and currents. Thus the Antarctic ice has a greater maximum and a smaller minimum ice extent. Because the Arctic and Antarctic time series are six months

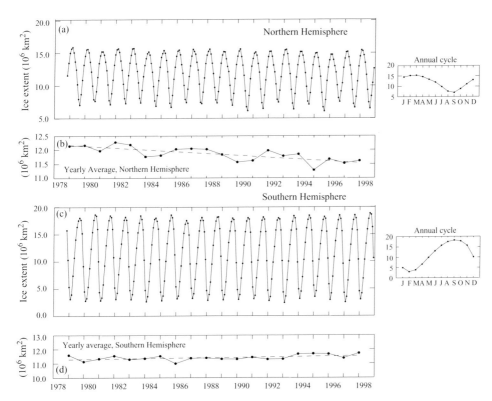

Figure 9.17. Time series of the Northern and Southern Hemisphere ice extent for 1978–1998. (a) The Arctic monthly averaged time series; (b) the Arctic annual averaged time series with the line of least-square fit; (c) the Antarctic monthly time series; (d) the Antarctic annual series with the line of least-square fit. The smaller inserts show the hemispheric annual cycles (Arctic figures courtesy D. J. Cavalieri, P. Gloersen, C. L. Parkinson, J. C. Comiso, and H. J. Zwally, used with permission; Antarctic figures from Figure 2 of Zwally *et al.*, 2002, © 2002 American Geophysical Union, reproduced/modified by permission of AGU).

out of phase, the global sea ice extent lies between 19 and 25×10^6 km^2 with the maximum in the austral winter.

Examination of the yearly averaged data shows an asymmetry between the two hemispheres. For the Northern Hemisphere, the annual average ice extent decreases linearly at a rate of about 3% per decade, while for the Southern Hemisphere, the extent shows no significant trend. A possible explanation for this difference may be that the position of the oceanographic Antarctic Convergence controls the southern ice extent, while atmospheric heating and cooling control the northern extent (Peter Rhines, private communication, 2001). These and similar regional time series illustrate the power of the passive microwave sea ice retrieval.

9.8 Further reading and acknowledgements

The website for Remote Sensing Systems (RSS), which is Frank Wentz's company, is at http://www.remss.com. From this site, Wentz and Meissner (1999) and other related papers can be downloaded from http://www.remss.com/support/publications.html. Also at this site, SSM/I and TMI data are available in both viewable and downloadable forms. The National Snow and Ice Data Center (NSIDC) website discusses polar applications of passive microwave data (http://nsidc.org/index.html). It also contains movies of Arctic and Antarctic monthly mean sea ice concentrations beginning in 1978 (http://nsidc.org/gallery/index.html), has files of sea ice concentrations and brightness temperatures on their ftp site, and also distributes them on CD. At RSS, I thank Chelle Gentemann, Thomas Meissner and Frank Wentz for helpful conversations, reprints and figures. I also thank Donald Cavalieri, Josephino Comiso, Gary Lagerloef and Simon Yueh for their help and contributions to this chapter.

10

Introduction to radars

10.1 Introduction

A radar is an active microwave device that transmits short directional pulses of energy, then operates as a sensitive receiver to measure the returned energy or radar *echo*. The term radar is an acronym for *radio detection and ranging*. The oceanographic value of radar is due to its response to different surface conditions. When the pulse interacts with a surface that is strongly reflective, the return is strong or bright; when the surface is non-reflective, the return is weak or dark. The properties of the reflected and scattered pulse are called backscatter. Because of the large variety of oceanic surface and wind phenomena that modulate the backscatter, radars can retrieve wind speed and direction, ocean swell properties and the presence of heavy rain. They can also make precise measurements of distance, and observe internal waves, sea ice, oil and biological slicks, and man-made structures such as ships and oil platforms.

Two specialized radars discussed in this and the following chapters are the *scatterometers* and *imaging radars*. A scatterometer makes quantitative measurements of the surface backscatter from small surface areas. It does this by receiving the return, then correcting it for atmospheric interference and instrument noise. With these corrections, any radar that measures backscatter can serve as a scatterometer (Ulaby *et al.*, 1981, pp. 9–10). In oceanography, scatterometers are used to retrieve the vector wind speed. In contrast, an imager receives the return from a large surface area, uses a variety of techniques to sub-divide the return into the contributions from many smaller areas, and displays the relative changes in backscatter in an image format.

This chapter describes how radars work, the methods used to subdivide an FOV and the backscatter properties of the ocean surface. Specifically, Section 10.2 derives the radar equation and describes its application to the ocean surface. Section 10.3 describes four different antenna configurations used in remote sensing. Sections 10.4 and 10.5 discuss two techniques used to sub-divide the FOV, called range and Doppler binning. Section 10.6 summarizes aircraft observations of backscatter from a wind-roughened ocean surface, and its dependence on incidence and azimuth angle. In particular, at near vertical incidence, the backscatter is specular, decreases with increasing wind speed and is independent of azimuth angle. In contrast, at oblique incidence angles, the backscatter depends on resonant

Bragg scattering, which causes the backscatter to increase with wind speed and vary with azimuth angle relative to the wind direction. The section also discusses the backscatter dependence on V- and H-polarization, and briefly describes its dependence on the other Stokes parameters.

10.2 Radar equation

The discussion of the radar equation follows Ulaby *et al.* (1982, Section 7.1) and Elachi (1987, Chapter 6). Specifically, Section 10.2.1 derives the radar equation for a perfectly transmissive atmosphere and an isolated nonemitting object, then extends this discussion to scattering from an extended surface. Section 10.2.2 discusses radar polarization and Section 10.2.3 describes how an absorbing and emitting atmosphere, an emitting surface and the instrument noise individually affect the radar return.

10.2.1 Radar backscatter from an isolated object and an extended surface

For the scattering of an electromagnetic pulse from an isolated object or target, Figure 10.1 shows the configuration of the antenna and target; the inset shows the transmitted pulse and the reflected return. The target is an isolated irregularly shaped object located in the radar far field at a range or distance R_0 from the radar, the atmosphere through which the pulse travels is perfectly transmissive, and the blackbody radiation emitted by the target is negligible compared to the magnitude of the backscattered energy. The antenna boresight is assumed to point at the target. The antenna has an aperture A, a gain $G(\theta, \phi)$ and a maximum boresight gain G_0. At some time t_0, the radar transmits a pulse of duration τ; the pulse then interacts with the target to reflect a fraction of the incident energy back to the antenna. Because the magnitude of the scattered energy depends on the target shape, composition and conductivity, the material properties of the target contribute to the nature of the backscatter. In summary, the radar transmit/receive cycle divides into three steps:

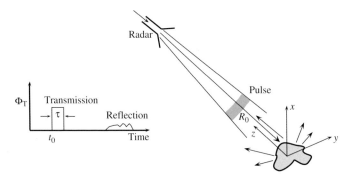

Figure 10.1. A pulse of energy incident on an isolated target. See text for definition of variables.

transmission, scattering at the target, and reception of the reflected pulse, which are discussed in this order.

The antenna transmits a pulse with a radiant flux Φ_T. From Equations (8.8) and (8.13), in the boresight direction, the transmitted power has a radiant intensity I_0, given by

$$I_0 = G_0 \Phi_T / 4\pi \tag{10.1}$$

The pulse interacts with the target in four possible ways. It can be transmitted through the target, absorbed within it, scattered away from the antenna, and backscattered toward it. For the case when no energy is transmitted through the target, the target area A_T is the cross-sectional area of the target at right angles to the boresight direction, so that the solid angle $\Delta\Omega$ subtended by the target at the radar is

$$\Delta\Omega = A_T / R_0^2 \tag{10.2}$$

Combining Equations (10.1) and (10.2) and assuming that $\Delta\Omega \ll 1$ and that the target is located in the boresight direction, the power Φ_{RS} incident on the target can be written as

$$\Phi_{RS} = \Phi_T G_0 A_T / 4\pi R_0^2 \tag{10.3}$$

As the pulse encounters the target, the incident energy excites eddy currents that are either absorbed within the target or generate a re-radiated power. For a particular target orientation relative to the antenna, if f_A is the fraction of incident power absorbed and dissipated by the target, then the magnitude of the re-radiated power is $\Phi_{TS} = \Phi_{RS}(1 - f_A)$. Further, if the gain at the target of the re-radiated power in the antenna direction is G_{TS}, and I_{TS} is the radiant intensity in the antenna direction, then

$$I_{TS} = \Phi_{TS} G_{TS} / 4\pi \tag{10.4}$$

Because the solid angle subtended by the antenna at the target is $\Delta\Omega_A = A / R_0^2$, the power Φ_R received at the antenna is

$$\Phi_R = A \Phi_{TS} G_{TS} / 4\pi R_0^2 \tag{10.5}$$

Combination of the transmission, target interaction and reception of the reflected pulse described in Equations (10.3) through (10.5) shows that the ratio of the received to transmitted power can be written as

$$\Phi_R / \Phi_T = \left[G_0 / 4\pi R_0^2 \right] \left[A_T (1 - f_A) G_{TS} \right] \left[A / 4\pi R_0^2 \right]$$
$$\text{(a)} \qquad\qquad \text{(b)} \qquad\qquad \text{(c)} \tag{10.6}$$

In (10.6) the term (a) is proportional to the transmitted power measured at the target; (b) gives the target properties; (c) is proportional to the power received at the antenna. Equation (10.6) shows that terms (a) and (c) are written in terms of the antenna properties and the range R_0, while (b) contains all of the target properties including its area, fraction of absorbed energy, and power pattern. Because the target properties are difficult to measure and not by themselves of much interest, they are grouped into one term called the *radar scattering cross section* σ with dimensions of m^2, where

$$\sigma = A_T(1 - f_A)G_{TS} \tag{10.7}$$

Substitution of (10.7) into (10.6) gives the following form of the radar equation:

$$\Phi_R/\Phi_T = \sigma G_0 A/(4\pi)^2 R_0^4 \tag{10.8}$$

Equation (10.8) shows that the ratio of received-to-transmitted power varies inversely as the fourth power of range, so that the radar must combine a powerful transmitter with a sensitive receiver. To eliminate the antenna aperture A from (10.8), Ulaby *et al.* (1981, Section 3-2.5 and Equation 3.133) show that for an antenna with no resistive losses, the gain can be written as

$$G_0 = 4\pi A/\lambda^2 \tag{10.9}$$

Substitution of (10.9) into (10.8) yields

$$\Phi_R/\Phi_T = \left[G_0^2\lambda^2/(4\pi)^3 R_0^4\right]\sigma \tag{10.10}$$

Rearranging terms in Equation (10.10) yields

$$\sigma = [\Phi_R/\Phi_T]\left[(4\pi)^3 R_0^4/G_0^2\lambda^2\right] \tag{10.11}$$

From Equation (10.11), σ is a function of λ, R_0, the ratio of the received-to-transmitted power and the known antenna properties.

In contrast to an isolated target, the half-power field-of-view A_{FOV} of an antenna pointed at the ocean contains an extended surface of scatterers and reflectors. If the ocean spatial properties are uniform within the FOV, then by analogy with Equation (10.7), σ is linearly proportional to A_{FOV}. Given this dependence of σ on area, a dimensionless cross section σ_0 can be defined from

$$\sigma = \sigma_0 dA_S \tag{10.12}$$

where dA_S is a differential element of surface area and σ_0 is the *normalized scattering cross section* or *normalized radar cross section* (NRCS). From this definition, σ_0 is dimensionless and, for constant surface properties, independent of surface area. For an extended surface and following Stewart (1985), Equation (10.10) can be rewritten in terms of σ_0 as

$$\frac{\Phi_R}{\Phi_T} = \frac{\lambda^2}{(4\pi)^3} \int_{A_{FOV}} \frac{G^2(\theta, \phi)\sigma_0}{R_0^4}dA_S \tag{10.13}$$

Consider the special case of a narrow beam scatterometer pointed at the ocean surface, with an FOV of area ΔA_{FOV} small enough that R_0, θ and ϕ are approximately constant over the field-of-view. For this case, Chelton *et al.* (2001b) show that $G(\theta, \phi)$ can be replaced by the boresight antenna gain G_0 so that Equation (10.13) yields the following algebraic equation for σ_0:

$$\sigma_0 = (\Phi_R/\Phi_T)[R_0^4(4\pi)^3]/\left[\lambda^2 G_0^2\Delta A_{FOV}\right] \tag{10.14}$$

If the antenna properties and the magnitudes of the transmitted and received power are known, then from (10.14), σ_0 can be calculated. With the additional assumptions of a noise-free instrument, a non-radiating environment and a non-attenuating atmosphere, Equation (10.14) is applicable to narrow or pencil beam instruments such as the altimeter. The magnitude of σ_0 depends on the ocean scattering properties, and additionally on the radar frequency, polarization, and the azimuth and look angles. Before discussion of the scattering properties, the next sections discuss polarization and how the return is affected by atmospheric attenuation and emission, instrument noise and the surface emissivity.

10.2.2 Polarization

The electromagnetic waves generated by radars are generally plane-polarized. For Earth remote sensing, the radars transmit and receive in either the V or H plane. Antennas that broadcast and receive in either V or H are called VV or HH antennas. The other less common options include broadcast in H and receive in V (HV), and broadcast in V and receive in H (VH). Because the return power is much smaller for VH and HV, at this time, the most common satellite radar modes are HH and VV. For a specific incidence angle and frequency, the measurement of all four polarization combinations or modes (HH, HV, VV, VH) completely specifies the surface reflection properties and is equivalent to determining the Stokes parameters for the reflected energy (Boerner *et al.*, 1998). As Chapter 13 describes for SARs, radars that operate in all four modes are called polarimetric.

10.2.3 Contributions of the ocean and atmosphere to the radar return

For a radar viewing the ocean through the atmosphere, the correction for atmospheric attenuation and the contributions from the various emission terms must be included in the σ_0 retrieval. For a real atmosphere and a reflecting and emitting ocean surface, the received power Φ_R can be written as

$$\Phi_R = \Phi'_\sigma + \Phi_{TN} \tag{10.15}$$

where Φ'_σ is the received power that is attenuated by the atmosphere and Φ_{TN} is the *thermal noise* (Spencer, Wu and Long, 2000). The term Φ_{TN} can be written

$$\Phi_{TN} = \Phi_N + \Phi_B \tag{10.16}$$

where Φ_N is the instrument noise and Φ_B is the sum of the environmental emissions. Specifically, Φ_B equals the sum of the surface emitted radiance, the upwelling atmospheric radiance and the reflection of the downwelling atmospheric and extraterrestrial radiances.

The instrument noise Φ_N sets the lower bound on the radar resolution. The noise is sometimes described in terms of a *noise floor*, which corresponds to the signal level at which σ_0 equals the noise. This occurs for a signal-to-noise ratio of 1, and is often described in terms of a *noise-equivalent sigma-zero* ($NE\sigma_0$) given in dB. The noise floor gives the optimum σ_0 resolution; the atmosphere, ocean and extraterrestrial blackbody radiances contribute additional terms that must either be estimated and removed, or be so small relative to the received σ_0 that they can be ignored. One of the advantages of radars over the passive microwave and VIR instruments is that, subject to available power, the designer sets the magnitude of the transmitted pulse, so that by making this pulse sufficiently large the noise is minimized.

The atmospheric attenuation affects the return as follows. If Φ_σ is defined as the received power that is corrected for atmospheric attenuation, then

$$\Phi_\sigma = \Phi'_\sigma / t^2 \tag{10.17}$$

In (10.17), t is the spatially and time-dependent atmospheric transmissivity that is adjusted for incidence angle; its square occurs because the radar pulse makes two passes through the atmosphere. Because most radars and scatterometers operate at frequencies less than 14 GHz, except for heavy rain, the transmissivities are close to 1 (Chapter 9). For the wind scatterometers, Chapter 11 describes the various methods used for determination of Φ_σ. Because SAR users are more concerned about relative rather than absolute changes in backscatter, the SAR operation generally ignores the attenuation and environmental blackbody terms. The nadir-viewing altimeter described in Chapter 12 is a special case, in that its primary purpose is to determine the time difference between the transmitted and received pulse, or the distance between the satellite and the surface. In summary, the absolute determination of σ_0 involves measurement of all the following terms: the radar return, the instrument and environmental noise and the atmospheric transmissivity.

10.3 Determination of σ_0 within an FOV

There are several ways to retrieve the dependence of σ_0 on the surface conditions, polarization, and look and azimuth angle. The first is that in combination with Equation (10.14), a pencil-beam scatterometer is pointed at different areas of the surface and the σ_0 distribution is defined. The second is to use a slant-looking broad-beam radar, then to subdivide the surface footprint into many smaller areas. This subdivision takes place in at least two ways, called range and Doppler binning. In range binning, the return is placed into bins based on the range or time delay between pulse generation and reception. In Doppler binning, the surface footprint is subdivided according to the Doppler shift of the return. As the following shows, scatterometers and imagers use both techniques. This section describes several common satellite radar antennas; the following sections describe the different kinds of binning.

Figure 10.2 shows four antenna configurations used in radar remote sensing and their FOVs. Relative to the antennas, location within the FOVs is described by an x, y coordinate system oriented in the cross-track (x) and along-track (y) directions. In this coordinate system, along-track is parallel to the flight direction, cross-track is at right angles to it and the coordinate origin is at the sub-satellite point. The antennas have the following

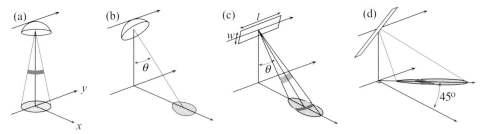

Figure 10.2. The configuration of four different antennas used in remote sensing. (a) The nadir-viewing altimeter parabolic antenna, (b) the side-looking parabolic antenna, (c) the side-looking rectangular antenna, (d) the scatterometer stick antenna, oriented at 45° to the flight path in a plane parallel to the surface. For each case, the light gray area on the surface is the FOV, while the dark gray swaths within the FOVs are in (c), a contour of constant range, and in (d), a contour of constant Doppler shift.

characteristics. First, Figure 10.2a shows the altimeter case of a nadir-pointing parabolic antenna and its circular FOV, where from Equation (8.5), the antenna half-power beamwidth $\Delta\theta_{1/2}$ is

$$\Delta\theta_{1/2} \sim \lambda/D \qquad (10.18)$$

Second, Figure 10.2b shows the same parabolic antenna pointed off-nadir, where the FOV is an ellipse described by the intersection of $\Delta\theta_{1/2}$ with the surface. The SeaWinds scatterometer described in Chapter 11 uses this antenna configuration.

Third, Figure 10.2c shows a rectangular side-looking radar with its long axis parallel to the flight direction. The configuration is used with the SAR; a typical antenna has a length $l = 10$ m and width $w = 2$ m. Equation (10.18) applies to each axis, so that the half-power beamwidth generated by l in the along-track direction is $\Delta\phi_{1/2}$, and that generated by w in the cross-track direction is $\Delta\theta_{1/2}$, where

$$\Delta\phi_{1/2} = \lambda/l, \qquad \Delta\theta_{1/2} = \lambda/w \qquad (10.19)$$

Thus as Figure 10.2c shows, a rectangular antenna generates a wide beam in the cross-track direction and a narrow beam in the along-track direction. Within the surface footprint, the dark gray curve is a contour of constant range or time delay. Fourth, Figure 10.2d shows the scatterometer case of a high-aspect-ratio stick antenna. This generates a long narrow FOV, with the long FOV axis at right angles to the length, the short axis at right angles to the width. Within the FOV, the dark gray curve is a contour of constant Doppler shift.

10.4 Range binning

This section first describes how range binning works, and shows that its resolution varies inversely with pulse length. Section 10.4.1 discusses the constraints on the generation of

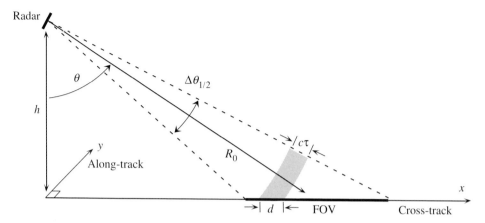

Figure 10.3. The interaction of a single pulse with the surface, where $c\tau$ is the pulse length and d is its projection on the surface.

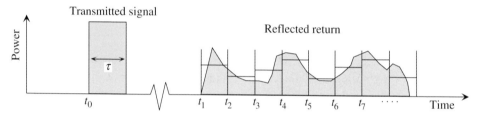

Figure 10.4. The binning of the radar return by time delay or range. The horizontal lines within each bin represent the average received power.

short pulses and a method called *chirp* that synthesizes short pulses from long pulses. Section 10.4.2 describes the constraints on the *pulse repetition frequency* (*PRF*).

In range binning, the backscattered energy received at a side-looking radar from within the surface footprint is binned according to the time delay between transmission and reception of the pulse. Figure 10.3 shows the interaction of a single pulse with the surface. If d is the length of the surface projection of the pulse, c is the speed of light and τ is the pulse duration, then d is approximately given by

$$d = c\tau \cos\theta \tag{10.20}$$

For range binning to work, d must be much smaller than the swathwidth. Figure 10.4 shows a schematic drawing of a transmitted pulse and the binning of its return into equally spaced time intervals. If the average time delay corresponding to each bin is converted into a cross-track distance, then for each pulse the average binned power can be plotted versus distance. If the radar moves at a uniform velocity, generates multiple pulses and is oriented so that it looks in the cross-track direction, this procedure generates a two-dimensional image in the along- and cross-track directions.

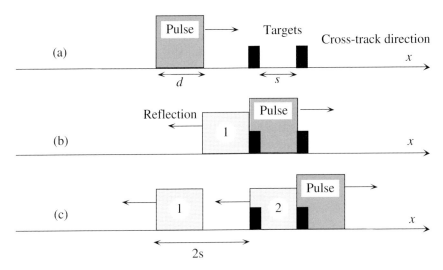

Figure 10.5. A radar pulse of length d incident on two targets, and its subsequent reflections. The two targets are separated by a distance s. (a) The incident pulse, (b) the reflection 1 of the pulse from the first target, and (c) the reflection 2 from the second target.

For this procedure, the cross-track resolution Δx depends on τ and is calculated as follows. Figure 10.5 shows the interaction of an incident pulse with two targets separated by a distance s. When the pulse reaches the first target, some of its energy is reflected. The remainder of the transmitted pulse propagates to the second target, where a second reflection occurs. For the second reflection to reach the first target, it must travel a distance s, which means that the echoes generated by reflection from the two targets are separated by $2s$. Therefore, as long as the distance between the two targets is greater than half the projected pulse length, so that $2s > d$, the two targets generate separate and identifiable returns. This means that the optimum cross-track resolution Δx is

$$\Delta x = d/2 \tag{10.21}$$

Thus for a given pulse length, even if the time bins are very small, the surface resolution cannot be less than that specified in Equation (10.21). Because the resolution improves as the length of the pulse decreases, the next section describes the chirp technique for the generation of short pulses.

10.4.1 Chirp

There are two restrictions on the generation of very short pulses. First, for a center or carrier frequency f_0 and from Fourier transform considerations, a pulse of duration τ has a frequency bandwidth $\Delta f_B \sim \tau^{-1}$. For example, a 10-cm pulse length corresponds to $\tau = 0.3$ ns, so that $\Delta \phi_B = 3$ GHz. Given the large number of users within the 1–14 GHz radar frequencies, because of EMR leakage into adjacent frequency bands, such short pulses cannot be used.

Second, even though a 3-ns pulse with its 1-m length and $\Delta f_B = 0.3$ GHz stays within its assigned bandwidth, for the return to have a satisfactory signal-to-noise ratio requires a large peak power, which for short pulses is expensive and difficult to generate. Following Ulaby *et al.* (1982, Chapter 7-5.3), many radars replace short pulses with long frequency-modulated pulses, which have the same integrated power and bandwidth as the desired short pulses. Within each pulse, the frequency increases linearly with time, producing the name chirp. When the reflection of the chirped pulse is received, the signal is run through a filter that reconstitutes the short pulse. Consequently, the chirped pulse has the same frequency bandwidth as the short pulse, but is longer, of much lower power and can be reconstituted into the desired short pulse. In the following, because the short and long chirped pulses have equivalent properties, the chirped pulse is treated as if it were a short, single frequency pulse.

10.4.2 Pulse repetition frequency

Radars are often designed such that their pulses repeat at a regular interval τ_p. The pulse repetition frequency (PRF) is defined as

$$\mathrm{PRF} = 1/\tau_p \tag{10.22}$$

From (10.22), a rate of 90 pulses per second corresponds to a PRF of 90 Hz. For most satellite instruments, it is desirable to make the PRF as large as possible, since by obtaining multiple looks of the same region, the signal-to-noise ratio can be increased by averaging the returns. The swathwidth in the cross-track direction, however, sets an upper limit on the PRF. To calculate the maximum PRF, Figure 10.6 shows a characteristic swath geometry. For a single pulse, the reflection occurs first from the near edge of the swath, then from the far edge. If the PRF is too great, then for successive pulses the echo from the second pulse returns from the near edge before the echo from the first pulse returns from the far edge. These overlapping echoes generate ambiguity in the return, making the data worthless and setting an upper limit on the PRF.

Calculation of the maximum PRF proceeds as follows. On Figure 10.6, $d_p = c\tau_p$ is the distance between successive pulses, R_1 is the range to the near edge of surface swath, and R_2 is the range to the far edge. The requirement that the first pulse reflection returns from the far edge before the second pulse reflection returns from the near edge yields the inequality

$$d_p = c\tau_p > 2(R_2 - R_1) \tag{10.23}$$

In terms of the PRF, (10.23) becomes

$$\mathrm{PRF} < c/2(R_2 - R_1) \tag{10.24}$$

For a hypothetical satellite at an altitude of 800 km with $\theta_1 = 21°$ and $\theta_2 = 45°$, the pulse separation must be greater than about 560 km or 2 ms, yielding a maximum PRF of 530 Hz.

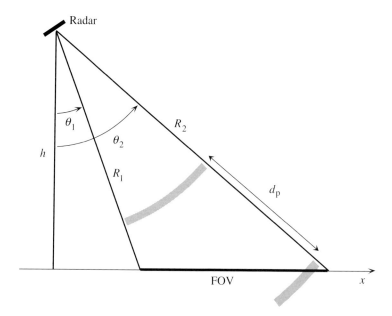

Figure 10.6. Determination of the maximum pulse repetition frequency (PRF) for a side-looking radar. To simplify the figure, the first pulse is allowed to propagate below the surface. See text for further description.

The restriction in (10.24) does not prohibit the interleaving of pulses or bursts of pulses. For example, the TOPEX altimeter generates a burst of pulses, followed by a period when it receives their echoes in sequence. Also, the twin pencil-beam SeaWinds instrument described in Chapter 11 alternates between the inner beam pulse, outer beam echo, outer beam pulse, inner beam echo. To avoid ambiguity in any of these schemes, the PRF of any sequence of pulse transmission and reception must satisfy (10.24).

10.5 Doppler binning

For a radar or scatterometer pointed in an arbitrary direction relative to the spacecraft trajectory, the return signal can also be binned according to its Doppler shift. The reason this is possible is that the surface velocity relative to the spacecraft, or equivalently the Doppler shift, depends on the antenna view angle relative to the trajectory. For real aperture radars, Doppler processing involves the generation of a long pulse at a constant f_0, then Doppler binning the return. Instruments such as the SEASAT and NSCAT scatterometers employed this technique, and as Chapter 13 describes, the SARs obtain their resolution from a combination of range and Doppler binning. In the following, Section 10.5.1 reviews the concept of Doppler shifts, and for a flat surface derives the location of the lines of constant Doppler shift, called *isodops*. Section 10.5.2 describes the spatial resolution of the Doppler binning and Section 10.5.3 shows how the rotating Earth alters the isodop locations.

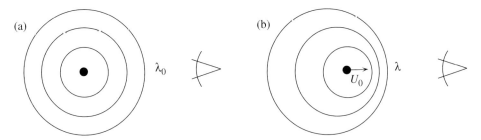

Figure 10.7. The change in wavelength associated with electromagnetic waves generated from (a) a stationary source; (b) a moving source, where the source velocity is uniform and non-relativistic.

10.5.1 Dependence of the Doppler shift on view angle

Figure 10.7 compares the difference between waves radiated from an idealized stationary and a moving source of electromagnetic waves, where in both cases the source radiates spherical waves at a constant frequency f_0 and wavelength λ_0. For the stationary case in Figure 10.7a, the time between crests is $\Delta t = 1/f_0$. Figure 10.7b shows the same radiating source moving at a velocity U_0 toward a stationary observer at the right. During Δt, the transmitter moves a distance $U_0 \Delta t$, so that the received λ is shortened by an amount

$$\lambda = \lambda_0 - U_0 \Delta t = \lambda_0 - U_0/f_0 \qquad (10.25)$$

$$\text{or} \quad \Delta \lambda = \lambda - \lambda_0 = -U_0/f_0 \qquad (10.26)$$

Because $c = \lambda f$, if λ changes by $\Delta \lambda$, f changes by

$$\Delta f/f_0 = -\Delta \lambda/\lambda_0 \qquad (10.27)$$

From Equations (10.26) and (10.27), $\Delta f = U_0/\lambda_0$. If the transmitter and receiver move together at a uniform velocity toward a stationary reflecting surface, the Doppler shift is doubled, so that

$$\Delta f = 2U_0/\lambda_0 \qquad (10.28)$$

Now suppose that the scatterometer views a flat surface at a constant oblique view angle γ relative to the spacecraft trajectory (Figure 10.8). The component of the spacecraft velocity in the γ-direction is $U_0 \cos \gamma$, so that the Doppler shift received at the spacecraft is

$$\Delta f = 2U_0 \cos \gamma /\lambda_0 \qquad (10.29)$$

From (10.29), Figure 10.8 shows a characteristic isodop along which Δf is constant.

Because the location of a point on the surface relative to the spacecraft trajectory is more commonly described by the look angle θ and the azimuth angle ϕ, for a flat surface, the following derives the relation between γ and θ, ϕ. Figure 10.9 shows the configuration of the antenna relative to a flat non-rotating Earth. The satellite is at altitude h, the distance from the antenna to the FOV is R_0, the along-track distance of the FOV from the antenna

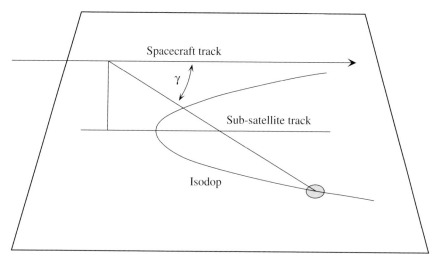

Figure 10.8. The spacecraft and sub-satellite tracks, the surface isodop, the scatterometer FOV and its view angle γ relative to the spacecraft track.

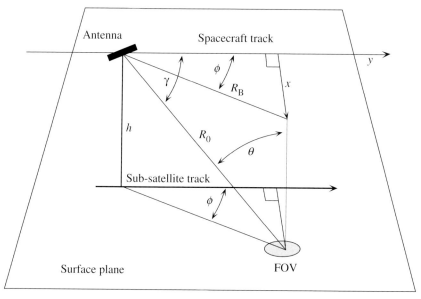

Figure 10.9. The geometry of a Doppler scatterometer above a horizontal plane, where ϕ is the azimuth angle, θ is the incidence angle and γ is the view angle relative to the spacecraft track. See text for further description.

is y and the projection of R_0 into the plane of the spacecraft is R_B. For γ, ϕ and θ, the following relations hold:

$$\cos \gamma = y/R_0 \quad \cos \phi = y/R_B \quad \sin \theta = R_B/R_0 \tag{10.30}$$

From (10.30), $\cos \gamma = \cos \phi \sin \theta$. Equation (10.29) can then be written as

$$\Delta f = (2U_0 \cos \phi \sin \theta)/\lambda_0 \tag{10.31}$$

giving the dependence of Δf on U, ϕ and θ.

Consider a scatterometer similar to the NSCAT discussed in Chapter 11 that is traveling to the right above a plane surface at an altitude of 800 km and a velocity $U_0 = 6.7$ km s^{-1}. From Equation (10.31), Figure 10.10 shows the pattern of isodops and the circles of constant range. These show that the combination of range and Doppler binning allows the entire surface to be gridded into unevenly shaped cells. Both the figure and Equation (10.31) show that the largest values of Δf occur in the fore and aft directions at $\phi = 0$ and π where $\Delta f = \pm U_0/\lambda$, while at right angles to the spacecraft trajectory Δf is zero. For the NSCAT carrier frequency of $\phi_0 = 14$ GHz, the maximum Doppler shift in the forward direction is

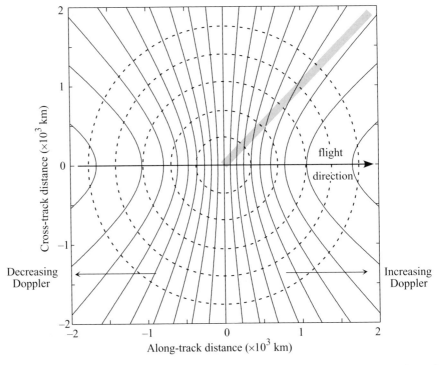

Figure 10.10. The solid lines are the isodops contoured at equal intervals of $0.1\Delta f_{max}$ on the surface plane for a satellite in horizontal motion above the plane; the dashed circles are lines of constant range (derived from Ulaby *et al.*, 1982, Equation 7.46). The origin lies directly beneath the spacecraft; the gray bar shows a typical FOV. See text for further description.

$\Delta f_{max} = 6.4 \times 10^5$ Hz; in the aft direction, $\Delta f_{max} = -6.4 \times 10^5$ Hz. The gray bar inclined at 45° shows the idealized FOV of a stick antenna. Examination of this FOV shows that near the origin the isodops are closely spaced; further from the origin they are further apart. Consequently, the size of the spatial cells defined by isodops that are equally spaced in frequency increase with distance from the origin.

A special case relevant to the SAR concerns the behavior of the isodops at look angles close to the cross-track direction. Equation (10.31) and Figure 10.10 show that if $\phi = \pi/2 - \delta$, where δ is the azimuth angle relative to the cross-track direction and defined so that it is positive in the forward direction, then for small δ,

$$\Delta f = 2U_0 \sin \delta \sin \theta / \lambda_0 \cong 2U_0 \delta \sin \theta / \lambda_0 \qquad (10.32)$$

Equation (10.32) shows that even though $\Delta f = 0$ in the cross-track direction, for azimuth angles close to this direction, Δf varies linearly with δ. Chapter 13 uses this relation in the derivation of the SAR resolution.

10.5.2 Doppler surface resolution

Doppler binning is used in two different ways. First, for certain of the wind scatterometers, the surface distribution of σ_0 is determined from Doppler binning of relatively long pulses. Second, the SAR processing uses both Doppler and range binning to combine the returns from a sequence of many short pulses in a computationally intensive procedure that yields a high resolution in both the along- and cross-track directions. This section discusses the binning of a single long return; Chapter 13 discusses the SAR case.

Section 10.4 shows that for range binning, the cross-track resolution improves as the pulse length decreases. In contrast, this section shows that the Doppler resolution improves as the pulse length increases. From Fourier transform considerations and for a pulse of length τ, the smallest resolution Δf_{min} to which the Doppler shift can be resolved is

$$\Delta f_{min} = 1/\tau \qquad (10.33)$$

Equation (10.33) shows that the longer the pulse, the smaller Δf_{min} and the better the Doppler resolution. Because the Doppler resolution or equivalently the surface spatial resolution improves with increasing pulse length, one advantage of Doppler binning over range binning is that the radars can utilize longer, lower power pulses.

The Doppler determination of the surface properties proceeds by transmission of a long pulse with carrier frequency f_0, reception of the return, and removal of the carrier. The modified return is then passed through a series of filters with bandwidths corresponding to the desired Doppler shifts, where the return is averaged within each filter. Figure 10.11 shows a schematic of the transmitted and received pulse, where the received power is placed into a series of bins defined by the filter bandwidths. For example, for the NSCAT pulse length of $\tau = 5$ ms, $\Delta f_{min} = 200$ Hz. For this case, Naderi, Freilich and Long (1991) show that to obtain a 25-km resolution in the cross-track direction, the first bin in the near swath

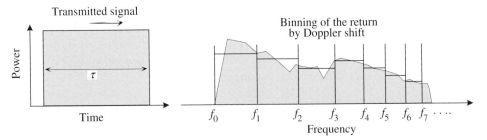

Figure 10.11. Binning of the return pulse by Doppler shift, where the width of each Doppler bin is proportional to a uniform displacement in range. The horizontal lines within each bin are the averaged return.

has a bandwidth of $\Delta f_{bin} = 15\,000$ Hz; the last has a bandwidth of 2000 Hz. Given that $\Delta f_{min} = 200$ Hz, the accuracy of the spatial resolution decreases from 1 to 10% across the swath. In contrast, if the pulse length is reduced by a factor of 5 to $\tau = 1$ ms, the accuracy across the swath decreases to values of 7 to 50%.

10.5.3 Rotation of the Earth

Because the Doppler shift responds to surface velocity, the processing must also consider the relative surface motion induced by the Earth's rotation. At the equator, the longitudinal velocity of the Earth is 0.5 km s^{-1}. If a spacecraft moving north crosses the equator at right angles, the cross-track direction is no longer a line of zero Doppler shift. Instead, for a spacecraft velocity of 6.5 km s^{-1}, the isodops are tilted by an angle equal to the tangent of the Earth and spacecraft velocities, or by 4°. Similarly, if the spacecraft crosses the equator moving south, the angle is reversed so that the total angular shift between the satellite traveling north and south is 8°. As Chapter 11 shows, in order that scatterometers compensate for this Doppler shift, either the spacecraft is rotated or the return is adjusted numerically.

10.6 Oceanic backscatter

Ocean backscatter divides into two cases, that from the open ocean and that from sea ice and from objects such as ships, oil rigs and icebergs. Although the open ocean σ_0 response depends on surface roughness and not directly on wind speed, wind speed is generally used as a proxy for roughness. In the following, Section 10.6.1 describes reflection from objects, Section 10.6.2 describes the difference between oceanic specular and Bragg scatter, and Section 10.6.3 discusses aircraft observations of backscatter.

10.6.1 Specular and corner reflectors

Unlike the ocean surface, objects such as ships, icebergs and oil rigs present reflective walls to the incident radiation. When these surfaces are perpendicular to the incident radiation, the

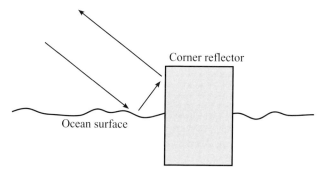

Figure 10.12. Corner reflection from a floating or bottom-mounted object with reflective vertical walls.

reflection is specular and the return is strong. When the surfaces are vertical, Figure 10.12 shows that the energy reflects off the ice or ocean surface, then off the walls, forming a corner reflector so that the antenna again receives a strong return. As the examples in Chapter 13 show, such reflectors appear bright in radar imagery.

10.6.2 Two kinds of oceanic backscatter

As Sections 5.2 and 9.4.1 discuss, the nature of the reflection or backscatter from the ocean surface depends on the wavelength distribution of the surface waves relative to the radiation wavelength λ. In general, the incident energy is scattered from short waves and reflected from long waves, where the long waves satisfy the radius-of-curvature condition in Equation (5.9). For radar backscatter from an ocean surface, Plant (1990) discusses *composite surface models* that are similar to the two-scale model described in Section 9.4.1. The composite surface consists of a large scale surface that satisfies the radius-of-curvature restriction, and a small scale surface with an rms amplitude much less than λ. The effect of the large scale surface is to advect and tilt the wind-generated patches of small scale roughness. As Plant describes, the wavelength separating the two scales is of order λ with a complicated dependence on incidence angle and the rms roughness.

For no winds and a flat surface, specular reflection occurs with its properties governed by the Fresnel coefficients described in Chapter 5. As the wind speed and roughness increase, coherent specular reflection decreases and incoherent scatter increases. Figure 10.13 shows the reflection and scattering of a radiance that is normally and obliquely incident on specular and wave-covered ocean surfaces. For normal incidence on a specular surface, Figure 10.13a shows that all of the incident radiance is returned to the antenna. For normal incidence on a rough surface, Figure 10.13b shows that the area of those facets perpendicular to the incident radiation decreases, so that the incident energy is in part specularly reflected back to the antenna and in part reflected and incoherently scattered away from the antenna. For normal incidence, this means that σ_0 decreases with increasing U. Because as Section 2.2.1 describes, the maximum ocean wave slope rarely exceeds 15°, for incidence angles $\theta <$ 15°, σ_0 continues to decrease as θ and U increase.

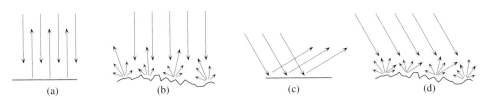

Figure 10.13. The specular reflection and incoherent scattering of a radiance incident on a surface. (a) Normal incidence, specular surface; (b) normal incidence, wave-covered surface; (c) oblique incidence, specular surface; (d) oblique incidence, wave-covered surface. See text for further description.

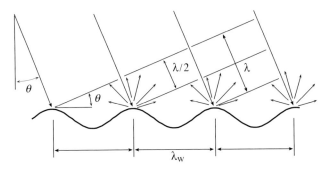

Figure 10.14. A schematic drawing of Bragg scatter modeled after the ERS-1 SAR. For the numbers given in the text, $\lambda_w = 72$ mm. See text for further description.

For $\theta > 15°$, Figure 10.13c shows that for specular reflection there is no return, while for a wave-covered surface, Figure 10.13d shows that only incoherent backscatter occurs in the antenna direction. In spite of the lack of specular reflectors at large look angles, early radar experimenters observed strong oceanic backscatter even for θ as large as 70° (Barrick and Swift, 1980). The source of this large angle backscatter is called *Bragg scatter*, which is named after William Bragg's investigation of the backscatter generated by radiation incident on a regular crystal structure (Ulaby *et al.*, 1982, p. 842). Bragg found that for specific incidence angles and frequencies, or when the crystal lattice spacing equalled half the projection of the incident wavelength on the lattice, the backscatter exhibited strong resonances. For the ocean, if the surface wave spectrum contains a wavelength component with a similar relation to the incident radiation, Bragg resonance also occurs.

Figure 10.14 shows a schematic drawing of the Bragg scatter generated by the interaction between an incident radiance and a specific water wavelength. In this example, the incident radiance is based on the 23° look angle of the ERS-1 5.3 GHz SAR, corresponding to $\lambda = 56$ mm. For this geometry, Bragg resonance occurs if there exists a surface wave component with a λ_w equal to half the surface projection of the radar wavelength λ, or when

$$\lambda_w = \lambda/2\sin\theta \tag{10.34}$$

If Equation (10.34) is satisfied, then the power reflected back to the antenna from two adjacent water wavecrests is in phase, so that radiances that are incoherently backscattered from the waves add coherently at the antenna, explaining the strong return observed for $\theta >$ 15°.

From examination of the figure, a more general form of this relation is

$$(2\lambda_w/\lambda)\sin\theta = n, \quad n = 1, 2, 3, 4, \ldots \tag{10.35}$$

so that there are whole families of Bragg-scattering solutions. Given that the wind generates a continuous spectrum of short ocean waves, resonant waves are generally present. Because, as Section 2.2.4 shows, the mean-square wave slope and surface roughness increase with wind speed, Bragg scatter increases with U. Further, observations and modeling results show that Bragg scatter also occurs from short waves riding on long ocean swell. In summary, for near nadir incidence angles, σ_0 decreases as U increases, while for oblique angles, σ_0 increases with U.

10.6.3 Aircraft observations

A series of aircraft experiments beginning in the 1940s and culminating in the 1970s investigated the dependence of σ_0 on U, θ, azimuth angle and polarization (Jones and Schroeder, 1978). Using an aircraft-mounted pencil-beam scatterometer set at a variety of incident and azimuthal angles, Jones and his colleagues carried out three kinds of experiments. First, during periods of steady offshore winds, the aircraft flew for distances up to 45 km away from the coast and back while observing the surface at a variety of fixed incidence angles. For this case where the ocean swell amplitudes and wavelengths increase with distance from the coast, Figure 10.15 shows that the VV and HH backscatter is approximately constant and independent of fetch. This and similar observations established that the wind-generated small scale surface roughness and resultant σ_0 are generated locally and are independent of fetch and swell height.

Second, for several different wind speeds, and with the scatterometer set at fixed incidence angles between 0° and 50°, the aircraft flew along flight lines in different directions relative to the wind. From observations made with the scatterometer pointed downwind, Figure 10.16 shows the VV and HH dependence of σ_0 on θ and U. The results for the scatterometer looking upwind and crosswind are similar (Jones *et al.*, 1977). The figure shows that for θ less than about 10°, specular reflection is dominant, and as expected σ_0 decreases as U increases. In contrast, for $\theta > 15°$ where Bragg scatter applies σ_0 increases with U. Between these two cases at an incidence angle of 10–15°, the effects of Bragg and specular scatter cancel, so that σ_0 is independent of U. Also at any constant U, σ_0 decreases as θ increases, so in all cases the return power decreases with increasing look angle. Finally, the figure shows that for $\theta > 20°$, the VV return is significantly greater than the HH. In summary, as θ increases, the nature of the backscatter changes from a process dominated by specular

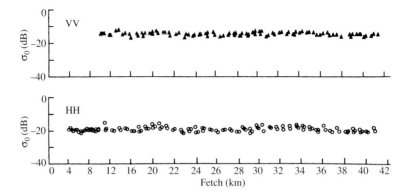

Figure 10.15. The dependence of the VV- and HH-scattering cross section on fetch, for an aircraft scatterometer flown downwind and away from a lee shore. The scatterometer frequency is 13.9 GHz, the look angle is 53°. Ross and Jones (1978) do not specify whether the scatterometer pointed upwind or downwind. On the horizontal axis, the scale change at 20 km occurs on the original figure. The wind speed varied between 10 and 13 m s^{-1} during the flight (Figure 5 from Ross and Jones, 1978, © 1978 Kluwer Academic Publishers, used with permission).

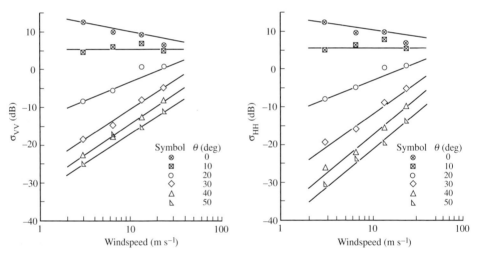

Figure 10.16. The VV- and HH-scattering cross section versus wind speed and θ. The scatterometer is pointed downwind. The reason σ_0 exceeds 0 dB at 0° and 10° is due to the factor of $(4\pi)^3$ in Equation (10.14) (Figures 7c and 7d from Jones, Schroeder and Mitchell, 1977, © 1977 IEEE, used with permission).

reflection where the return decreases as U increases, to one dominated by Bragg scatter where the return increases with U.

Third, the aircraft flew in a series of highly banked circles while observing the surface at different fixed values of θ. Similar to the investigation of the passive microwave emission from the sea surface described in Section 9.4.5, this maneuver conically scanned the antenna

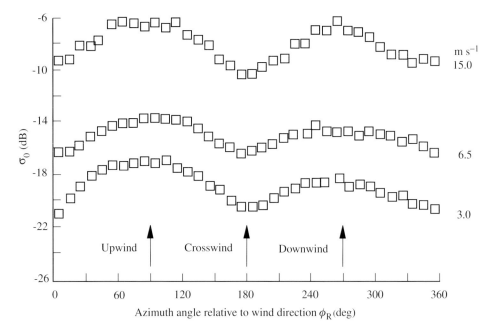

Figure 10.17. The VV-scattering cross section as a function of azimuth angle for three wind speeds and $\theta = 30°$. The upwind direction is at an azimuth angle of 90° (Figure 2 from Jones, Wentz and Schroeder, 1978, © 1978 American Institute of Aeronautics and Astronautics, used by permission of AIAA).

over 360° of azimuth. The difference between the two cases is that in Chapter 9 the aircraft measured the azimuthal dependence of the passive emissivity, here it measures the azimuthal dependence of the backscatter. For three wind speeds and a VV antenna at $\theta = 30°$, Figure 10.17 shows the dependence of σ_0 on U and on the azimuth angle relative to the wind direction ϕ_R defined in Equation (9.6). At each wind speed, the curves are symmetric around the wind direction, with their maxima in the upwind and downwind directions and minima in the crosswind directions, so that σ_0 varies approximately as $\cos 2\phi_R$. The source of this variability is the azimuthal distribution of wind-generated short waves and roughness described in Section 2.2.4. The figure also shows that as U increases, σ_0 increases, and that the upwind maxima are slightly larger than the downwind maxima, where this difference is called the *upwind/downwind asymmetry*. The source of this asymmetry is the preferential growth of parasitic capillary waves and foam formation on the downwind faces of the longer waves.

In an extension of this discussion to polarimetric observations, Yueh *et al.* (2002) compare a two-scale theoretical model with aircraft scatterometer observations of backscatter at VV, HH and HV polarizations. Because for symmetry reasons VH backscatter is identical to HV, they omit the VH case. Figure 10.18 shows the results of their two-scale scattering model. On the figure, the curves for each of the backscatter combinations (VV, HH, HV) appear similar and are symmetric about the wind direction, with maxima in the upwind and

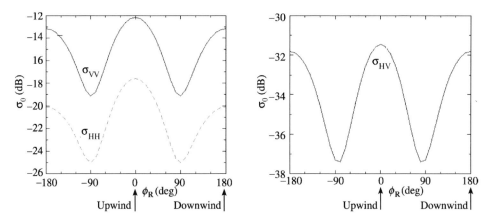

Figure 10.18. Theoretical dependence of σ_0 on ϕ_R for a polarimetric 14 GHz scatterometer at a 45° incidence angle and for a 10 m s^{-1} wind speed (Figure 3 from Yueh *et al.*, 2002, © 2002 IEEE, used with permission, courtesy Simon Yueh).

downwind directions, and minima in the crosswind directions. Also, as shown in Figure 10.16, σ_{VV} has the strongest response and is about 7 dB larger than the σ_{HH}. In contrast, the σ_{VH} and σ_{HV} responses are nearly 20 dB smaller than σ_{VV}. Because the HV response is small compared with VV and HH, it is not presently used in the vector wind retrieval. The next chapter describes the use of the VV and HH response curves in retrieval of the vector wind speed, and Chapter 13 describes how the modulation of the Bragg scatter by wind and waves allows radars to view a wide variety of ocean surface phenomena.

10.7 Further reading and acknowledgements

Ulaby, Moore and Fung (1981, 1982) and Elachi (1987) present the basics of radar operation and theory, and Donelan and Pierson (1987) review ocean surface properties and their effect on remote sensing. In more recent work, Yueh *et al.* (2002) compare a theoretical model with aircraft observations of the polarimetric radar response to waves. I thank Maria Colton, Gene Poe, Son Nghiem and Simon Yueh for their contributions to this material.

11

Scatterometer observations

11.1 Introduction

The primary function of the scatterometer is to use the backscatter dependence on azimuth angle described in the previous chapter to retrieve near global vector surface winds at intervals of 1–2 days. In addition, the scatterometer measures the backscatter from the polar pack and land ice. The importance of the winds is that they drive the ocean currents, modulate the air/sea fluxes of heat, moisture and gases such as carbon dioxide, and influence regional and global climate. Specifically, surface winds are the largest source of momentum for generation of surface waves and basin-scale ocean currents. The distribution of vector wind speeds determines the height distribution and propagation direction of ocean swell, and allows prediction of the effects of this swell on ships, offshore structures and seacoasts. Because surface observations of ocean winds by island weather stations, moored meteorological buoys and ships are sparse in the Southern Hemisphere, without scatterometers large areas of ocean would have no observations of vector winds.

Liu (2002) summarizes the current state of scatterometry and its applications to atmospheric and oceanic studies. In the atmosphere, timely assimilation of the high spatial and temporal resolution scatterometer data contributes to improved numerical weather prediction (NWP) forecasts in the Southern Hemisphere (Atlas *et al.*, 1999), to better descriptions of mid-latitude marine storms (Zierden, Bourassa and O'Brien, 2000), and to improved forcing of numerical ocean circulation models (Kelly, Dickinson and Yu, 1999). For Atlantic hurricanes, scatterometer winds have extended the lead time in which potential hurricanes can be identified, especially for small atmospheric vortices close to Africa that lack clouds and are too small to be identified in numerical models (Katsaros, Forde and Liu, 2001). Scatterometers have improved the understanding of winds in the Pacific Inter-Tropical Convergence Zones (ITCZ) (Liu, 2002) and of the coupling between winds and SST in the ITCZ (Chelton *et al.*, 2001a). They have also led to the discovery of a long wake downwind of the Hawaiian Islands (Xie *et al.*, 2001), to an improved understanding of the Asian and African monsoons (Liu, 2002), and to the nature of coastal wind jets and eddies in the Gulf of Tehuantepec off Central America (Bourassa, Zamudio and O'Brien, 1999).

Although discussions of the importance of winds frequently emphasize strong winds in the form of storms, hurricanes or typhoons, wind speeds less than 5 m s^{-1} account for nearly

40% of the hourly averaged winds (our Figure 2.1; Shankaranarayanan and Donelan, 2001). Because weak winds are concentrated in the tropics and subtropics where the majority of the ocean-to-atmosphere heat flux occurs, shifts in their patterns affect the global heat flux balance. Shankaranarayanan and Donelan show that the ability to track these weak winds contributes to the forecasting of El Niño.

This chapter describes three kinds of wind scatterometers, discusses their accuracies and presents examples of retrieved winds. Each scatterometer works by taking multiple looks at the same sea surface area from either different directions or at different polarizations. As Chapter 10 shows, because the wave properties and surface scattering cross section σ_0 are functions of both the wind speed and the azimuthal difference ϕ_R between the wind direction and the scatterometer look angle, such multiple looks can retrieve the wind speed and direction. In addition to the multiple look requirement, there are two additional considerations. First, because the wind retrieval requires precise measurement of σ_0, the received backscatter must be corrected for noise and atmospheric attenuation; second, the wind solutions must be corrected for possible ambiguities.

For the noise correction, as Section 10.2.3 describes, an accurate measurement of σ_0 requires that for each transmitted pulse or group of pulses the scatterometer must measure not only the received power Φ_R, but also two additional quantities. These are the atmospheric transmissivity t and the total thermal noise Φ_{TN}, which is the sum of the instrument noise and the natural emissions from the sea surface and atmosphere. These quantities must either be determined simultaneously or, for the case of t, be provided from other satellite instruments or from climatological lookup tables. As the following sections show, the different scatterometers employ a variety of methods to measure Φ_R, Φ_{TN} and t. For the ambiguity correction, the scatterometer-derived wind product also depends in part on coincident surface wind observations. As Section 11.3.2 shows, the reason for this is that, depending on the number of scatterometer looks, the scatterometer produces one, two or four different wind vector solutions, separated from each other by azimuth angles of $\pm 90°$ or $\pm 180°$. The way in which the false wind solutions, called *ambiguities*, are removed is to use the NWP wind directions as a best guess. This procedure is similar to the derivation of the global infrared SST data sets described in Chapter 7, where the final product depends on correction of the satellite SSTs by buoy observations.

In the following, Section 11.2 summarizes past, present and some of the proposed future scatterometer missions and gives their requirements for accuracy and coverage. Section 11.3 describes how scatterometers retrieve the vector wind speed. Sections 11.4 through 11.7 describe three kinds of operational scatterometers and their relative advantages and disadvantages. Section 11.8 describes the accuracy of the retrievals, and Section 11.9 discusses several examples of retrieved winds and of Antarctic pack ice observations.

11.2 Background

Table 11.1 lists the past, present and some of the proposed future scatterometer missions. The scatterometers divide into three classes. The first includes the NASA SEASAT-A Satellite

Table 11.1. *List of satellite scatterometer missions in order of their launch dates*

Satellite	Agency	Instrument	Frequency/operation	Launch date	Status/end date
SEASAT	NASA	SASS	14.6 GHz, 4-antennas, Doppler bin, left, right	June 1978	October 1978
ERS-1	ESA	AMI	5.3 GHz, 3-antennas, range bin, right side	July 1991	June 1996
ERS-2	ESA	AMI	5.3 GHz, 3-antennas, range bin, right side	April 1995	January 2001
ADEOS-1	NASA/NASDA	NSCAT	14 GHz, 6-antennas, Doppler bin, left, right	August 1996	June 1997
QuikSCAT	NASA	SeaWinds-1	13.4 GHz, two rotating pencil beams	June 1999	Operational
ADEOS-2	NASDA/NASA	SeaWinds-2	13.4 GHz, two rotating pencil beams	December 2002	October 2003
METOP	ESA	ASCAT	5.3 GHz, 3-antennas, range bin, left, right	2005	Pending
GCOM-B1	NASDA/NASA	OVWM (AlphaSCAT)	SeaWinds successor	2008	Planned

Adapted from Plant, 2001.

273

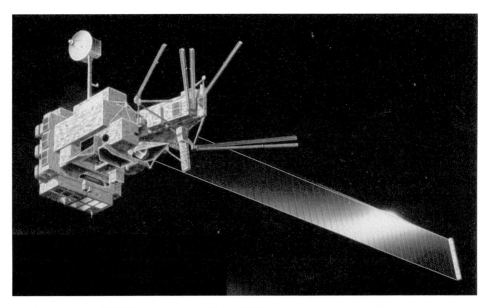

Figure 11.1. The ADEOS-1 spacecraft with the NSCAT 3-m stick antennas. The antennas and the solar panel dominate the spacecraft configuration. (Figure 1 from Naderi *et al.*, 1991, © 1991 IEEE, used with permission).

Scatterometer (SASS) and the NASA Scatterometer (NSCAT) that use stick-like antennas and Doppler bin the returns. NSCAT flew on the Japanese Advanced Earth Observing Satellite-1 (ADEOS-1) (Figure 11.1). The second includes the Advanced Microwave Instrument (AMI) scatterometer on the European Remote-sensing Satellites ERS-1 and ERS-2 (Figure 11.2) and the 2005 European Advanced Scatterometer (ASCAT) on the forthcoming European MÉTéorologie OPérationnelle (METOP) satellite; these use three long rectangular antennas and range bin the returns. The third is the SeaWinds scatterometer mounted on the dedicated QuikSCAT satellite and on ADEOS-2. SeaWinds uses a rotating dish antenna to generate a pair of conically scanned pencil beams each at a different incidence angle, then range bins the return from each beam. As Chapter 14 describes, the successor instrument to SeaWinds is the Ocean Vector Winds Mission (OVWM) instrument, called AlphaScat, which is proposed for the Japanese Global Change Observation Mission-B1 (GCOM-B1) successor to ADEOS-2.

The NASA scatterometers operate at K_u-band (14 GHz); the European scatterometers at C-band (5.3 GHz), where $\lambda = 6$ cm at C-band, and 2 cm at K_u-band. As Chapter 9 shows, C-band has the advantage that at this frequency the atmospheric transmissivity is almost identically equal to 1, while at K_u-band it is nearly 1. In contrast, because the short capillary-gravity waves are more responsive to changes in wind speed than the longer waves, K_u-band has a greater response to changes in wind speed and a greater dynamic range than at C-band. Because of the shorter radiation wavelength, however, K_u-band is also more responsive to raindrop roughening of the sea surface.

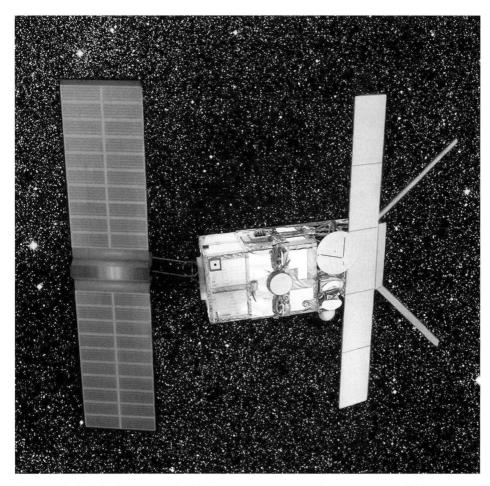

Figure 11.2. The ERS-1 spacecraft with the solar panel to the left, the rectangular SAR antenna to the right, and the three narrow aspect ratio rectangular scatterometer antennas at the far right. The main payload structure measures 2 m × 2 m × 4 m. (Courtesy ESA, used with permission).

As Table 11.1 shows, the first scatterometer mission was the 1978 NASA SASS stick scatterometer (Johnson *et al.*, 1980). This was followed by the European AMI fan-beam scatterometers on ERS-1 and -2; these observations began in 1991 and terminated in 2001. The SASS follow-on was the NSCAT launched in August 1996 on the Japanese ADEOS-1 satellite. Because of the catastrophic failure of a solar panel, the ADEOS-1 mission ended on June 30, 1997, so that NSCAT lasted less than a year (Wentz and Smith, 1999). Because of the gap in satellite wind coverage created by this loss, the SeaWinds scatterometer was launched on 19 June 1999 on the dedicated US QuikSCAT satellite (Figure 11.3). Consequently, during 2001 and as Table 11.1 shows, SeaWinds was the only functioning wind scatterometer. In December 2002, it was joined by an identical scatterometer on ADEOS-2,

Figure 11.3. The SeaWinds scatterometer mounted on the QuikSCAT satellite. The QuikSCAT antenna has a 1-m diameter (Courtesy NASA/JPL/Caltech, used with permission).

which failed in October 2003, and in January 2003 by the polarimetric passive microwave WindSat instrument discussed in Chapters 9 and 14. The next European scatterometer will be the 2005 launch of ASCAT on METOP-1.

Table 11.2 lists the mission requirements for the QuikSCAT scatterometer; Naderi *et al.* (1991) give similar requirements for NSCAT. For the scatterometer wind measurements to be of global or regional value to the meteorological community, accurate wind measurements must be obtained from the entire ice-free ocean at daily intervals. For wind speeds of 3–20 m s^{-1}, the rms speed accuracy must be better than 2 m s^{-1}; for 20–30 m s^{-1}, the rms

Table 11.2. *QuikSCAT scatterometer mission requirements for areal coverage*
and wind accuracy

Quantity	Requirements	Applicable range
Wind speed	2 m s^{-1} (rms)	3–20 m s^{-1}
	10%	20–30 m s^{-1}
Wind direction	20° (rms) for the closest ambiguity	3–30 m s^{-1}
Spatial resolution	25 km	σ_0 cells
	25 km	Wind cells
Location accuracy	25 km (rms)	Absolute
	10 km (rms)	Relative
Coverage	90% ice-free ocean every day	—
Mission duration	36 months	—

Perry, 2001.

accuracy must be within 10% of the wind velocity. The directional accuracy of the best wind solution must have an rms error of no more than 20°. The location of each σ_0 measurement must have an rms accuracy of 25 km, and the winds must be determined within cells with characteristic dimensions of 25 to 50 km. The system must produce useful products within hours after the data are acquired and the scatterometer should be able to acquire data for at least three years. As Sections 11.4 through 11.6 show, these requirements for accurate winds and global coverage dictate the scatterometer orbit and swathwidth.

11.3 How scatterometers derive the wind velocity

Each of the instruments listed in Table 11.1 retrieves the vector wind speed by taking multiple looks at the same surface area at different azimuth angles and polarizations. Because, as Figure 10.18 shows, the retrieved values of σ_{HV} and σ_{VH} are much smaller than σ_{VV} and σ_{HH}, the scatterometers presently operate in HH or VV. The SEASAT SASS made only two such looks, the AMI made three looks, and NSCAT made four looks at three different look angles and two polarizations from each side of the satellite. For the SeaWinds rotating beams, the number of looks varies from two to four, depending on position within the swath. For these multiple looks, the following describes the technique used to retrieve wind speed and direction.

Figure 11.4 shows the conceptual scatterometer design. For a steady wind, each scatterometer retrieves the backscatter from the same FOV at two or four different times, azimuth angles and polarizations. For the three beams shown on the figure, the antenna look angles relative to the satellite trajectory are 45° ahead, at right angles to the trajectory and 45° behind. For directional wind retrieval to be possible, Bragg scattering must dominate, so that the antenna incidence angles must be greater than 15°–20°. Assume that the satellite is at an altitude of 800 km, with a surface velocity of about 7 km s^{-1}. If the FOV in question

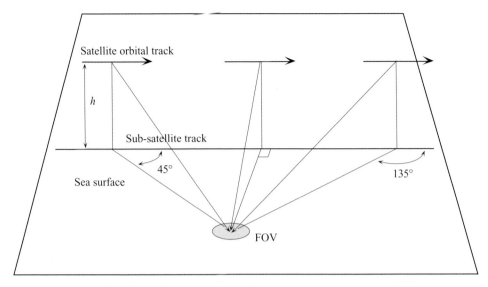

Figure 11.4. Example of several looks by a scatterometer at the same FOV.

is at a perpendicular distance of 500 km from the sub-satellite track, then the mid-beam observes it about 70 s after the forward beam and the aft beam observes it an additional 70 s later. This procedure gives three measurements of σ_0 over a period of about two minutes. From these observations and assuming steady winds, the next section shows that the vector wind speed can be derived through use of the dependence of σ_0 on azimuth angle shown in Figures 10.17 and 10.18.

11.3.1 Geophysical model function

Determination of the wind velocity from multiple measurements of σ_0 requires knowledge of the functional relation between σ_0 and the near surface winds, where this relation is called the *geophysical model function* or simply the *model function*. Because the look and azimuth angle-dependent backscatter is proportional to sea surface roughness and not necessarily to the 10-m wind speed U, scatterometer measurements of wind velocity are indirect. The wind velocity used in scatterometry is the *neutral stability* wind velocity measured at a 10-m height, where neutral stability means in the absence of atmospheric stratification. Although the following refers to the neutral stability wind speed as either the wind speed or 10-m wind speed, as next discussed, the scatterometer and observed winds differ slightly.

The importance of stratification is that it modifies the momentum transport through the surface boundary layer. When the ocean surface is warmer than the atmosphere, the boundary layer is unstable, so that momentum is more easily transferred from the 10-m winds to the surface. For this unstable atmosphere, a specific U produces a greater surface roughness and more backscatter than a stably stratified atmosphere, so that unstable stratification makes the scatterometer-inferred neutral winds larger than the observed, while a stable stratification makes them smaller. Consequently, before comparison of scatterometer and buoy winds and depending on the observed atmospheric stratification, the buoy winds must be adjusted to be greater or smaller than their observed values. This adjustment has typical values of $0.1–0.2 \, \mathrm{m \, s^{-1}}$ (Dudley Chelton, private communication, 2003). Other factors such as organic or inorganic slicks that increase surface tension and reduce the ocean surface roughness also produce an apparent smaller wind speed.

The most general form of the model function gives σ_0 as a function of the polarization P, where P represents a VV or HH antenna, the incidence angle θ, the wind speed U and the relative wind direction ϕ_R. This relation can be written as

$$\sigma_0 = F(P, U, \theta, \phi_R) \tag{11.1}$$

Based on aircraft and satellite observations similar to those described in Section 10.6.3, and for a constant U and fixed θ and polarization, the σ_0 dependence on ϕ_R is commonly described by an empirically derived truncated Fourier series called the *two-cosine function* (Wentz, Peteherych, and Thomas, 1984; Wentz and Smith, 1999; Brown, 2000):

$$\sigma_{0P} = A_{0P}(1 + A_{1P} \cos \phi_R + A_{2P} \cos 2\phi_R + \cdots) \tag{11.2}$$

In (11.2), the subscript P again represents polarization. Although Wentz and Smith (1999) state that the contributions to Equation (11.2) of higher order terms such as $\cos 3\phi_R$, $\cos 4\phi_R$ and so forth do not exceed 4% of the first three terms, Equation (11.2) is sometimes supplemented with higher order harmonics.

The coefficients in Equation (11.2) are empirically derived by comparison of the scatterometer winds with surface and other satellite wind data sets. Examples include comparison with NDBC buoy winds (Freilich and Dunbar, 1999), or with SSM/I wind magnitudes at small wind velocities and ECMWF NWP winds at larger velocities (Wentz and Smith, 1999). The model functions are continually updated; the current SeaWinds model function is the QuikSCAT function; the ERS-1, -2 model function is the CMOD-4 (Liu *et al.*, 1997). The reason that the model functions are derived empirically and not from theoretical models of surface wave roughness is that a sufficiently accurate theory does not yet exist to describe the response of short wind waves to changes in U.

Difficulties exist with the accuracy of the buoy winds at both small and large U (Zeng and Brown, 1998). At small U, the buoy and scatterometer winds both have poor directional accuracies, and at these velocities, ocean currents can bias the scatterometer winds relative to the buoy measurements. At large U, problems occur with buoy tilt, the effect of spray on the buoy anemometer, and for heavy swell conditions with the uncertainty about the anemometer height relative to the sea surface.

As an example of a geophysical model function and for the SeaWinds incidence angles and polarizations, Figure 11.5 shows the dependence of σ_0 on U and direction for the QuikSCAT model function. As Section 10.6.3 describes, the maxima of these curves approximately occur in the upwind and downwind directions, the minima in the crosswind directions. Three factors characterize the model curves: a general increase in σ_0 with U, an upwind/crosswind

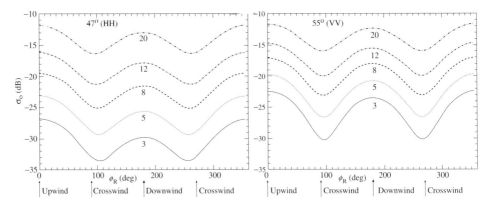

Figure 11.5. The SeaWinds geophysical model function for the SeaWinds incidence angles and polarizations of 47° HH and 55° VV. The curves are lines of constant wind speed; the numbers below each curve give the wind speed in m s^{-1}. Upwind is at 0°; downwind at 180° (Courtesy Michael Freilich, used with permission).

difference and an upwind/downwind asymmetry. First, for fixed incidence and azimuth angles, σ_0 increases approximately as log U (Freilich, 2000). The increase in σ_0 with U at a constant ϕ_R means that σ_0 measurements generally become more accurate with increasing U, although this trend breaks down in the presence of severe wave breaking. Second, the difference in the magnitude of the upwind and crosswind σ_0, called the upwind/crosswind ratio, and the associated dependence of σ_0 on ϕ_R permits the retrieval of the wind direction. The figure shows that this ratio is largest at small U, then decreases with increasing U.

Third, the upwind/downwind asymmetry occurs because the upwind maximum is slightly larger than the downwind; this occurs because of the presence of foam and the enhanced growth of short capillary-gravity waves on the downwind faces of longer waves (Section 2.2.1). As the next section shows, this asymmetry makes possible the retrieval of a unique wind solution. In general, the magnitude of this asymmetry increases with incidence angle, is larger for HH than for VV, and is largest for small U (Freilich, 2000). As Freilich shows, the sensitivity of σ_0 to changes in U, and the magnitudes of the upwind/crosswind ratio and the upwind/downwind asymmetry increase with incidence angle. Also, Section 10.6.3 shows that for the same incidence angle, σ_0 is about 7 dB larger at VV than HH. This difference explains why the SeaWinds outer beam is VV and the inner is HH, so that the two beams have about the same return power.

The model functions described by Equation (11.2) are given in lookup tables for A_{0P}, A_{1P} and A_{2P} and if necessary for higher order coefficients as functions of wind speed and direction, look angle and polarization. The fan-beam scatterometers such as the NSCAT and AMI require that the model function be specified for range of look angles from approximately 15° to 65°, where the NSCAT model function was also specified for VV and HH. Because these scatterometers have approximately 20 observational cells in the cross-track direction, they require relatively complicated lookup tables. In contrast, because the

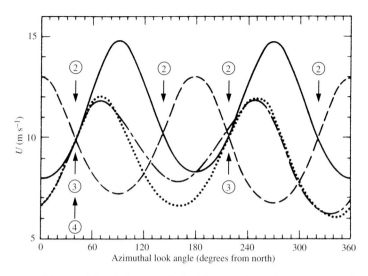

Figure 11.6. Loci of the possible wind vectors derived from co-located measurements from different antennas. —, 45° antenna azimuth angle (VV); — — — 135° (VV); ······, 65° (HH); - — · — -, 65° (VV). The arrows marked by the ② show the four solutions derived from two looks; the arrows marked by the ③ show the two solutions derived from three looks, and the arrow marked by the ④ shows the single solution derived from four looks (Figure 5 of Naderi *et al.*, 1991, © 1991 IEEE, used with permission).

SeaWinds model functions are only specified at two look angles, they are easier to upgrade to a greater resolution in wind speed and direction than the fan-beam functions.

11.3.2 Derivation of the vector wind speed from the model function

Based on NSCAT, Figure 11.6 shows how the σ_0 derived from these multiple looks is used to estimate the vector wind speed (Naderi *et al.*, 1991). On the figure, the curves are not the constant wind speed contours shown in Figure 11.5, rather they are contours of constant σ_0 that give the dependence of U on ϕ_R. The curves are discussed in the order listed in the caption. First, the solid curve is derived from a single VV observation of σ_0 at a 45° azimuth angle relative to the flight direction. For this observation, the curve shows that the wind speed solutions lie between 6 and 15 m s^{-1} with no directional information. Second, the dashed curve is the solution for a VV σ_0 observation at right angles to the first. The solid and dashed curves representing these two looks intersect at the four points marked by the arrows and the ②. Each of these intersections represents possible wind solutions called ambiguities. At these points, the wind speeds are about 10 m s^{-1} with four choices of direction separated from one another by approximately 90°. This case of two looks and four ambiguities corresponds to the outer SeaWinds swath and the entire SASS swath.

Third, the dotted curve shows the solution for an observation at 65° and HH, where the first three curves have two common intersections marked by the arrows and the ③. Because these two wind solutions are identical in magnitude and approximately 180° apart in direction, three looks yield the correct wind magnitude, but do not reveal whether the wind is blowing *to* or *from* a specific direction. Finally, the short-dash–long-dash curve is the solution for an observation at 65° and VV. The four curves intersect at the single point marked by the arrow and the ④, corresponding to a scalar wind velocity of 10 m s^{-1} and a wind direction of 40°. Examination of the two intersections marked by the ③ and ④ shows that there is only a small difference between the correct and the 180° ambiguous solution, where the cause of this difference is the upwind/downwind asymmetry. Without this asymmetry, the 180° ambiguity could not be eliminated. Because this small asymmetry is easily obscured by noise, many scatterometers use only three looks and accept the two ambiguities. These are reduced through use of the NWP co-located wind directions as a first guess, then by applying the median filter iteration described in Freilich and Dunbar (1999) and Gonzales and Long (1999).

11.4 NSCAT scatterometer

The NSCAT scatterometer was a K$_u$-band (13.995 GHz) scatterometer that was launched on August 17, 1996 on ADEOS-1. The satellite was in a sun-synchronous orbit at an altitude of 795 km, a period of 101 minutes and a sub-satellite velocity of 6.7 km s^{-1}.

Naderi *et al.* (1991) give a detailed description of NSCAT. It consisted of six identical dual-polarization stick antennas, measuring about 3 m in length, 6 cm in width, and 10–12 cm in thickness. Each antenna produced a fan beam with incidence angles of $20° < \theta < 55°$ in the along-beam direction, and a 0.4° beamwidth in the cross-beam direction. Figure 11.7 shows the NSCAT illumination pattern, where the left-hand swaths are at angles of 45°, 65° and 135° relative to the flight direction, and the right-hand swaths are at angles of 45°, 115° and 135°. The lack of Doppler response at right angles to the spacecraft explains this asymmetric choice of angles. There are three swaths on each side of the spacecraft, and because the antennas at 65° and 115° operate at both VV and HH, the antennas on each side make four different measurements. In the cross-track direction, the swathwidths are 600 km. Directly beneath the spacecraft in a region measuring ±165 km from the nadir track, specular backscatter dominates the return so that there is no directional wind data from within this *nadir gap*, which occurs for all fan-beam scatterometers. Outside of this gap, each swath is divided into 24 Doppler cells, measuring 25 km in the cross-track direction.

To obtain a 25-km resolution in the along-track direction, each antenna was sampled every 3.74 s, during which time the spacecraft traveled 25 km. During this 3.74 s, because the NSCAT had a single transmitter/receiver that rotated among the eight different beams, each beam was sampled within a subperiod of 468 ms. Within this subperiod and for each cell, the scatterometer measured Φ_R and Φ_{TN}. To do this, the subperiod was further divided

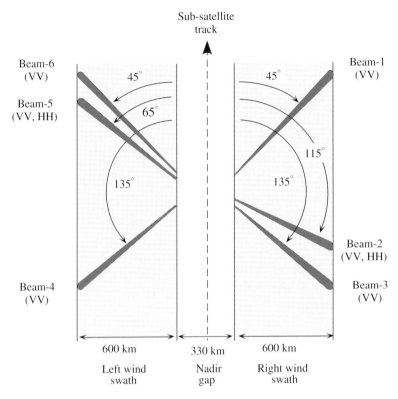

Figure 11.7. The NSCAT surface swath is shown in gray, the antenna surface footprints are outlined, and the nadir gap is shown in white (Adapted from Figure 7 of Naderi *et al.*, 1991).

into 29 observational cycles of 16 ms duration that consisted of 25 transmit/receive cycles and 4 observations of noise. A transmit/receive cycle consisted of a 5-ms transmitted pulse, and 11 ms of receive. For the received pulses, Section 10.5.2 describes the Doppler binning procedure. The four noise measurements consisted of 5 ms with *no* transmission followed by 11 ms of receive, which within each footprint provide a measurement of Φ_{TN}. To obtain σ_0, the Φ_R and Φ_{TN} were averaged over their respective observational periods, then Φ_{TN} was subtracted from Φ_R to obtain Φ'_σ in Equation (10.15). The backscattered power corrected for attenuation Φ_σ was then obtained from Equation (10.17), where the transmissivity comes from a climatological lookup table.

As Section 10.5.3 shows, the Doppler shift observed by the scatterometer is also a function of the Earth's rotation, so that in the NSCAT processing, the center frequency and bandwidth of each Doppler cell were adjusted as a function of distance from the equator so that the size and positions of the surface bins relative to the satellite did not change. In contrast, SASS had only four stick antennas at $45°$ and $135°$ to right and left of the spacecraft trajectory, and its onboard Doppler filters were fixed (Johnson *et al.*, 1980). This created difficulties near the equator, where the Doppler cells observed by the fore and aft antennas had very different lengths, which reduced the cell overlap in the two-look intercomparison of σ_0.

11.5 AMI scatterometer

The Advanced Microwave Instrument (AMI) flew on the ERS-1 and ERS-2 satellites. These satellites were in a sun-synchronous circular orbit at an altitude of 785 km, a nominal period of 100 minutes and a 1030 local equator crossing time (Attema, 1991). The AMI was a vertically polarized C-band scatterometer that combined a high resolution SAR with a low resolution wind scatterometer, using a common transmitter and receiver and two separate antennas (Attema, 1991). The SAR used the large rectangular antenna shown in Figure 11.2; the scatterometer used the three large-aspect-ratio rectangular antennas. The system operated in three modes: a high resolution SAR image mode that was only used when the satellite was within range of a ground station so that the data could be directly downloaded; a low resolution SAR mode used for wave observations; and the scatterometer mode. The data from the wave and scatterometer modes were recorded onboard for later downloading. Because the scatterometer and SAR used the same electronics, wind data were not always taken in the vicinity of ground stations.

Figure 11.8 shows the footprints of the scatterometer antennas. The three rectangular antennas generate beams to the right of the spacecraft at azimuth angles of 45°, 90° and 135°, where the central antenna measures 2.3 m × 0.35 m and the fore and aft antennas measure 3.6 m × 0.25 m. The central antenna beamwidth measures 24° in elevation and 1.4° in azimuth; the fore and aft antenna beamwidths measure 26° in elevation and 0.9° in azimuth. For the fore and aft antennas, the receiver center frequencies were adjusted to account for their respective Doppler shifts. To minimize the effect of the Earth's rotation

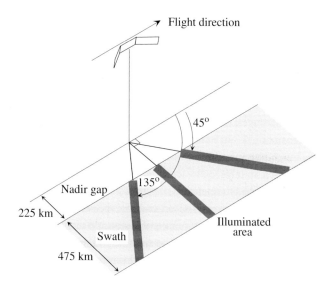

Figure 11.8. The surface swaths of the ERS scatterometer antennas.

on the scatterometer, the satellite was actively rotated about its nadir axis in what is called *yaw* steering, so that the Doppler shift observed by the mid-antenna beam was always zero.

Because of the nadir gap, the swath begins at 225 km from nadir and has a cross-track width of 475 km. The AMI used range binning to observe σ_0 within cells with cross-track and along-track widths of about 50 km. For the central antenna, the pulse duration was 70 µs; for the fore and aft antennas, the duration was 130 µs, where the fore and aft pulse lengths were longer because of the oblique antenna azimuth angles. For the central antenna, the PRF was 115 Hz, and for the fore and aft antennas the PRF was 98 Hz, so that for each antenna, the time between pulses was about 10^4 µs. This relatively long inter-pulse interval was used in three ways: to receive Φ_R, to record an internal calibration pulse and to make a passive observation of Φ_{TN}. Because this procedure was followed for every pulse, calibration of the return and removal of Φ_{TN} was straightforward.

For each pulse, σ_0 was calculated by applying the calibration, removing the system and environmental noise, and correcting for atmospheric transmissivity from a climatological lookup table. The σ_0 from each antenna was then resampled onto a 25-km square grid, with 19 data points across the swath. The individual σ_0 were then resampled to a 50-km resolution to improve their signal-to-noise ratio, where across the swath, the noise was nearly constant at about 6% of the signal (Ezraty and Cavanie, 1999). The three looks yield two wind speed estimates where the best wind solution was determined by comparison with NWP solutions. Additional external calibration and a check on instrument drift and degradation were provided by surface transponders and by observations of the Amazon rain forest with its relatively uniform backscatter.

The AMI scatterometer observations terminated in January 2001; its replacement will be the European C-band Advanced Scatterometer (ASCAT) on the ESA METOP-1 mission currently scheduled for 2005, which Chapter 14 discusses further (Rostan, 2000). Unlike the ERS spacecraft, METOP will not include a SAR. Figure 11.9 shows the configuration of the ASCAT antennas; the instrument is a C-band range-binned scatterometer with three

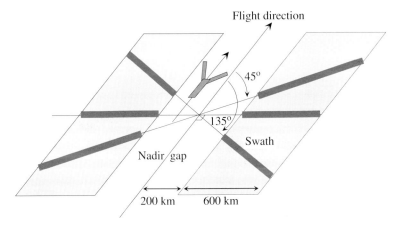

Figure 11.9. The surface swaths of the METOP-1 Advanced Scatterometer (ASCAT) scheduled for launch in 2005 (Adapted from Rostan, 2000).

rectangular antennas similar in design to AMI, except that they look off to both sides of the spacecraft, doubling the effective swathwidth while keeping the nadir gap.

11.6 SeaWinds scatterometer

The SeaWinds scatterometer is on the QuikSCAT and ADEOS-2 satellites; Spencer, Wu and Long (1997, 2000) describe its design and operation. Both satellites are in sun-synchronous orbits at an altitude of 803 km and a period of 101 minutes. QuikSCAT has a 0600 ascending equator crossing time; ADEOS-2 had a 1030 descending crossing time. SeaWinds consists of a 1-m rotating parabolic antenna, with two offset feeds that generate two 13.4-GHz pencil beams at different incidence angles (Figure 11.10). The inner beam operates at HH, an off-nadir angle of 40° and an incidence angle of 47°; the outer beam operates at VV, an off-nadir angle of 46° and an incidence angle of 55°. The antenna rotates at 18 revolutions per minute; the surface footprints have diameters of approximately 25 km. The return from this footprint can either be binned in its entirety or, as shown below, divided into a number of range-dependent cells. As Figure 11.10 shows, the SeaWinds swath has an overall width of 1800 km with no nadir gap. The swath divides into two parts: in the dark gray areas, the winds are determined from four looks; in the light gray areas, from two looks. Part of the two-look region occurs at distances for which only the outer beam takes data, and part occurs adjacent to nadir.

Figure 11.11 shows the rotating pattern of a single SeaWinds footprint; during one rotation, the satellite advances about 25 km. For the four-look region, Figure 11.12 shows that the FOV is viewed twice by the outer beam looking forward at time t_1 and backward at t_4,

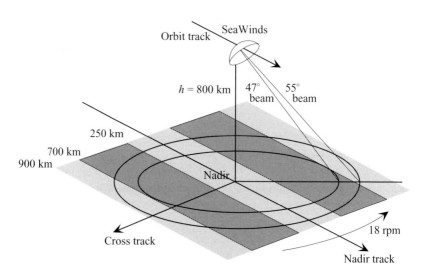

Figure 11.10. The SeaWinds conceptual design and scan coverage for the listed incidence angles. In the dark portion of the swath, the winds are determined from four looks; in the light portion, from two looks (Adapted from an unpublished figure of Michael Freilich).

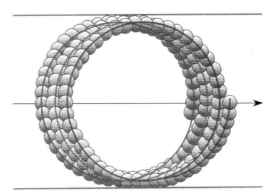

Figure 11.11. The surface scanning pattern of a single SeaWinds beam. The diameter of a single FOV is about 25 km (Figure courtesy Michael Freilich, used with permission).

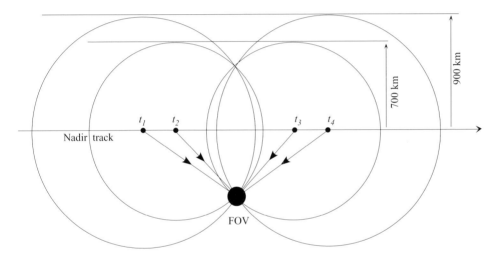

Figure 11.12. An example of how two looks by the outer beam and two looks by the inner beam generate four looks at the same FOV (Redrawn from an undated NASDA publication on ADEOS-2).

and twice by the inner beam looking forward at t_2 and backward at t_3. Spencer *et al.* (1997) show that the wind retrieval performance of the SeaWinds varies with distance from the nadir track, where the best wind retrievals occur when the azimuthal differences between measurements are close to 90°, so that even in the four-look region, the quality of the retrieved winds is not uniform.

Table 11.3 lists some properties of the SeaWinds beams. The transmit/receive cycle alternates between the two beams as follows: inner beam transmit, outer beam receive, outer beam transmit, inner beam receive, so that each echo returns after the following transmit pulse. For both beams, the overall PRF is about 192 Hz, corresponding to a transmit/receive cycle of about 5.2 ms, within which the antenna rotates about half a beamwidth.

Table 11.3. *SeaWinds parameters*

Parameter	Inner beam	Outer beam
Rotation rate	18 rpm	
Polarization	HH	VV
Zenith angle	40°	46°
Surface incidence angle	47°	55°
Slant range	1100 km	1245 km
3-dB footprint (along-scan × cross-scan)	24 × 31 km	26 × 36 km
Pulse length (unchirped)	1.5 ms	
Pulse length (chirped)	Programmable, > 2.7 µs	
Along-track spacing	22 km	22 km
Along-scan spacing	15 km	19 km

Adapted from Spencer *et al.*, 2000, Table 1.

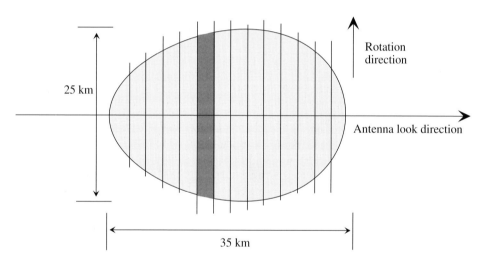

Figure 11.13. The division of the SeaWinds footprint into range slices. See text for additional description.

The footprint of either beam is an ellipse measuring approximately 25 km in azimuth and 35 km in range. Within this footprint, which is sometimes called an egg, range binning is used to improve the resolution (Perry, 2001). Figure 11.13 shows the egg footprint, and the division of the egg into different range bins, called *slices*. There are 12 slices per footprint, and in the analysis σ_0 is calculated for both the full footprint and the inner eight slices. This means that the scatterometer measures σ_0 at a variety of resolutions, including the full beam footprint, the different slice footprints and a variety of footprints made up of different combinations of slices. For each of these, ground processing locates the geographic center of the egg and its slices.

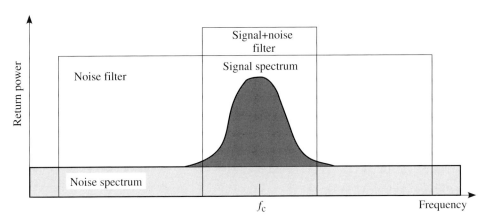

Figure 11.14. The SeaWinds filters used for detection of signal and noise. The figure shows the signal and noise spectra, and the filters used to discriminate between the two (Adapted from Figure 6 of Spencer *et al.*, 1997).

11.6.1 Calibration and removal of noise

At every half rotation, SeaWinds generates an internal calibration pulse as a check on the scatterometer gain. For noise removal and unlike the other scatterometers, SeaWinds measures Φ_{TN} and Φ_R simultaneously (Spencer *et al.*, 2000). From Figure 11.14 this measurement is done as follows. The return signal has a center frequency f_c that is adjusted for the azimuth-dependent Doppler shift. Around f_c, Φ_R has a symmetric peaked spectrum that lies within an 80 kHz bandwidth; in contrast, Φ_{TN} has a flat, very broad spectrum that overlaps Φ_R. To recover the signal and noise, for each pulse, the return is passed through two filters centered on f_c, one with a broad 1-MHz bandwidth, the other with an 80-kHz bandwidth. The broad filter primarily measures noise, the narrow filter measures signal and noise. To a good approximation, subtraction of one from the other yields the corrected signal. The advantage of this simultaneous measurement of signal and noise is that it can account for conditions where the surface or atmospheric properties change rapidly, such as near an ocean front or at the ice edge. The disadvantage is that unlike the AMI and NSCAT, the measurements of Φ_{TN} and Φ_R are not independent. Spencer *et al.* (2000) show that compared to NSCAT, the noise increase associated with this method is small.

11.6.2 Atmospheric transmissivity and rain

As Equation (10.17) shows, the radar backscatter from each wind cell must be corrected for the atmospheric transmissivity t. Because QuikSCAT does not carry a passive microwave radiometer, t is provided from a global monthly mean SSM/I-derived climatology that is interpolated in time and space to the surface wind cell, then calculated for the scatterometer look angle (Perry, 2001). For SeaWinds on short-lived ADEOS-2, the AMSR radiometer provided co-located observations of columnar water vapor, liquid water and rain rate. When rain is present, the σ_0 measurements experience two problems. First, the decrease in transmissivity with increasing rain rate shown in Figure 9.5 means that the return is greatly attenuated. Second, raindrops roughen the sea surface, affect the σ_0 measurement and alter the wind solution. Because at the shorter wavelengths the scatterometer is more sensitive to small scale surface roughness and the atmospheric attenuation has a greater variability, both problems are more serious at K_u-band than at C-band. For SeaWinds on ADEOS-2, AMSR served as a rain detector; for SeaWinds on QuikSCAT, even though the scatterometer antenna is a less than optimal radiometer, its observations of the attenuation caused by heavy rain are used to set a rain flag (Perry, 2001).

11.7 Relative advantages and disadvantages of AMI, NSCAT and SeaWinds

The greatest disadvantage of the fan-beam scatterometers is the nadir gap in the wind coverage. In contrast, SeaWinds provides a broader swath with no nadir gap, yielding a greatly improved daily coverage. For the ERS-2 AMI, NSCAT and QuikSCAT and from data respectively taken in 1997, 1998 and 2000, Figure 11.15 shows the daily global swath coverage and the improvement in this coverage since the ERS-1 launch in 1991. The AMI has a daily coverage of 41% of the world ocean and a 50-km resolution; NSCAT has a 77% coverage and a 25-km resolution; QuikSCAT has a near global 93% coverage and a 25-km resolution that with some loss in accuracy can be processed to 12.5 km.

Other SeaWinds advantages are as follows. First, because the 2–3 m long large-aspect-ratio rectangular or stick antennas required by the fan-beam scatterometers require unobstructed fields-of-view from the spacecraft, they cannot be accommodated on all vehicles. Also, these antennas must be designed to fold into a compact package to fit into the launch vehicle, so that their deployment in space is more difficult and subject to problems. In contrast, the SeaWinds dish antenna is easier to accommodate and deploy. Second, the SeaWinds geophysical model function need only be known at the two discrete incidence angles, rather than at the broad range of incidence angles associated with the NSCAT and AMI scatterometers. Third, because all of the SeaWinds energy is incident at a specific angle onto a small surface footprint, the backscatter also avoids the fourth-power decrease with range that occurs across the fan-beam swath.

The SeaWinds disadvantages are first, because the antenna constantly rotates, there is less integration time available for averaging of adjacent measurements and noise reduction. In contrast, the entire fan-beam swath is illuminated with each pulse, so that the available averaging time is greater. Second, the fan-beam observations of a particular FOV are always made at the same azimuth angles, while the SeaWinds observations occur at a variety of azimuth angles. Because the wind algorithms work best for azimuth angles separated by 90° and have greater errors for azimuth angles separated by 180°, unlike the fan-beam instruments, the SeaWinds observational accuracy varies across the swath.

Figure 11.16 gives a schematic and tabular comparison of the different antenna and swath configurations of the five wind scatterometers. The third row shows the surface patterns of the different instrument antennas; the seventh row compares the instrument swaths, where

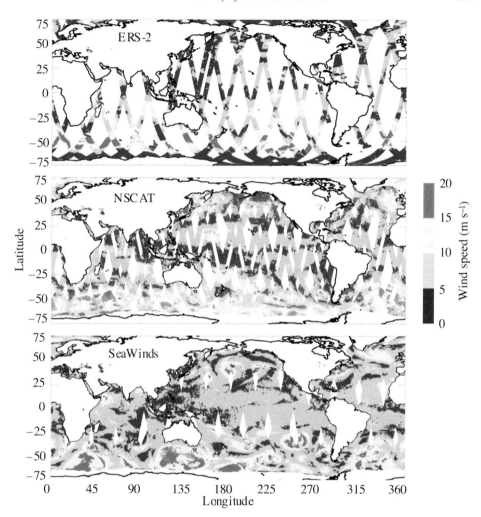

Figure 11.15. Comparison of the daily global coverage of the ERS-2 (July 8, 1997), NSCAT (May 6, 1998) and QuikSCAT (May 25, 2000) scatterometers over the ocean (The images were obtained from the NASA/NOAA sponsored data system Seaflux at JPL through the courtesy of W. Timothy Liu and Wenqing Tang, used with permission).

the variable width of the SASS swath is caused by its fixed Doppler filters. The figure shows that NSCAT, SeaWinds and the planned ASCAT provide the best coverage and that only SeaWinds lacks the nadir gap.

11.8 Accuracy of the wind retrievals

From their comparison of scatterometer and buoy winds, Freilich and Dunbar (1999) derive the accuracy in speed and direction of the scatterometer winds. Their analysis is

Instrument	SASS	AMI (ERS-1, 2)	NSCAT	SeaWinds	ASCAT
Frequency	14.6 GHz	5.3 GHz	13.995 GHz	13.402 GHz	5.3 GHz
Scan pattern		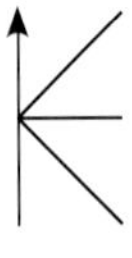	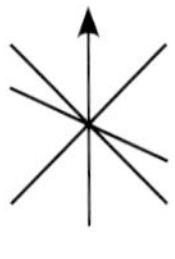	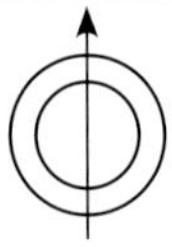	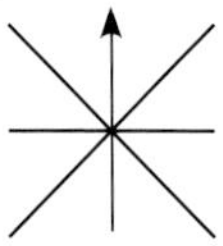
Incidence angle	22°–55°	20°–50°	20°–50°	47°, 55°	20°–50°
Beam resolution	Fixed Doppler	Range binning	Variable Doppler	Scanning pencil	Range binning
Resolution	50 km	50 km	25 km	12.5, 25 km	25 km
Swath	500 km 500 km 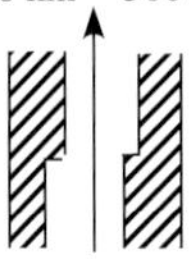	500 km 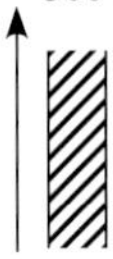	600 km 600 km 	1800 km 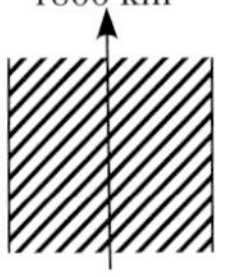	600 km 600 km 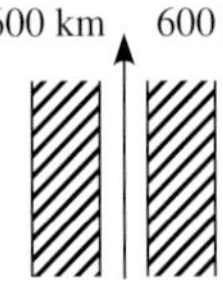
Daily coverage	Variable	41%	77%	93%	80% (estimated)
Operation dates	1978	1991–2001	1996–97	1999–, 2002–03	2005 (estimated)

Figure 11.16. Comparison of the properties of the five scatterometers (Adapted from Figure 3 from Atlas *et al.*, 2001, © 2001 American Meteorological Society, used with permission).

based on about 56 000 25-km resolution NSCAT wind measurements that are located within 50 km and 30 minutes of buoy anemometer wind measurements made at 30 US coastal NDBC buoys. Because the scatterometer winds represent neutral stability 10-m winds, the measured buoy winds were adjusted for anemometer height and atmospheric stratification.

In their scalar wind speed comparison, each buoy/scatterometer wind pair is placed into 0.5 m s^{-1} wide consecutive NDBC velocity bins, then averaged. Figure 11.17 compares the co-located averages. The triangles are averages of more than 100 NSCAT measurements; the stars are averages of 5–99 measurements. The light solid line is the 45° line of perfect agreement; the heavier offset solid line is the best fit to the triangles. The figure shows that for wind speeds less than 15 m s^{-1}, the comparison produces many co-located data points, but fewer at greater wind speeds. For wind speeds of 1–18 m s^{-1}, with the 3-standard deviation outliers removed from the data, the NSCAT winds have an rms accuracy of about $\pm 1.2 \text{ m s}^{-1}$ and a low bias of about 0.3 m s^{-1}. The largest deviations occur for buoy wind speeds less than 3 m s^{-1}. At these wind speeds, because the backscatter is determined from very small amplitude waves, the scatterometer wind errors are large, and because of anemometer errors and the neutral stability correction, the buoy wind errors are also large. For velocities greater than 15 m s^{-1}, problems occur with the paucity of data, heaving of the buoys and the presence of spray and foam.

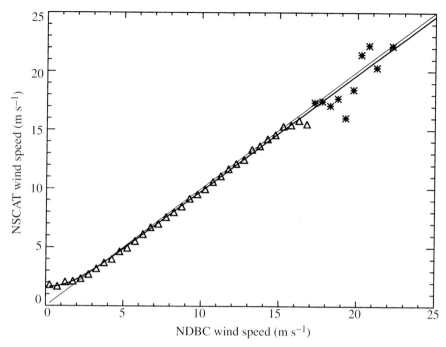

Figure 11.17. Comparison of co-located NDBC buoy and NSCAT wind magnitudes. See text for further description (Figure 5 from Freilich and Dunbar, 1999, © 1999 American Geophysical Union, reproduced/modified by permission of AGU).

In the directional comparison, Freilich and Dunbar (1999) sort the winds into 1 m s^{-1} wide consecutive NDBC wind speed bins, then additionally into 10° wide NDBC relative direction bins. For four different velocities, Figure 11.18 plots the frequency of occurrence of relative direction where the speeds listed on the plot are the center velocity of the bin. For each velocity, the figure shows that the maximum in relative direction occurs close to 0° and the directional distribution is approximately symmetric. Freilich and Dunbar determine the directional accuracy in two ways. The first estimates the number of gross ambiguities, defined as those directional differences greater than 60°. The second is the mean and rms error of those differences that are closer to 0°. The figure shows that in all but the 10.5 m s^{-1} curve, there are small peaks at ±180°, corresponding to ambiguity errors. The number of ambiguous solutions decreases with increasing wind speed; for wind speeds greater than 6 m s^{-1}, the ambiguity error is less than 3%. The figure also shows that the smallest 3.5 m s^{-1} curve has a broader directional distribution than the higher velocities. Ignoring the ambiguities and for wind speeds of about 2–17 m s^{-1}, Freilich and Dunbar find that the mean difference in wind direction is about 8° and the rms directional error is about ±18°. Smaller wind speeds have larger ambiguities and larger standard deviations. From an unpublished comparison, the SeaWinds retrievals have similar errors (Dudley Chelton, private communication, 2003).

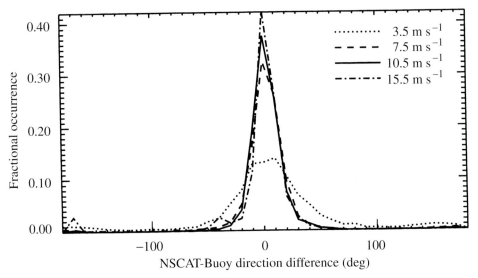

Figure 11.18. The distribution of the difference in wind direction between the NSCAT and buoy winds for four different wind speeds, as binned into $10°$ bins, as in $-5°$ to $+5°$, $5°$ to $15°$, and so forth. See text for further description (Figure 6a from Freilich and Dunbar, 1999, © 1999 American Geophysical Union, reproduced/modified by permission of AGU).

11.9 Applications and examples

This section gives three examples of scatterometer wind retrievals, then discusses a Sea-Winds backscatter image of Antarctic ice. The wind examples include a North Pacific weather front, a large scale example of daily winds from the Atlantic and Pacific, and a strong wind event off the Pacific coast of Central America.

11.9.1 A weather front in SeaWinds swath data

Plate 13 shows a single swath of SeaWinds wind vectors across a front in the North Pacific taken on September 2, 1999 at 1530 UTC. The swath is 1800-km wide and is centered just south of the Alaskan Peninsula. The black vectors are rain-free, the red vectors are rain-contaminated. The arrows outside of the figure at the top right and middle left mark the discontinuity associated with the SeaWinds transition from four to two looks. The line of discontinuity in wind direction running from the bottom left to the upper right is a weather front; the adjacent red vectors show a frontal rain band. At the lower right, the anomalously large wind vectors may be the result of improper rain-flagging. The plate shows that the large number of wind vectors retrieved within the swath provide a powerful tool for case studies of fronts and other wind and storm systems. A comparable collection of wind measurements would be impossible to obtain from ship observations.

11.9.2 Hemispheric winds

For the Atlantic and Pacific Oceans, Plate 14 shows the QuikSCAT winds for April 19, 2000. The image is from the same day as the cloud observations in Plate 2 and the passive microwave wind retrievals in Plate 10. On the plate, the lines and arrows show wind direction; the colors show wind speed, where yellow and green represent wind speeds greater than 13.5 m s^{-1}. In the North Atlantic, the strong winds southeast of Greenland are associated with a storm approaching the British Isles; in the North Pacific, a similar cyclonic storm occurs just south of the Aleutian Islands. In the South Pacific, strong winds occur south of Australia and east of New Zealand. In the South Atlantic, strong winds also occur between South America and Africa. Comparison with the passive microwave winds in Plate 10 shows that strong winds occur at the same locations, indicating that the passive microwave and QuikSCAT give similar results.

11.9.3 Gulf of Tehuantepec

The Sierra Madre mountains located along the Pacific coast of Mexico and Central America play an important role in the regional meteorology. For the winter cold fronts that propagate south across the Gulf of Mexico from North America, the mountains act as a barrier between the Gulf and the Pacific. Only two gaps in these mountains permit the flow of cold dense air into the Pacific: Chivela Pass in southern Mexico and the low lying terrain around the central Nicaraguan lake district. Within these gaps, the air flow can be strongly accelerated. Bourassa *et al.* (1999) report observations of wind velocities in Chivela Pass as large as 60 m s^{-1}. For December 1, 1999 at 00 UTC, Plate 15 shows a QuikSCAT-derived example of these winds into the Gulf of Tehuantepec, where such a wind event is called a *Tehuanos*. On the plate, land topography is shown in color according to the scale below; over the ocean, the shades of blue and the white arrows show the wind magnitude and direction, where darker shades of blue indicate stronger winds. In the Atlantic, the image shows the flow acceleration approaching the pass, and in the Pacific, the concentration of large velocities adjacent to the coast. The image is derived from 25-km gridded QuikSCAT winds following the scheme described in Pegion *et al.* (2000). In the Gulf of Tehuantepec, these winds generate the cold coastal upwelling that Plate 9 shows at a small scale.

11.9.4 Polar ice studies

As an example of SeaWinds observations of land and sea ice, Figure 11.19 shows for July 19, 1999, a 24-hour average of the SeaWinds σ_0 measurements of the Antarctic continent and the surrounding sea ice. To remove the wind signal, the open water surrounding the pack ice has been masked. Because of the strong radar returns from continental snow and ice, the Antarctic continent has a large range of backscatter with several very bright regions. In contrast, the pack ice surrounding the continent is darker with a smaller dynamic range. Within the pack ice and in Drake Passage, the bright objects are icebergs, where the source

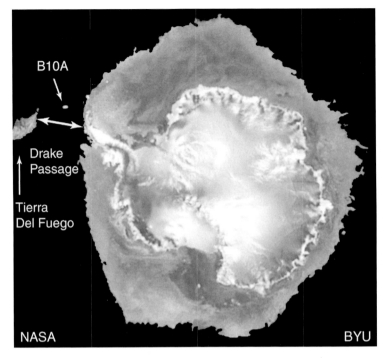

Figure 11.19. Backscatter properties of the Antarctic continent and the surrounding sea ice derived from a 24-hour average of SeaWinds σ_0 measurements on July 19, 1999 (Courtesy David Long, Brigham Young University (BYU), used with permission).

of their brightness is corner reflection from their vertical walls. The bright rectangular object just outside of the passage at the left of the image is a 21 × 42 km large iceberg called B10A. The B10A iceberg broke off the end of the West Antarctic Thwaites ice shelf in about 1992 and since that time has drifted around the continent. As Long and Drinkwater (1999) describe, daily scatterometer images of the pack ice show its extent, circulation patterns and response to winds, without the concerns about water vapor at the ice edge that occur in the passive microwave sea ice algorithms. The imagery is also used by the National Ice Center (http://www.natice.noaa.gov) to track large icebergs and to post their positions as hazards to navigation. (Figure and descriptive material courtesy David Long.)

11.10 Further reading and acknowledgements

O'Brien (1999) is a special issue of *Journal of Geophysical Research* on NSCAT and scatterometer wind measurements. The NASA repository for NSCAT and QuikSCAT data is at http://podaac.jpl.nasa.gov/quikscat/. At this site, files may be browsed and individual data days can be ordered via ftp and in some cases ordered on CD-ROM. Perry (2001) can be downloaded from the same website at http://podaac.jpl.nasa.gov/quikscat/qscat_doc.html/.

Also at JPL, many images and movies of winds, hurricanes and sea ice are at http://winds.jpl.nasa.gov/scatdata/dataindex.html/. The Freilich (2000) document is at http://eospso.gsfc.nasa.gov/eos_homepage/for_scientists/atbd/. The Microwave Earth Remote Sensing laboratory website at Brigham Young University (BYU) directed by David Long (http://www.ee.byu.edu/ee/mers/) contains publications and information about scatterometry and its applications to sea ice. The site also contains movies of scatterometer winds and of pack ice and iceberg motion. The author acknowledges the help of Mark Bourassa, Robert Brown, Josh Grant, Michael Freilich, Timothy Liu, Ellen Lettvin, David Long, James J. O'Brien and William Plant in the preparation of this chapter.

12

The altimeter

12.1 Introduction

The radar altimeter transmits short pulses of energy vertically downward toward the ocean surface, then receives the reflected signal. The return yields information on the global distribution and variability of sea surface height, ocean swell amplitude and scalar wind speed. Specifically, the time difference between the transmitted and received signal gives the distance or range between the satellite and the sea surface, the shape of the return yields the significant wave height and the magnitude yields the scalar wind speed. If the satellite orbit is precisely determined and the range is corrected for a variety of atmospheric, ocean surface and solid Earth factors, then these measurements allow determination of changes in sea surface height (SSH) due to tides, geostrophic currents and other oceanic phenomena to an accuracy of 2–3 cm.

This chapter describes how the altimeter works, discusses its sources of uncertainty and describes some of its oceanographic applications. Wunsch and Stammer (1998) and the collection of papers edited by Fu and Cazenave (2001) contain more detailed and extended discussions of the TOPEX altimeter results, and within this collection Chelton *et al.* (2001b) describe the physics of the TOPEX altimeter and its associated error budget. Wunsch and Stammer (1998) show that the altimeter provides near global coverage of the temporal and spatial scales of ocean variability, the meridional transports of heat and the distribution and properties of Rossby waves. Altimeters contribute in at least two additional ways. First, because the sea surface responds to the ocean bottom, Smith and Sandwell (1997) show how altimeter measurements have contributed to an improved seafloor topography. Second, because the altimeter serves as a precision tide gauge, Wunsch and Stammer (1998) show that it has also enhanced our knowledge of deep ocean tides and their dissipation.

One difficulty with altimeter measurements is that sea surface height must be measured relative to the geoid, which is defined as the shape of the ocean surface in the absence of all external forcing and internal motion. Currently, except for a few specific regions, the geoid is only known at scales greater than about 500 km (Wunsch and Stammer, 1998). In the future, the results of recent gravity missions such as the German CHAllenging Minisatellite Payload (CHAMP), the US/German Gravity Recovery and Climate Experiment (GRACE) and the forthcoming European Gravity Field and Steady-State Ocean Circulation Explorer (GOCE) will determine the geoid at shorter length scales. At scales where a reference geoid is lacking, studies of sea surface height are presently done in terms of the variability of the observed SSH around its mean.

For potential users, archives such as the Physical Oceanography Active Archive Center (PODAAC) at JPL (http://podaac.jpl.nasa.gov) contain the data in a variety of browse images and downloadable files that are corrected for the biases and uncertainties discussed in this chapter. This permits a user to work with the data without understanding, for example, how to correct for the inverse barometer effect. In spite of this data availability, the descriptions of the trade-off between orbit period and equatorial spacing, the factors that determine the surface footprint size and resolution, and the contributions of the ionosphere, atmosphere and solid Earth to the error budget should help the reader understand both the potential of the altimeter and its limitations.

In the following, Section 12.2 describes the satellite orbit, the shape of the Earth and the geoid, and defines the variables used in the SSH retrieval. Section 12.3 summarizes the historical ocean altimeter missions; Sections 12.4 and 12.5 describe the TOPEX/POSEIDON altimeter mission and its successor, JASON-1. Sections 12.6 and 12.7 describe how the altimeter works, discuss the interaction of the altimeter pulse with the sea surface and show how ocean swell and sea surface roughness alter the reflected pulse. Section 12.8 discusses the altimeter error budget, Section 12.9 gives examples and Section 12.10 gives further references and data sources.

12.2 Shape of the Earth

Figure 12.1 shows the variables used to describe the shape of the Earth and the sea surface. Along the radial line between the satellite and the Earth's center of mass, the altimeter measures the height or range $h = h(\chi, \psi, t)$ of the satellite above the sea surface, where χ is latitude and ψ is longitude. The other radial variable is $H(\chi, \psi, t)$, which is the height of the satellite above the Earth's center of mass. As Section 12.4.4 discusses under the topic of *precision orbit determination* (POD), H is precisely measured through a variety of means. The difference between H and h is the oceanographic variable of interest $h_S(\chi, \psi, t)$, which is the height of the sea surface above the Earth's center of mass, where

$$h_S = H - h \qquad (12.1)$$

The goal of the altimeter is to determine h_S to within the 2–3 cm accuracy necessary to resolve geostrophic flows. Equation (12.1) shows that this determination depends on the accurate measurement of two quantities, the satellite height H above the Earth's center of mass and its height h above the sea surface. These measurements are equally important and, as this chapter shows, are made in very different ways.

The sea surface height h_S is described in terms of three successive approximations (Wunsch and Stammer, 1998; Chelton *et al.*, 2001b). The first is the *reference ellipsoid*

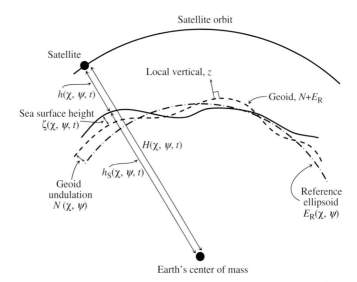

Figure 12.1. The altimeter geometry showing the satellite orbit, the reference ellipsoid, geoid undulation and height of the sea surface above the Earth's center of mass, where χ is latitude and ψ is longitude. See text for further description (Adapted from Figure 3, Wunsch and Stammer, 1998).

$E_R(\chi, \psi)$, which is the shape of the time-independent uniform distribution of the Earth's mass generated by gravitational and centrifugal forces. The short axis of the ellipsoid runs through the poles, the long axis runs through the equator and the ellipsoid is symmetric about the polar axis. The length of the equatorial axis is such that at the equator the ellipsoid surface corresponds to mean sea level. The TOPEX ellipsoid has a polar radius of about 6359 km, an equatorial radius of 6380 km and accounts for about 90% of the geoid.

Problems with the ellipsoid occur because of the uneven distribution of the Earth's mass. At a range of horizontal scales of 10 to 1000 km, lateral gravity forces determine the surface topography, so that a region of excess mass at the sea bottom such as a continental ridge attracts water to produce a topographic rise, while mass deficits generate valleys (Figure 12.2). The sea surface produced by this uneven mass distribution is the equipotential surface corresponding to the sea level in the absence of external forces such as winds and tides. This surface is defined relative to the ellipsoid and is called the *geoid undulation N*(χ, ψ), where the corresponding *geoid* is the sum $N + E_R$ (Wunsch and Stammer, 1998). The common practice that this book will follow is to refer to N as the geoid.

The geoid is derived from expansions of spherical harmonics fitted to the altimeter data, where the ability of the geoid to resolve spatial features depends on the number of harmonics. Relative to the ellipsoid, N has an amplitude of about ±100 m. Figure 12.3 shows the marine geoid derived from the University of Texas UTGF26 model with 26 spherical harmonics and a spatial resolution of about 1500 km (Bindschadler *et al.*, 1987, Figure 6a). The geoid has a topographic low south of India and a high north of New Guinea. A more recent geoid

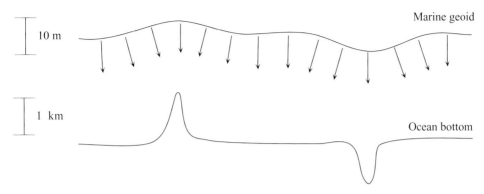

Figure 12.2. The effect of a rise and depression in the seafloor topography on the marine geoid. The horizontal scale is of order 10 to 1000 km. The arrows show the local gravitational accelerations, which are normal to the geoid (Adapted from Figure 10 of Chelton, 1998).

is the Earth Geopotential Model 96 (EGM96) with 360 spherical harmonics and a spatial resolution of about 500 km (Wunsch and Stammer, 1998). At the scale of Figure 12.3, EGM96 and UTGF26 are identical.

The geoid in Figure 12.3 is heavily smoothed. For an example of the geoid variability at shorter length scales, Figure 12.4 shows an example of the geoid response to changes in the seafloor topography over distances of 10–1000 km. The upper part of the figure gives the altimeter response in meters; the lower part gives the bottom topography in kilometers. The figure shows that the subsurface ridges associated with the Line Islands and the Hawaiian Ridge generate a sea surface response of 1 to 5 m. The Murray Fracture Zone also generates a topographic depression, while the seamounts do not have a noticeable effect on the surface, probably because of their small geographic extent. Such responses have led to the development of bottom topographic charts derived from the combination of depth soundings from ships and altimetric measurements (Smith and Sandwell, 1997).

Relative to the geoid, the third surface is the sea surface height $\zeta(\chi, \psi, t)$, defined as

$$\zeta(\chi, \psi, t) = h_S(\chi, \psi, t) - N(\chi, \psi) - E_R(\chi, \psi) \qquad (12.2)$$

The height ζ describes the sea level variability relative to the geoid induced by a wide variety of atmospheric and oceanographic phenomena. These include geostrophic flows, tides, atmospheric pressure changes, and seasonal heating and cooling. The geoid describes the ocean at rest; ζ describes the nonequilibrium surface of a dynamic ocean.

The sea surface height has both steady and variable components. The steady components include features such as the mean flows of the Gulf Stream and Kuroshio; the variable components include tides, fluctuations associated with the weight of the atmosphere, the surface response to seasonal oceanic heating and cooling, planetary waves, and variable currents and eddies. As Table 2.1 shows, the presence of long period planetary waves, eddies and currents means that relative to the geoid, ζ has a variability of about 1 m. The

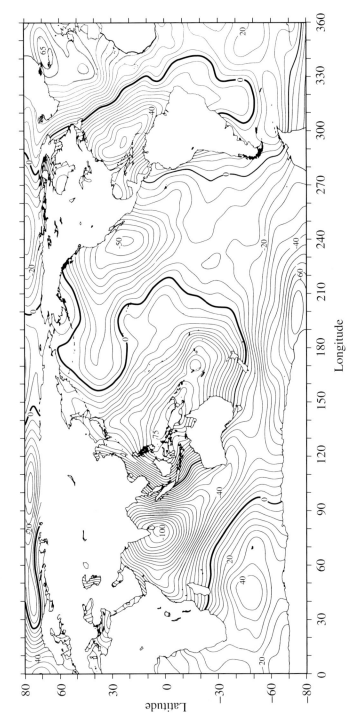

Figure 12.3. The undulations of the oceanic geoid at contour intervals of 5 m, where the 0-m contour is darkened. The geoid is the University of Texas UTG26 model; the rms accuracy is about 1 m (Figure 6a from Bindschadler *et al.*, 1987; not subject to US copyright).

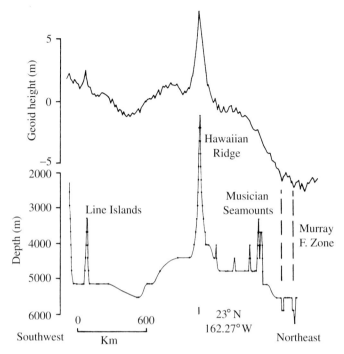

Figure 12.4. Geoid height from the GEOS-3 altimeter and the corresponding bottom profile along the sub-satellite track. The long-wavelength regional geoid is subtracted from this profile to emphasize bottom topography (Figure 3 from Watts, 1979, © 1979 American Geophysical Union, reproduced/modified by permission of AGU; discussion from Wunsch *et al.*, 1981).

purpose of the altimeter is to measure ζ and determine the sea surface response to a variety of geophysical forcing.

For measurements of ζ to be of oceanographic value, both H and h must be determined to an accuracy of 2–3 cm. As the following sections show, the satellite orbital position H is determined to a centimeter accuracy from a combination of laser and radio ranging and Global Positioning System (GPS) measurements; the range h is determined from the pulse travel time. Many factors contribute to the uncertainty in h. For example, variable columnar concentrations of V and ionospheric free electrons alter the electromagnetic phase velocity and change the apparent range between the satellite and the surface. Variations in V produce apparent changes in sea level of as much as 30 cm, and the diurnal and annual cycles of the ionospheric free electrons create changes of order 1 m. Rain and variations in the mass of atmosphere along the altimeter path also contribute to the range uncertainty. Finally, the ocean is not a quiescent specular reflector, but is covered by waves ranging from capillaries to large amplitude ocean swell, where all these scales affect the range retrieval and error budget.

12.3 Past, present and future altimetric satellites

The past, present and proposed altimeter missions include an experimental altimeter on
Skylab in 1973, the single frequency altimeters launched on GEOS-3 in 1975, SEASAT in
1978, GEOSAT in 1985–1990, and the ERS-1 and ERS-2 altimeters that operated from 1991
to 2000. The more accurate dual-frequency altimeters began with TOPEX/POSEIDON in
1992, continued with JASON-1 in 2001, ENVISAT in 2002 and, as Chapter 14 describes,
will continue in 2005 with the Ocean Surface Topography Mission (OSTM) replacement for
JASON-1. TOPEX and JASON-1 occupy orbits that are specifically designed for topography and are not sun-synchronous, while the ENVISAT Radar Altimeter-2 (RA-2) occupies
a less desirable, sun-synchronous orbit. Because of concerns about atmospheric drag and
the need for precise orbit determination, the ideal altimeter satellite would have the size and
shape of a cannon ball. Consequently, the most successful altimeter missions are compact
low-drag satellites that only carry an altimeter with supporting measurements of atmospheric water vapor, liquid water and ionospheric free electrons. As the next two sections
discuss, this condition is nearly satisfied by the two current ocean topography missions,
TOPEX/POSEIDON and JASON-1.

12.4 TOPEX/POSEIDON

This section describes the TOPEX/POSEIDON altimeter, and its supporting instruments
and surface measurements. Specifically, Section 12.4.1 describes the satellite and its choice
of orbit, and Section 12.4.2 describes the TOPEX Microwave Radiometer. Section 12.4.3
describes the determination of the free electron concentration, and Section 12.4.4 describes
three techniques in the precise determination of the satellite orbit. Section 12.4.5 concludes
with a description of the surface calibration sites.

12.4.1 The satellite and its orbit

Figure 12.5 shows the TOPEX/POSEIDON satellite, which is a joint project between NASA
and the French space agency *Centre National d'Études Spatiales* (CNES). TOPEX is an
acronym for *TOPography EXperiment*; POSEIDON is a dual French and English acronym:
Premier Observatoire Spatial Étude Intensive Dynamique Océan et Nivosphère, and *Positioning Ocean Solid Earth Ice Dynamics Orbiting Navigator*. The cumbersome name reflects
the difficulties of international collaboration (Wunsch and Stammer, 1998). TOPEX was
launched on August 10, 1992 into an exact repeat orbit (Section 1.3), and began taking data
in September 1992.

A number of considerations determined the TOPEX choice of orbit. First, for a single
satellite mission, the temporal and spatial resolution compete with one another. The temporal
resolution is determined by how long it takes for the satellite to repeat a particular orbit; the
spatial resolution by the equatorial spacing between successive orbits. A short repeat period
yields a large spatial separation; a long repeat period yields a small separation. Second,

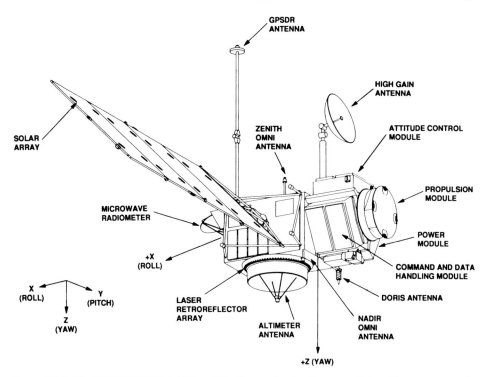

Figure 12.5. The TOPEX/POSEIDON satellite. For scale, the diameter of the altimeter antenna is 1.5 m (Figure 1 from Fu *et al.*, 1994, © 1994 American Geophysical Union, reproduced/modified by permission of AGU).

TOPEX is not in a sun-synchronous orbit, rather it is at a higher altitude. This choice of altitude reduces atmospheric drag; its disadvantage is that because of the fourth power dependence on range of the transmitted-to-received power, the altimeter must supply more power to achieve a satisfactory signal-to-noise ratio. The choice of a non sun-synchronous orbit was to avoid aliasing the 24-hour or diurnal tides, which generate a spurious mean displacement (Wunsch and Stammer, 1998). Third, the orbit was designed so that in the subtropics, the ascending and descending paths cross each other at nearly right angles. At the crossover points, this choice of crossing angle permits accurate retrievals of the two geostrophic velocity components. Fourth, the TOPEX exact repeat orbit repeatedly samples the same locations at about 10-day intervals.

The result of these considerations is that TOPEX is in a circular orbit at an altitude of 1336 km, with a period of 112 minutes and an inclination providing surface coverage between ±66° of latitude. The satellite makes approximately 14 orbits per day, for a ground speed of about 6 km s^{-1}. The orbits exactly repeat at a period of 9.916 days, referred to as a 10-day repeat cycle. Figure 12.6 shows the TOPEX ground track over a single 10-day cycle and shows that with the exception of the Atlantic north of Iceland, TOPEX covers most

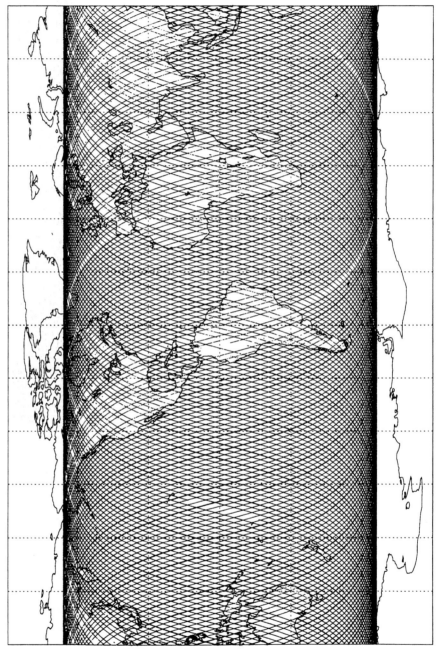

Figure 12.6. The orbit tracks of TOPX/POSEIDON for the 10-day orbit cycle 013. Gaps in the trajectories indicate missing data (Courtesy Robert Benada, used with permission)

of the ice-free ocean. For this orbit, the equatorial separation between adjacent tracks is 320 km, and away from the northern and southern extremes, the ascending and descending tracks cross each other at large angles. The restrictions on temporal sampling imposed by this orbit mean that the highest frequencies observable in the data have a 20-day period. Also, because of the relatively large separation between successive orbit tracks and the 10-day separation between exact repeat tracks, the altimeter only provides information on mesoscale ocean features.

The advantage of the exact repeat orbit is that, over a sufficiently long averaging period and at any point along the track, a time average $\overline{h}_S(\chi, \psi)$ of the sea surface height relative to the Earth's center of mass can be defined as an approximation to the geoid. This average includes both the geoid and the spatial changes in sea surface height associated with steady currents. Relative to this average, the variability in sea surface height $\Delta\zeta(\chi, \psi, t)$ is defined as

$$\Delta\zeta(\chi, \psi, t) = h_S(\chi, \psi, t) - \overline{h}_S(\chi, \psi) \tag{12.3}$$

where h_S is the instantaneous height measurement from Equation (12.1). Over many 10-day cycles and at spatial scales greater than the track separation, Equation (12.3) allows derivation of the variable flow properties.

TOPEX contains two separate altimeters that share a single 1.5-m diameter parabolic antenna. These are the NASA dual-frequency altimeter (ALT or TOPEX) operating at C-band (5.3 GHz) and Ku-band (13.6 GHz) with wavelengths of 6 and 2 cm, and the CNES experimental low weight, low power, single-frequency solid state altimeter (SSALT or POSEIDON) operating at 13.65 GHz. The advantage of a dual-frequency altimeter is that at each frequency the altimeter responds differently to variations of ionospheric free electrons and to rain. Because of this response, these altimeters can measure the columnar electron density and determine its effect on the electromagnetic phase speed, as well as identify regions of heavy rain. At C-band, the ALT has a half-power beamwidth of 2.7°, a gain of 35.9 dB and a PRF of 1220 Hz; at Ku-band, it has a beamwidth of 1.1°, a gain of 43.9 dB and a PRF of 4200 Hz (Zieger *et al.*, 1991). In both cases, the instrument uses the chirp technique described in Section 10.4.1 to generate a pulse with bandwidth $\Delta f_B = 320$ MHz, so that the pulses have an effective duration of 3.125 ns or a length of 1 m. The TOPEX and POSEIDON altimeters alternate in their observations, such that over eleven cycles TOPEX operates for ten cycles, POSEIDON for one. As Section 12.5 shows, the successful operation of SSALT on TOPEX is the reason that a dual-frequency POSEIDON altimeter is the only altimeter on JASON-1.

TOPEX carries four additional instruments, the TOPEX Microwave Radiometer (TMR) used for determination of L, V and U, and three instruments used for POD, where one of the POD instruments also measures free-electron density. For both TOPEX and JASON-1, the three POD instruments and ground systems are the NASA Laser Retroreflector Array (LRA) on the satellite and its associated laser ground stations, the CNES Doppler tracking system (Doppler Orbitography and Radiopositioning Integrated by Satellite or DORIS), and a NASA GPS receiver. The following sections discuss the TMR, the ionospheric correction, the POD systems and the surface calibration sites.

12.4.2 TOPEX Microwave Radiometer (TMR)

The purpose of the TMR is to measure V, L and to identify or flag regions with a heavy rain rate R_R. As Chelton *et al.* (2001b) show, variations in the tropospheric concentrations of V and L alter the real part of the index of refraction and thus the electromagnetic phase speed. These changes are sometimes grouped under the name *refraction*; because they reduce the phase speed, they create an apparent additional distance to the sea surface called *range delay*. The range delays induced by these changes are of order 1 m; for example, a change in V from 0 to 70 mm of columnar water equivalent yields a range delay of 50 cm (Wunsch *et al.*, 1981). For cloud liquid water, the effect of L on range is 1–2 orders smaller than V and is generally ignored. Because the rain rate R_R strongly affects the transmissivity, the electromagnetic phase speed and the scattering surface, pixels containing heavy rain must be masked.

TMR is a nadir-viewing radiometer that was built almost entirely from SMMR spare parts and operates at the 18, 21, and 37 GHz SMMR frequencies described in Section 8.6.1 (Janssen, Ruf and Keihm, 1995). Because TMR is nadir-viewing, the vertical and horizontal polarizations are identical, so that TMR has only three channels. These are used to solve for V, L and U in a manner similar to the SSM/I algorithm described in Section 9.6.1, where SST is provided from a lookup table. The retrieved V and L are then used to calculate the transmissivity, phase speed and range delay along the two-way altimeter path. TOPEX sets a rain flag in two ways. First, if L exceeds a preset threshold, rain is assumed and the pixel is masked. Second, because the rain-induced attenuation is an order greater at K_u- than at C-band, a rain flag is also set if the difference in attenuation between the two frequencies exceeds a threshold (Chelton *et al.*, 2001b). The TMR footprint is also spatially constrained. At 21 GHz, TMR has a 35-km diameter footprint, but because of the sidelobe interference described in Section 8.5, its observations cannot be used within 50 km of land (Ruf and Giampaolo, 1998).

12.4.3 Ionospheric free electrons

As Section 4.2.5 describes, the density of ionospheric free electrons has a strong diurnal and interannual cycle, where their density retards the electromagnetic phase velocity at a rate proportional to their columnar concentration (Chelton *et al.*, 2001b, Section 3.1.3). Specifically, if Δh_{ion} is the range delay in cm and ρ_{ion} is the electron columnar density in TECU, the dependence of Δh_{ion} on f and ρ_{ion} is

$$\Delta h_{ion} \sim \rho_{ion}/f^2 \tag{12.4}$$

Equation (12.4) shows that the range delay decreases with increasing f. If the effects of V and R_R are removed from h, then measurements at the two altimeter frequencies yield two equations for Δh_{ion} and ρ_{ion}, so that Δh_{ion} can be calculated and removed from h. At $f = 5.3$ GHz, $\Delta h_{\mathrm{ion}} = 1.45$ cm/TECU; at $f = 13.6$ GHz, $\Delta h_{\mathrm{ion}} = 0.22$ cm/TECU. Since ρ_{ion} varies between 10 and 120 TECU, at 5.3 GHz, Δh_{ion} varies from about 10 to 160 cm; at 13.6 GHz, Δh_{ion} varies from 2 to 30 cm (Chelton *et al.*, 2001b). As the next section describes, the ionospheric correction can also be determined from the two DORIS frequencies.

12.4.4 Precision orbit determination (POD)

As stated above, the satellite height H is the second critical variable in the determination of SSH. This height is measured by a variety of methods that are grouped under the name of precision orbit determination (POD). POD is defined as the precise determination of the satellite position at regular time intervals and in three dimensions relative to the Earth's center of mass, where the resultant time series is the orbit ephemeris (Chelton *et al.*, 2001b). For the SSH retrieval to be of oceanographic value, H must have an accuracy of 1 to 2 cm. As Tapley *et al.* (1994), Bertiger *et al.* (1994) and Chelton *et al.* (2001b) describe in detail, the POD systems on TOPEX and JASON-1 include laser and DORIS tracking and GPS positioning. The laser and DORIS measurements determine the spacecraft position and velocity at irregular intervals; the GPS measurements continuously determine satellite position. Combination of these measurements with numerical orbit models allows calculation of a precise orbit.

Following Chelton *et al.* (2001b), the POD systems work as follows. First, on TOPEX, the NASA-provided LRA is mounted around the base of the altimeter antenna and is used in combination with a network of manned satellite laser ranging (SLR) stations, where the station locations are known to within a centimeter. For 1994, Figure 12.7 shows the station locations and their visibility masks; at any given time there are as many as 30 stations available for tracking a specific satellite. The laser measurements of spacecraft range are used to determine the three components of spacecraft position. Although the lasers require cloud-free conditions and a human operator, because their optical wavelengths are not affected by water vapor or ionospheric refraction, the measurements have a 1-cm precision. In the future, NASA plans to automate all of the SLR sites, with several such sites operational in 2003.

Second, the DORIS tracking system determines the spacecraft velocity using an onboard receiver in combination with a global network of about 50 ground beacons (Figure 12.8). The DORIS ground beacons broadcast continuously and omnidirectionally at 0.4 and 2.04 GHz. When these signals are received at the spacecraft, its velocity is determined from the observed Doppler shifts at 10-s intervals to a precision of 0.5 mm s^{-1}. These measurements determine the changes in satellite velocity due to radiation pressure and drag. Measurements at the two DORIS frequencies also determine the free-electron concentrations along the DORIS slant paths, which permits their removal from the velocity estimates. For the POSEIDON altimeter, the electron measurements are extrapolated to the nadir path for removal of the ionospheric phase delay, although with less accuracy than the dual-frequency ALT measurements. DORIS is also used by JASON-1 and ENVISAT for orbit determination and

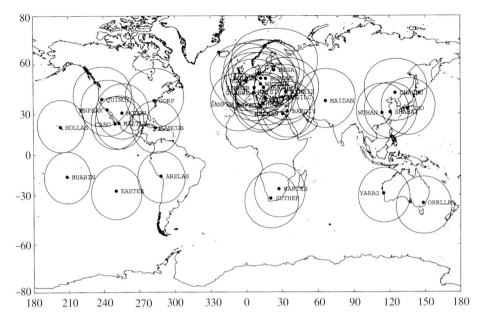

Figure 12.7. The 1994 distribution of the satellite laser ranging (SLR) tracking stations and their 20° elevation visibility masks (Figure 6 from Tapley *et al.*, 1994, ©1994 American Geophysical Union, reproduced/modified by permission of AGU).

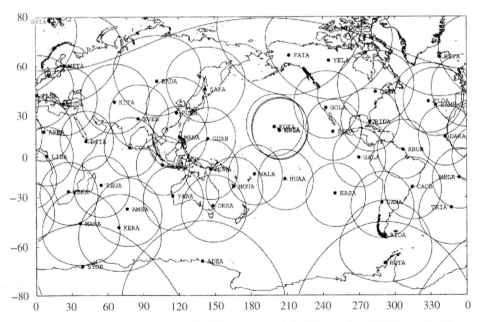

Figure 12.8. The 1994 locations of the DORIS beacons and their 15° elevation visibility masks (Figure 7 from Tapley *et al.*, 1994, ©1994 American Geophysical Union, reproduced/modified by permission of AGU).

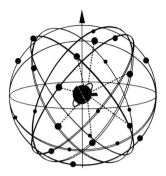

Figure 12.9. The constellation of GPS satellite orbits. The small inner orbit corresponds to TOPEX; the dotted lines show an example of the satellites used in a specific determination of position (Figure 3 from Bertiger *et al.*, 1994, ©1994 American Geophysical Union, reproduced/modified by permission of AGU).

geolocation. Because the SLR and DORIS measurements provide oblique estimates of position and velocity, they must be combined with a numerical orbit model to produce a precise orbit. Chelton *et al.* (2001b) state that DORIS is the primary contributor to the orbit accuracy, while the SLR contribution is to align the orbit center with the Earth's center of mass.

Third, the NASA GPS Demonstration Receiver (GPSDR) on TOPEX continuously tracks the satellite position (Bertiger *et al.*, 1994). The GPS space system consists of 24 satellites orbiting at an altitude of 20 200 km with a 12-hr period, where the satellites are distributed into six orbit planes (Figure 12.9). The satellites broadcast at 1.58 and 1.23 GHz, where the use of two frequencies allows for correction of ionospheric delays. At any time, the GPSDR collects navigation data from between five and nine of these satellites. When the GPS is not encrypted, data from a minimum of four satellites allow TOPEX to determine its position with an rms radial accuracy of ± 2 cm (Fu *et al.*, 1994). The advantage of GPS is that it continuously determines satellite position with potentially better accuracy than SLR and DORIS and without their spatial and temporal gaps. The continuous GPS tracking also means that its orbit solution is less dependent on a numerical model. The disadvantage of the GPSDR was that beginning in 1994 the GPS data were encrypted, which degraded the position accuracy from 2 to about 5 cm. As shown below, however, the JASON GPS can use these encrypted data to obtain the original position accuracy (Chelton *et al.*, 2001b, Section 4.2).

12.4.5 Surface calibration

TOPEX has two surface calibration sites, the US Platform Harvest, which is a Texaco oil platform located in the Pacific, 12 km west of Point Conception on the southern California coast, and the French Lampedusa Island/Lampione Rock site between Sicily and Tunisia (Christensen *et al.*, 1994). Harvest lies directly beneath an ascending orbit path; Lampedusa, beneath a descending path. Figure 12.10 illustrates the measurements made at Harvest;

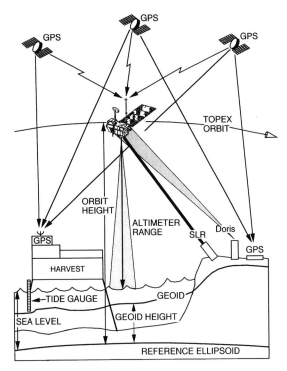

Figure 12.10. Examples of the TOPEX and JASON-1 surface calibration measurements made at Platform Harvest (Figure 1 from Christensen *et al.*, 1994, © 1994 American Geophysical Union, reproduced/modified by permission of AGU).

similar measurements are made at Lampedusa. Harvest contains three separate tide gauges, an upward-looking water vapor radiometer, and instruments for measuring ionospheric free electron density, sea state, and meteorological parameters such as relative humidity and atmospheric pressure. Harvest also has a GPS receiver so that the Harvest and TOPEX sea level measurements are in a common reference frame. When TOPEX passes over Harvest, the satellite position is measured by four SLR stations, TOPEX GPS and DORIS. The combination of the sea level and satellite position measurements means that the satellite height above sea level is determined independently of the altimeter, providing a check on the accuracy and drift of the altimeter range measurement. One problem with Harvest is that it is so close to the coast that sidelobe contamination makes the TMR water vapor measurement unusable; instead, the values of *V* at Harvest are extrapolated from offshore measurements. The use of the Harvest calibration site continues with JASON.

In summary, the TOPEX improvements over earlier altimeters are as follows:

1. The three POD instruments provide a better orbit determination.
2. By measuring the free-electron concentrations and by setting a rain flag, the dual-frequency radar produces a better range determination.

3. The TMR radiometer simultaneously measures atmospheric water vapor, liquid water and the presence of rain.

12.5 JASON-1

The successor to TOPEX is the US/French JASON-1 satellite launched on December 7, 2001 (Figure 12.11). Although JASON-1 is similar in design to TOPEX, because of advances in electronic miniaturization, its mass is only 500 kg compared with 2400 kg for TOPEX. Unlike TOPEX, JASON-1 carries only the POSEIDON-2 altimeter. This is a solid state dual-frequency K_u- and C-band (5.3 and 13.6 GHz) instrument based on SSALT, where the dual frequencies are again used to measure range and correct for ionospheric delays. For POD, JASON-1 uses DORIS and the Laser Retroreflector Array (LRA), which as Figure 12.11 shows is mounted on a truncated cone adjacent to the radar antenna. JASON-1 also determines position from two independent Turbo Rogue Space Receiver (TRSR) BlackJack GPS receivers that, regardless of GPS encryption, determine the satellite position to an accuracy of 2–3 cm (Haines *et al.*, 2002).

   For atmospheric correction, JASON-1 carries the JASON Microwave Radiometer (JMR), which is a three-frequency nadir-looking microwave radiometer similar to TMR. JMR operates at 18.7, 23.8 and 34.0 GHz, which differ slightly from the TMR frequencies. The radiances received at these frequencies are used to solve for V, L and U, and to set a rain flag. The reasons for the frequency changes are as follows. First, the changes from 21.0 to 23.8 GHz and from 37.0 to 34.0 GHz were made to reduce the possibility of interference from the higher harmonics of the 5.3-GHz altimeter. Second, the change from 18 to 18.7 GHz was a political decision made to support the proposed adoption of a remote sensing band at that frequency (L. Fu, private communication, 2001). The algorithms have been modified

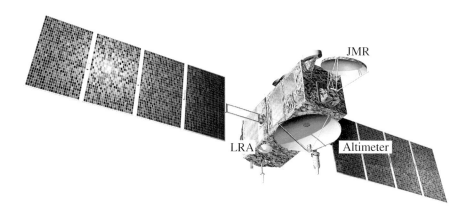

Figure 12.11. The JASON-1 satellite with its large solar panels. JMR is the Jason Microwave Radiometer, Altimeter is the 2-m radar antenna, and LRA identifies the truncated cone containing the Laser Retroreflector Array (Courtesy CNES/NASA).

to accommodate these frequency changes; the expected rms accuracy of the water vapor retrieval remains identical to TMR at 1.2 cm.

JASON-1 has the same orbit and ground track as the original TOPEX orbit, providing for continuity of the original TOPEX observations. Initially, JASON-1 was positioned in the same orbit behind and within 60 s or 500 km of TOPEX. In this common orbit and for a period of about six months after launch, JASON-1 and TOPEX cross-calibrated their instruments by taking near simultaneous measurements of the same sea surface conditions. At the end of this calibration period, TOPEX was moved into a parallel orbit that lies midway between two adjacent JASON-1 orbits. Until TOPEX fails, operation of this altimeter pair doubles the surface resolution. Because the JASON-1 orbit is identical to the original TOPEX orbit, the JASON-1 surface calibration sites are the existing US Harvest site and a different French site at a Mediterranean location called Corsica-Capraia.

12.6 Altimeter pulse interaction with a specular sea surface

This and the following section discuss the precise measurement of the distance h between the satellite and the sea surface. This section considers the case of pulse reflection from a specular surface; Section 12.7 describes the complications that arise for reflection from a wave-covered surface. Within the present section, Section 12.6.1 shows that for small off-nadir look angles, the range retrieval is independent of look angle, Section 12.6.2 derives the beam footprint and Section 12.6.3 describes the retrieval of the round trip travel time.

12.6.1 Effect of variable pointing angle on range retrieval

The altimeter boresight direction unavoidably varies about nadir. For example, Figure 12.12 gives the early time history of the averaged off-nadir look angle of the TOPEX boresight direction and shows that this angle settled to a value of about 0.05°. With this pointing accuracy and from the TOPEX altitude of 1340 km, the projection of the altimeter boresight onto the surface lies within a 1.2-km radius circle centered on nadir. Simple trigonometry shows that the range variation along the boresight direction associated with this uncertainty is 0.5 m, or of the same order as the height variations associated with geostrophic currents. In spite of this variation, because the altimeter generates spherical waves, the following shows that for small off-nadir look angles the measured range is independent of θ.

For nadir- and slant-looking antennas, Figure 12.13 shows schematic diagrams of the radiating wavefronts. In both cases the antennas are at a height h above the sea surface, and have a half-power beamwidth of $\Delta\theta_{1/2}$. Figure 12.13b shows that as long as $\theta < \Delta\theta_{1/2}$ and for a spherical wavefront, the pulse from the tilted antenna has a component propagating in the nadir direction, so that its round trip travel time is the same as for the nadir-look case. This independence of the range measurement from small off-nadir look angles is a major reason for the success of the altimeter.

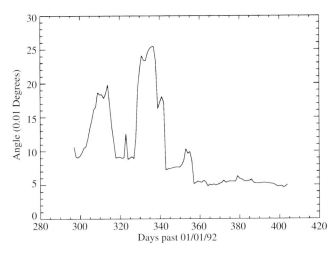

Figure 12.12. Time history of the daily averaged off-nadir pointing angle for cycles 4–14 (Figure 4 from Fu *et al.*, 1994, © 1994 American Geophysical Union, reproduced/modified by permission of AGU).

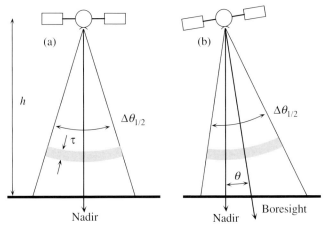

Figure 12.13. The propagation of a spherical wavefront for cases of (a) a nadir-looking and (b) a slant-looking antenna both with a greatly exaggerated beamwidth.

12.6.2 Pulse-limited footprint

Because the altimeter generates short pulses, the resultant footprint is smaller than the beamwidth-limited footprint described in Section 8.2.1. This smaller FOV is called the *pulse-limited footprint* and has an area proportional to the pulse duration. The analysis proceeds as follows. For a specular surface and a nadir-looking antenna, the time t_0 for the

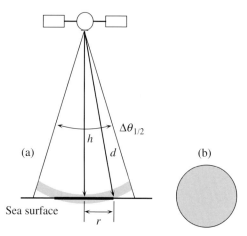

Figure 12.14. The encounter of the radar pulse with a specular surface, in (a) side view and (b) top view. The dark solid line at the surface in (a) shows the diameter of the illuminated area corresponding to the circle in (b).

pulse leading edge to travel from the antenna to the sea surface is

$$t_0 = h/c \tag{12.5}$$

Figure 12.14 shows the pulse encounter with the sea surface and the footprint size, where to simplify the figure the pulse is shown passing through the surface without reflection. From Figure 12.14, if $t' = t - t_0$ and $0 \le t' \le \tau$, the footprint radius r is written

$$r^2 = (d^2 - h^2) = (ct)^2 - (ct_0)^2 = c^2 \left[(t_0 + t')^2 - t_0^2 \right] \tag{12.6}$$

For $t' \ll t_0$, Equation (12.6) becomes

$$r^2 = 2c^2 t_0 t' = 2hct' \tag{12.7}$$

Equation (12.7) shows that for $0 \le t' \le \tau$, the footprint is a disk with its area increasing linearly with t'. Assume that the altimeter has a narrow beam nadir-pointing antenna of constant gain G_0 as described in Equation (10.14), and that the surface conditions within the instantaneous footprint are uniform. For this case, both the power backscattered to the antenna and σ_0 also increase linearly with t'. From Figure 12.14 and Equation (12.6), the maximum radius of the illuminated disk is proportional to τ and is given by

$$r^2 = 2hc\tau \tag{12.8}$$

As the wavefront continues to propagate and for $t' > \tau$, Figure 12.15 shows that the surface footprint becomes an annulus described by

$$r_2^2 = 2hc(t - t_0) \qquad r_1^2 = 2hc[t - (t_0 + \tau)] \tag{12.9}$$

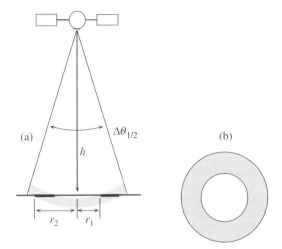

Figure 12.15. The annular area illuminated by the pulse for $t' > \tau$, in (a) side view and (b) top view.

so that $r_2^2 - r_1^2 = 2hc\tau$, and the area of the illuminated footprint remains constant at A_{max} $= 2\pi hc\tau$. In summary, for $0 \le t' \le \tau$, the illuminated area increases linearly with time; for $t' > \tau$, the area remains constant until r_2 extends beyond the halfpower beamwidth, at which time the return power falls off to zero.

For a specular surface, the above arguments show that the maximum disk and annulus areas are equal and proportional to τ. For TOPEX, $\tau = 3.125$ ns for a pulse length of 0.9 m, so that $r = 1.6$ km and $A_{max} = 8$ km^2. In contrast, the C-band beamwidth-limited footprint has a diameter of about 60 km; the corresponding K_u-band footprint has a diameter of about 26 km, so that for a specular surface, the pulse-limited footprint is much smaller than the beamwidth-limited footprint. As Section 10.4.1 discusses, to avoid interference with other spectrum users, the minimum pulse length is restricted to about 1 m, so that these are the minimum altimeter footprints. Given this interaction with the surface, the calculation of the round trip travel time is next described.

12.6.3 *Determination of the round trip travel time*

Figure 12.16 shows the return from an idealized interaction of the pulse with a specular surface. It divides into four parts. First, before the return pulse arrives, the instrument observes only the noise floor. Second, as the leading edge of the return arrives at the antenna, Φ_R increases linearly with time, proportional to the increase in footprint area. Third, when the footprint becomes an annulus, the return power is constant so that Φ_R reaches a plateau. Fourth, at the trailing edge of the return where the annulus becomes greater than the half-power beamwidth, Φ_R drops off in what is called plateau droop. Given this interaction, the round trip time t_{RT} for the midpoint of the pulse is defined as that time when the received

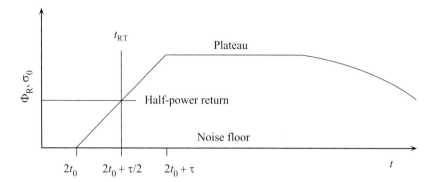

Figure 12.16. The time dependence of the backscattered energy and scattering cross section for the reflection received from a specular surface. The horizontal line shows the height at which the backscatter return equals half its plateau value; the vertical line marks the corresponding time. See text for further description.

reflection of the illuminated footprint equals half of its maximum size or when

$$t_{RT} = 2t_0 + \tau/2 \tag{12.10}$$

From this discussion, determination of t_{RT} becomes a question of finding the midpoint of the region of linear slope. The estimation of t_{RT} is done by an onboard tracking algorithm that determines when the return power equals half the difference between the plateau level and the noise floor.

Two factors complicate this determination: off-nadir pointing angles and ocean waves and surface roughness. Off-nadir pointing angles have two effects. First, slightly more energy is reflected away from the antenna, reducing the plateau level so that it must be adjusted for look angle. Second, if the angle is large enough, then in addition part of the circle or annulus falls outside of the beamwidth-limited footprint. This means that even with the plateau level adjusted for look angle, the plateau droop occurs earlier, making it more difficult to determine the plateau level.

12.7 Effect of waves on the altimeter return

When waves are present, three factors alter the radar return: the small scale surface roughness, the large scale ocean swell and the random nature of the sea surface. First, as the wind speed increases, the increase in roughness and mean-square surface slope scatters and reflects more energy away from the antenna, so that the plateau level decreases with increasing U. Second, because of the nature of scattering from surfaces with randomly distributed slopes, the idealized signal shown in Figure 12.16 has a large random component that must be removed by averaging (Chelton *et al.*, 2001b). Third, an increase in the ocean swell height reduces the slope of the leading edge of the return in Figure 12.16,

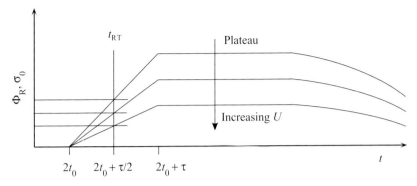

Figure 12.17. The decrease in plateau level with increasing wind speed.

and yields an algorithm for retrieval of $H_{1/3}$. The next three sections discuss each of these factors.

12.7.1 Small scale roughness and the determination of U

As Figure 10.16 shows for a nadir-oriented radar, the increase in sea surface roughness and mean-square slope with U causes σ_0 to decrease. Ignoring for the moment the random signal component, Figure 12.17 shows for wind-induced roughness that the decrease in σ_0 with increasing U reduces the plateau level while leaving the rise time unchanged. Because this response to surface roughness also occurs when ocean swell is present, this U-dependence forms the basis for a wind speed algorithm (Chelton *et al.*, 2001b, Section 7). Because off-nadir pointing angles also reduce the plateau level, to retrieve U the return must be adjusted for pointing angle. Also, because the surface roughness associated with rain cells attenuates the return and generates a false wind speed signal, rain must be identified and masked. Further, the dependence of the plateau level on U means that the sensitivity and linearity of the altimeter electronics and the range-retrieval accuracy are also functions of U. As the next section shows, because of the advantages of operating within the linear region of the altimeter electronics, the gain of the return is adjusted by an onboard function called Automatic Gain Control (AGC) so that when measured in digital counts, the plateau level is constant and independent of U.

12.7.2 Automatic Gain Control (AGC) and averaging of the return

For a random wave field, the return from any individual pulse is very noisy. To reduce the noise, the AGC carries out the following sequence of operations. First, it adjusts the individual returns for off-nadir pointing angles, then averages the returns over a sufficient period of time such that the mean signal dominates. Figure 12.18 shows for a simulated return that, as the number of averaged pulses increases, the return approaches the idealized shape in Figure 12.16. The AGC then adjusts the plateau level of the averaged return so that

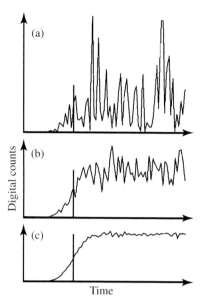

Figure 12.18. The effect of averaging a simulated return from a Gaussian distribution of wave heights with $H_{1/3} = 10$ m. (a) Single return; (b) average of 25 returns; (c) average of 1000 returns. The lower axis represents the noise floor (Figure 1b from Townsend, McGoogan and Walsh, 1981, © 1981 Kluwer Academic/Plenum, used with permission).

when measured in digital counts it is constant. This AGC adjustment value is transmitted to the ground for estimation of σ_0 and U; the half-power point and t_{RT} are determined by the difference between the plateau level and the noise floor. For the TOPEX altimeter, which transmits and receives 4000 pulses per second, the data are averaged at the satellite over 50 ms or 200 pulses. For oceanographic purposes and on the ground, the data are further averaged over 1 s (Chelton *et al.*, 2001b, Section 2). For a specular surface, the averaged surface footprint measures about 9 km in the along-track direction and 3 km in the cross-track. As the next section shows, the footprint size also increases with the ocean swell amplitude.

12.7.3 Effect of ocean swell

The long period ocean swell has two effects: it increases the footprint size and the rise time of the return power. For the altimeter, the swell amplitude is described in terms of the significant wave height $H_{1/3}$ (Section 2.2.3). From TOPEX observations, a characteristic value of $H_{1/3}$ is 3 m; the largest mean monthly value is about 12 m; the largest instantaneous value is 15–20 m (Lefèvre and Cotton, 2001).

Figure 12.19 illustrates the pulse encounter with ocean swell. The presence of swell means that the first pulse reflection instead taking place at t_0, now takes place at about

$$t_1 = t_0 - H_{1/3}/2c \tag{12.11}$$

Table 12.1. *The dependence on the significant wave height $H_{1/3}$*
of A_{max} and its diameter for a single pulse, and the along-track
and cross-track dimensions of the 1-s averaged footprint.

$H_{1/3}$ (m)	A_{max} (km²)	Diameter (km)	Footprint (km × km)
0	8	3.2	9 × 3
3	34	6.5	12 × 6
6	59	8.7	15 × 9
15	134	13	19 × 13

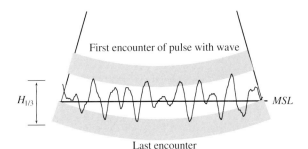

Figure 12.19. The encounter of the wavefronts with ocean swell; *MSL* is mean sea level. The curvature of the wavefronts is greatly exaggerated.

Similarly, at nadir the last pulse reflection takes place at

$$t_2 = t_0 + H_{1/3}/2c + \tau \tag{12.12}$$

Similar to the specular case and for $t_1 < t \le t_2$, the footprint is a disk with its area increasing linearly with time. For $t > t_2$, the footprint again becomes an annulus, so that the maximum illuminated area A_{max} can be written

$$A_{max} = 2\pi h(c\tau + H_{1/3}) \tag{12.13}$$

Equation (12.13) shows that A_{max} increases linearly with $H_{1/3}$. For TOPEX, $c\tau$ is about 1 m, so that for $H_{1/3} = 3$ m, A_{max} is four times its specular value. Table 12.1 shows the dependence on $H_{1/3}$ of A_{max}, its corresponding diameter and the along-track and cross-track dimensions of the 1-s averaged surface footprint. As $H_{1/3}$ increases from 0 to 15 m, the A_{max} diameter increases from 3 to 13 km, which is still less than the 26-km diameter of the K_u-band altimeter footprint. This increase in area with $H_{1/3}$ is called *defocusing*. For an $H_{1/3}$ of 3 m, the footprint measures 12 km × 6 km, while in regions with heavy swell, such as the Antarctic Convergence, the footprint size approaches 20 km × 15 km. This shows that the presence of swell increases the size of the surface footprint and limits the altimeter spatial resolution.

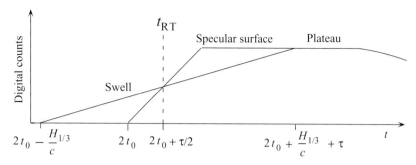

Figure 12.20. Comparison of the time dependence of the backscattered signal from a specular surface and from ocean swell.

Figure 12.20 compares the behavior of the return signal in both the presence and absence of ocean swell and shows that when swell is present the rise time is longer and the slope of the response is reduced. In spite of this change, for the AGC-adjusted plateau level, the half-power point occurs at the same distance from the altimeter. Consequently, for the wave-covered surface, the technique described for the specular surface can also be used for retrieval of the round trip travel time. The inverse dependence of $H_{1/3}$ on this slope permits retrieval of the global fields of $H_{1/3}$ and allows for description of their seasonal variation (Lefèvre and Cotton, 2001).

12.8 Errors and biases in retrieval of sea surface height

In the following, Sections 12.8.1–12.8.4 show that the errors and biases in the h_S retrieval have four sources: altimeter noise, atmospheric errors, sea state biases and uncertainties, and errors in orbit position. Section 12.8.5 combines these errors into a total error budget, and shows that for a single altimeter pass the uncertainty in SSH is 4.1 cm for TOPEX and 2.5 cm for JASON-1. Section 12.8.6 discusses two physical phenomena, the inverse barometer effect and ocean tides, both of which create real changes in sea surface height and generate additional uncertainties in the retrieval of geostrophic height. Because geoid uncertainties are an additional source of error, Section 12.8.7 concludes with a discussion of the recent and planned gravity missions designed to determine the geoid at shorter wavelengths.

12.8.1 Altimeter noise

Fu *et al.* (1994) describe how the TOPEX altimeter noise is determined from spectral analysis of 10-s sequences of the altimeter return. The resultant noise varies with significant wave height, such that at an $H_{1/3}$ of 2 m the rms ALT noise is 17 mm. As $H_{1/3}$ increases, the rms noise increases until for $H_{1/3} > 3$ m it reaches a stable value of 20–25 mm. The POSEIDON noise is slightly greater. From prelaunch data, the JASON-1 altimeter noise is ± 15 mm (JASON-1 noise estimates are from a prelaunch document at the Archiving, Validation and Interpretation of Satellite Oceanographic Data (AVISO) website: http://www.jason.oceanobs.com/html/portail/general/welcome_uk.php3).

12.8.2 Atmospheric sources of error

The atmospheric corrections and uncertainties divide into three categories: *dry troposphere*, *wet troposphere* and *ionosphere*. Dry troposphere refers to all tropospheric gases except water vapor and liquid water; wet troposphere to water vapor and cloud liquid water; ionosphere to free electrons.

Dry troposphere   The dry tropospheric range delay varies with the amount of atmospheric mass between the sensor and the surface or equivalently with sea level pressure, and equals 2.7 mm per 1-mbar change in pressure. Corrections for this delay use the surface pressure fields produced by the European Centre for Medium-range Weather Forecasts (ECMWF). For TOPEX and JASON-1, and based on the ECMWF rms pressure accuracy of 3 mbar, the associated range error is 7 mm (Chelton *et al.*, 2001b).

Wet troposphere   The wet troposphere range delay has contributions from V and R_R, where the contributions from L are sufficiently small that they are generally ignored. For V, comparison of TMR with ground-based radiometer and radiosonde measurements shows that the uncertainty in the TMR measurements of V corresponds to a range error of about 11 mm (Fu *et al.*, 1994). The JMR retrieval is assumed to have the same accuracy. As described in Section 12.4.2, both TMR and JMR as well as the dual-frequency altimeters can identify regions of heavy rain for later masking.

Ionospheric free electrons   For TOPEX and JASON-1, the ionospheric range correction derived from the dual-frequency altimeters has an error of about 5 mm. For the single-frequency POSEIDON altimeter on TOPEX, the ionospheric correction is determined from measurements made by the slant-range, two-frequency DORIS signals. Given the additional uncertainties associated with the adjustment of this slant measurement to a vertical path, the DORIS measurement uncertainty is 17 mm, or about three times the TOPEX uncertainty.

12.8.3 Sea state bias

Sea state bias is generated by ocean swell and can be divided into two parts. The first is the *electromagnetic* (EM) bias, which refers to the apparent depression of the mean sea level generated by the interaction of the radar pulses with the physical wave properties. The second is the *tracker* or *skewness* bias, which refers to the additional apparent surface depression generated by tracker determination of the half-power point. The sum of these two is called the *total sea state bias*. Skewness bias can be reduced by postprocessing; EM bias cannot be further reduced (Chelton *et al.*, 2001b).

Electromagnetic bias   In an ocean swell field, EM bias occurs because the troughs of even pure sinusoidal waves are better reflectors than the crests, so that the mean reflecting surface is depressed below mean sea level. Two factors enhance this bias: parasitic capillary formation on the wave crests and finite amplitude waves. For the first, preferential formation of parasitic capillaries on wave crests scatters energy away from the antenna and adds to the depression of the mean reflecting surface. For the second, as the wave slope ak_w increases, the waves develop broad troughs and narrow crests (Section 2.2.1). Because the broad troughs are better reflectors than the crests, the mean reflecting surface is further depressed.

From observational studies, the EM bias is negative and approximately linearly proportional to 2–3% of $H_{1/3}$, where the constant of proportionality depends on geographic region and on U (Chelton *et al.*, 2001b). Because waves of a given $H_{1/3}$ can have either small or large wave slopes, this bias has a large error. For the same $H_{1/3}$, the waves can consist entirely of long period sinusoidal waves of small ak_w, as occurs for waves generated by a distant storm, or of locally storm-generated troichoidal waves with large ak_w. Consequently, the wave slope or exact bias cannot be inferred from $H_{1/3}$, and can only be partially parameterized in terms of the TOPEX retrievable variables U and $H_{1/3}$ (Chelton *et al.*, 2001b). For both TOPEX and JASON-1, the rms error of the bias approximately equals 1% of $H_{1/3}$, so that, for $H_{1/3} = 2$ m and depending on geographic area, the bias is between 40 and 60 mm with an rms error of about ± 20 mm (Fu *et al.*, 1994). This error is difficult to reduce and is second in magnitude only to the errors in the POD.

Tracker or skewness bias   As Sections 12.6.3 and 12.7 describe, the function of the onboard tracker is to determine the midpoint of the region of linear rise in the altimeter return. In this calculation, the algorithm assumes that the wave amplitude has a Gaussian distribution. Because the actual waveform is non-Gaussian or skewed, the tracker generates an additional negative offset. The part of this offset that is proportional to $H_{1/3}$ is generally included in the EM bias. At $H_{1/3} = 2$ m the error in the TOPEX skewness bias is about ± 12 mm (Fu *et al.*, 1994); at $H_{1/3} = 10$ m the error reaches a maximum value of about ± 40 mm (Chelton *et al.*, 2001b). Because this is an instrument error, it can be reduced by postprocessing. In practice, because it is difficult to separate skewness from EM bias, the two are frequently grouped into a single sea state bias.

12.8.4 Errors in orbit determination

At short time scales, uncertainties in the satellite orbital position are the largest source of range error. Orbit errors divide into single pass errors, which are associated with a single

Table 12.2. *The single-pass rms measurement errors for the different components of the TOPEX and JASON-1 error budgets*

	TOPEX ALT	JASON-1
Range errors		
Altimeter noise	1.7 cm	1.5 cm
Atmospheric corrections		
Dry troposphere	0.7 cm	0.7 cm
Wet troposphere	1.1 cm	1.0 cm
Ionosphere	0.5 cm	0.5 cm
Sea state		
Electromagnetic bias	2.0 cm	1.0 cm
Skewness	1.2 cm	0.2 cm
Total range error (rss)	3.2 cm	2.25 cm
Single pass radial distance orbit error	2.5 cm	1.0 cm
Total rss sea surface height error	4.1 cm	2.5 cm
Wind/wave accuracies		
Wind speed	2 m s^{-1}	1.5 m s^{-1}
Significant wave height $H_{1/3}$	0.2 m	5% or 0.25 m[a]

[a] Whichever is greater

The TOPEX errors are estimated from altimeter data; the JASON-1 errors are the proposed goals of the JASON analysis. Except in the retrieval of wind speed and significant wave height, $H_{1/3}$ is set equal to 2 m.

TOPEX data adapted from Chelton *et al.*, 2001b, Table 11; JASON-1 data from AVISO at http://www.jason.oceanobs.com/html/donnees/precision_uk.html.

range estimate, and the error associated with monthly or greater time scale averages over spatial scales of a few hundred kilometers. For a single TOPEX orbit pass, the rms position error is about 2.5 cm, where both random and systematic errors contribute to this estimate (Chelton *et al.*, 2001b, Table 11). As the next sub-section shows, this single pass error is reduced when averaged over long periods and large spatial scales.

12.8.5 Summary and error discussion

Based on the above discussion, for a 1-s average of the height measured by a single satellite pass over the area of interest and an $H_{1/3}$ of 2 m, Table 12.2 gives the error budget for the TOPEX and JASON-1 altimeters. For TOPEX, the numbers correspond to the errors in the final Geophysical Data Records (GDR); for JASON-1, the numbers are the desired mission goals.

The table shows that there are two kinds of errors: first, the range errors generated by altimeter noise and by the atmosphere and sea state; second, the errors in the satellite radial position. The combination of these two yields the total measurement error. The table also lists the accuracies of the wind speed and significant wave height retrievals. For range errors alone, the root sum of the squares (rss) of the various contributions yields a range error of 3.2 cm for TOPEX and 2.25 cm for JASON. Addition of the single pass orbit error gives the total single pass rss error of 4.1 cm for TOPEX and 2.5 cm for JASON. The largest errors are those associated with orbit radial position, EM bias and instrument noise. For longer time and spatial averages over time scales of one month or greater and over spatial scales of a few hundred kilometers, comparison of the TOPEX range retrievals with the sea surface height measured at the several Pacific tide gauges shows that the total error drops to 2 cm (Cheney *et al.*, 1994).

12.8.6 *Environmental sources of uncertainty*

In addition to the height changes generated by geostrophic flows, sea surface height is also physically altered by ocean tides and the *inverse barometer* effect. Tides are generated by the relative motion of the Earth, moon and sun; the inverse barometer effect is the surface response to spatially variable changes in sea level pressure. Because these are real changes in sea surface height, they are not included in Table 12.2. For determination of geostrophic height, however, they must be removed from the altimeter signal.

Tides   Ocean tides occur at specific discrete frequencies, with components at semi-diurnal, diurnal, fortnightly, monthly, semi-annual and annual periods. Tides produce elevation changes of about 1–3 m, and except for very large ocean waves are the largest contributor to ocean surface variability (Wunsch and Stammer, 1998). Previous to TOPEX, tidal models were primarily based on observations made from a global network of coastal and island tidal stations. Because the altimeter measures tidal height in the interiors of ocean basins, the combination of TOPEX and surface observations with numerical tidal models means that the amplitudes of the major tidal components are now known to an error of 1 cm (Le Provost, 2001). Given these models, most of the tidal signal can be removed from the altimeter range retrieval, which greatly improves the accuracy of the geostrophic height retrieval.

Inverse barometer   The inverse barometer effect describes the hydrostatic response of the sea surface to spatially variable changes in sea level pressure at time scales greater than about two days. Because a spatially uniform change in sea level pressure does not affect sea level height, this pressure fluctuation must occur relative to the spatially averaged pressure. For pressure changes satisfying these conditions, a 1-mbar increase in pressure generates a 1-cm decrease in the surface elevation. The inverse barometer correction works well over the open ocean, but breaks down in small marginal seas and in the western boundary currents. Although the inverse barometer and dry troposphere corrections are both functions of sea level pressure, they differ fundamentally. The dry troposphere correction yields an electromagnetic range delay independent of surface displacement; the inverse barometer effect is a physical surface displacement. Similar to the dry troposphere correction, the inverse barometer effect is removed using ECMWF surface pressure fields. The error in this correction is about 3 mb or 3 cm in displacement (Chelton *et al.*, 2001b).

12.8.7 Gravity missions

Because a detailed knowledge of the ocean geoid and its accompanying gravity field is essential to the retrieval of absolute sea surface height, this section describes three gravity missions, two present and one forthcoming. These are the German CHAMP mission launched in July 2000, the US/German GRACE mission launched in March 2002 and the European GOCE mission scheduled for launch in 2005. The combination of these experiments should allow determination of the geoid to length scales of about 100 km.

CHAMP is a single satellite that was launched into a non sun-synchronous orbit at an initial altitude of 454 km, where because of atmospheric drag, the orbit is slowly decaying. The use of a non sun-synchronous orbit allows the satellite to observe the diurnal gravity components. CHAMP uses the BlackJack GPS receivers described in Section 12.5 for continuous orbit tracking and a three-axis accelerometer for measurement of surface forces. During its estimated five-year lifetime, the goal of CHAMP is to determine the long-wavelength variability in the Earth's gravitational field.

GRACE consists of a pair of satellites that are both in the same sun-synchronous orbit at an altitude of 500 km with an along-track separation of 200 km (described at http://www.csr.utexas.edu/grace/). GRACE also has a five-year lifetime. The position and separation of the satellites are determined from combination of GPS and a microwave ranging system that measures the distance between the satellites with an accuracy of ± 10 μm, or one-tenth the diameter of a human hair. These measurements allow determination of the gravity field and its variability, where the variability measurements may be able to observe surface and deep ocean currents through their effects on the gravity field.

GOCE is a single satellite that will operate in a sun-synchronous orbit at an altitude of 250 km. Within its two-year lifetime, GOCE will measure the Earth's static gravity field at length scales about 1 km. The combination of these three missions should yield an improved geoid with rms uncertainties of about 1 cm at scales of 100 km. Combined with altimetry data, this improved geoid will allow determination of the properties of such 100-km scale current systems as the Gulf Stream and the Antarctic Circumpolar Current. Once this improved geoid is available, all of the earlier altimeter data will be reanalyzed, so that these observations should lead to a great improvement in understanding the ocean circulation.

12.9 Applications and examples

Altimeter measurements provide information about the mesoscale ocean properties and their variability. For flows with length scales greater than 500 km where the geoid is known, they yield both the steady and variable geostrophic flow properties (Wunsch and Stammer, 1998). At shorter scales where the geoid is insufficiently determined, the measurements yield information on the variability of the flows around an altimeter-defined mean sea level. The sea surface height exhibits two kinds of variability: that associated with local changes in the water column density and volume, called *steric* changes; and that associated with redistribution of mass. Steric changes are caused by seasonal heating and cooling and by precipitation and evaporation; mass redistribution is caused by variability in ocean currents and by planetary waves.

12.9.1 Large scale geostrophic flow

Plates 6 and 7 illustrate the large scale properties of the global circulation. The first shows the sea surface height and the geostrophic flow determined relative to the geoid; the second shows the flow variability relative to the mean sea surface height. First, Plate 16 shows the four-year TOPEX average (October 12, 1992–October 9, 1996) of sea surface height ζ measured relative to the geoid as defined in Equation (12.2). The image is filtered to remove features with scales less than about 500 km. For this and the following examples, ζ is corrected for tides and the inverse barometer effect. On the plate, the colors show SSH; the arrows show the geostrophic flow. At the equator, the arrows are omitted because of the breakdown in geostrophy. Examination of the plate shows that the total range in SSH attributable to geostrophic flows is about 3 m. The smallest SSH values occur around Antarctica, where the northward increase in SSH corresponds to the Antarctic Circumpolar Current. The largest heights occur in the western Pacific and in the Indian Ocean off South Africa. In the Pacific, these are associated with the Kuroshio Current and with the western boundary current off Australia; in the Indian Ocean, with the Agulhas Current System. In the Atlantic, the height gradients associated with the Gulf Stream and the Brazil–Malvinas Confluence are also visible.

Second, as an example of the variability at shorter spatial scales and for the same four-year period as Plate 16, Plate 17 shows the rms variability of $\Delta\zeta$ relative to $\bar{\zeta}$ as defined in Equation (12.3). Because subtraction of $\bar{\zeta}$ removes all of the geoid undulations as well as the steady geostrophic currents, these SSH anomalies are valid for all length scales. Examination of this plate shows that the largest variability occurs in the regions of the western boundary currents, including the Gulf Stream, Kuroshio, the Agulhas Current, as well as in portions of the Antarctic Circumpolar Current, with an especially large variability south of Africa.

Because of the four-year time average, over much of the ocean, the variability approaches the ±2 cm TOPEX noise floor applicable to long-term averages.

12.9.2 Seasonal variations in sea surface height

The four panels of Plate 18 show the seasonal SSH anomalies relative to a nine-year TOPEX mean. The panels show the Northern Hemisphere autumn (September–November), winter (December–February), spring (March–May) and summer (June–August). Following Stammer and Wunsch (1994) and Wunsch and Stammer (1998), these anomalies have two sources: the steric changes caused by seasonal heating and cooling; and the dynamic changes in major current systems due to seasonal wind variations.

Examination of the plate shows that the Northern and Southern Hemisphere SSH anomalies are six months out of phase, with a complicated response in some regions due to seasonal winds. In the North Atlantic and Pacific, the anomalies are greatest in autumn and least in spring. Because of the intense winter heat exchanges, the largest annual change of about 20 cm occurs in the Gulf Stream and Kuroshio. Away from these current systems, the annual Northern Hemisphere change is about 12 cm. In the Southern Hemisphere, the response is reversed, in that the anomalies have a maximum in March–May and a minimum in September–November.

The Southern Hemisphere annual change is about 6 cm, or about half that of the Northern Hemisphere. The much larger Northern Hemisphere land area causes its greater response. Because winter land areas become much colder than the oceans, the northern winter offshore winds cool the ocean much more than the southern winds, so that the northern ocean experiences a larger annual temperature change than the southern. In both hemispheres, because cold sea water is relatively insensitive to changes in temperature, the largest steric changes occur at mid-latitudes instead of at high latitudes.

The plate also shows how the large scale current and wind systems contribute to the annual variability. The equatorial pattern is complicated because it is averaged over two El Niños, 1992–93 and 1997–98, and because of the complicated zonal structure of easterly and westerly flows. North of the equator, the plate shows the banded seasonal character of the North Equatorial Current and Counter Current. The Counter Current achieves its maximum eastward flow in September–November and its minimum flow in March–May, while the North Equatorial Current exhibits the opposite behavior. A similar but weaker system occurs south of the equator. In the northern Indian Ocean, the large change in seasonal amplitude is driven by the oceanic response to seasonal monsoon winds.

12.9.3 Rossby wave propagation

For the Indian Ocean, Plate 19 illustrates the propagation of long-period Rossby waves. Rossby waves are long-wavelength baroclinic waves that propagate to the west with large amplitudes on the pycnocline, but with variations of ocean surface height of only about 10 cm (Cushman-Roisin, 1994). Because of their relatively long wavelengths and slow phase

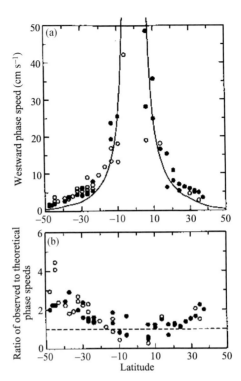

Figure 12.21. The Rossby wave phase speeds as a function of latitude. (a) comparison of the observed westward phase speeds (circles) with theory (solid line). The dark circles are Pacific observations; the open circles are Atlantic observations. (b) Comparison of the ratios of the observed and theoretical phase speeds. The dashed line is at a ratio of 1. See text for further description (Figure 5 from Chelton and Schlax, 1996, © 1996 American Association for the Advancement of Science, reprinted with permission, figure courtesy Dudley Chelton).

speeds, the surface manifestations of these waves are visible in TOPEX data. For TOPEX cycle 60, the left-hand figure shows the geographic distribution of the SSH anomaly in the Indian Ocean. The arrow marks a long rectangular strip of observations at 25° S. The right-hand figure, which is called a Hovmöller diagram, shows the SSH anomalies within this strip plotted against longitude in the horizontal and TOPEX cycle number in the vertical. Examination of the Hovmöller diagram shows the characteristic tilt of the SSH pattern from lower right to upper left, which corresponds to the westward propagation of the Rossby wave crests and troughs.

From construction of similar figures at many different latitudes for the Atlantic and Pacific Oceans, Figure 12.21 compares the latitude dependence of the observed Rossby wave properties with theory (Chelton and Schlax, 1996). For both sub-figures, the horizontal axis gives latitude between 50° S and 50° N. Figure 12.21a compares the observed westward phase speeds with theory. The figure shows that the westward phase speeds increase with decreasing latitude, where this increase continues until a singularity is approached close to the equator. Figure 12.21b compares the ratio of the observed and predicted wave speeds, and shows that for latitudes greater than 35° the observed speeds are about twice as large as predicted. As Wunsch and Stammer (1998) show, this disagreement between theory and observation has led to modifications of Rossby wave theory.

12.10 Further reading and acknowledgements

Wunsch and Stammer (1998) and the collection of papers in Fu and Cazenave (2001) review the TOPEX altimeter theory and results. In earlier work, there are two special issues of the *Journal of Geophysical Research* respectively concerning TOPEX geophysical evaluation and scientific results: Anon. (1994) and Cheney (1995). Information on TOPEX and JASON-1 is found at http://topex-www.jpl.nasa.gov/; the specific JASON-1 website is at http://topex-www.jpl.nasa.gov/mission/jason-1.html; the general JPL archive website is http://podaac.jpl.nasa.gov/. Altimeter data are also found at the Archiving, Validation and Interpretation of Satellite Oceanographic Data (AVISO) website: http://www.jason.oceanobs.com/html/portail/general/welcome_uk.php3. Haines *et al.* (2002) can be found at http://gipsy.jpl.nasa.gov/igdg/papers. The material in Sections 12.6 and 12.7 is based in part on unpublished notes of Dudley Chelton. I thank Dudley Chelton, Paolo Cipollini, Lee Fu, Meric Srokosz and Detlef Stammer for their contributions to this chapter.

13

Imaging radars

13.1 Introduction

Side-looking imaging radars provide a powerful way to retrieve ice and ocean surface backscatter properties at a high resolution and under nearly all weather conditions. Given that many geophysical processes modulate the Bragg scattering waves, the returns from these radars can be formed into images that display a wide variety of surface phenomena. Other advantages are that, depending on the processing, the resolution can be of order meters, and at the frequencies used by these radars, the atmosphere is transparent except for heavy rain.

There are two kinds of satellite imaging radars: *Synthetic Aperture Radar* (*SAR*) and the real aperture *Side-Looking Radar* (*SLR*). The SLR is a range-binned instrument with a surface resolution of about 1 km; the SAR is a more complicated instrument with a resolution that can be as high as 3 m. Because the radar pulses illuminate the surface, the instruments provide day and night coverage. SAR is the principal radar imager used in oceanographic research, where SARs have been flown by the US, Canada, Europe, Russia and Japan. Because the SLR shares some of the SAR characteristics, is used by Russia and Ukraine, and will be used as an interferometer on the US/France Ocean Surface Topography Mission (OSTM), this chapter covers both instruments, but emphasizes SAR.

SARs provide a variety of information about oceanographic and sea ice processes. For the ice-free ocean, SAR is used in the study of internal waves (Hsu and Liu, 2000), surface waves (Heimbach and Hasselmann, 2000), and ocean eddies (DiGiacomo and Holt, 2001). Other phenomena visible in SAR include shallow bottom topography, ocean currents, surface patterns of rain and wind, and the presence of oil and other surface-modifying substances. Specular reflectors such as ships, offshore structures and icebergs are also visible. For the polar pack ice, SARs observe the ice edge position and because of the general increase in surface roughness with ice thickness, SARs can also determine the areal extent of different ice types (Kwok, Rignot and Holt, 1992).

SARs operate in a variety of modes. The Standard mode has a 100-km swathwidth with a typical resolution of 25 m. The ScanSAR mode has a swathwidth of 350–500 km and a resolution of 75–150 m. ScanSAR is a mode of the present RADARSAT and ENVISAT SARs and of many future proposed SARs; it is of particular value to ocean and ice studies. In the Arctic, overlays of the RADARSAT ScanSAR swaths are used to construct the *Arctic snapshot*, which is a 3- to 6-day image of the entire ice cover. These data are analyzed by the RADARSAT Geophysical Processing System (RGPS) to derive the velocity, deformation and age of the offshore pack ice. For Antarctica, ScanSAR has been used to map the continental ice sheet. In the open ocean, Norway, the United States and Canada use ScanSAR for monitoring and managing of fishing fleets in national and adjacent international waters (Olsen and Wahl, 2000, Pichel and Clemente-Colón, 2000). NOAA uses ScanSAR in its CoastWatch program to monitor fisheries, oil pollution and extreme weather events, and in combination with numerical weather forecast models to produce high resolution maps of the vector wind speed (Pichel and Clemente-Colón, 2000; Thompson and Beal, 2000).

Both SAR and SLR depend on the relative motion of the spacecraft or aircraft to generate an image, have antennas that are generally much longer in the along-track than the cross-track direction and generate oblique fan beams at right angles to the spacecraft trajectory. In most cases, the fan beams operate at look angles greater than about $20°$. At these angles, the return avoids specular scatter and strongly depends on Bragg scatter. The SAR operation is complicated and data intensive; for any pixel in a SAR image, the brightness is derived from the phase and amplitude of the backscatter recorded from hundreds of pulses transmitted over a period of about 0.5 s. Because of the complications introduced by this procedure, understanding SAR imagery requires information on how SAR works and on the engineering constraints imposed by the solar array, antenna, electronics and associated ground system. Interpretation of the imagery also requires knowledge of the dependence of Bragg scatter on incidence angle and surface conditions.

In the following, Section 13.2 describes the general design of the SLR and SAR, and Section 13.3 derives the SLR resolution. Section 13.4 derives the SAR resolution and discusses the image constraints imposed by the PRF, system noise, and relative motions within the images associated with ocean currents, ships or surface waves. Using the currently operational RADARSAT SAR as an example, Section 13.5 discusses the SAR design, its imaging modes and the constraints on its operation. Section 13.6 shows examples of SAR open water and pack ice imagery. Finally, Section 13.7 describes the ENVISAT SAR and two proposed SAR missions.

13.2 Background

This section describes how SLR and SAR work, discusses the concept of resolution and how it differs from the VIR case, describes SARs that operate at multiple polarizations, discusses the interferometric SAR and concludes with a summary of the properties of past, present and future satellite SARs.

13.2.1 General description

A SAR or SLR satellite antenna has typical dimensions of about 10 m in the along-track direction, 2 m in the cross-track, and looks off to the side of the spacecraft at incidence angles

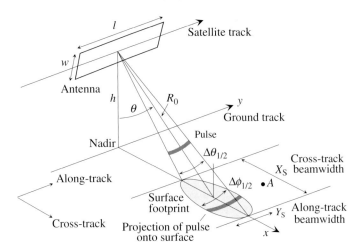

Figure 13.1. The viewing geometry for the SAR and SLR antenna: w is the antenna width; l is the length. For clarity, the along-track width of the surface footprint is greatly exaggerated relative to the cross-track; typical footprint dimensions are 3 km × 100 km. At the scale of the figure, the along-track beamwidth would be only slightly larger than the width of the line marked x.

of 20–50°. The antennas are made up of many distributed transmitter/receiver elements in what is called an active phased array (Luscombe *et al.*, 1993). Although the antennas described in this chapter are rectangular, a SAR can also consist of a parabolic antenna with a front feed, as illustrated by the 1990–1994 Magellan SAR Venus mapping mission (Anthony Freeman, private communication, 1999). Satellite SARs operate with PRFs of 1000–2000 Hz and frequencies of 1–10 GHz, corresponding to wavelengths of 3–25 cm. The reason for this choice of frequencies is that for $f < 1$ GHz, the radars are affected by reflection and absorption in the ionosphere, by terrestrial sources of radiation and by the galactic radiance described in Section 9.3; for $f > 10$ GHz, by atmospheric absorption.

   Figure 13.1 shows the viewing geometry and the half-power FOV or surface footprint of a rectangular side-looking radar antenna of width w and length l, that looks off at right angles to the spacecraft trajectory. For the RADARSAT SAR antenna, $w = 1.5$ m and $l = 15$ m. The size of the surface footprint follows from the definition in Equation (10.19) of the half-power beamwidths $\Delta\theta_{1/2}$, $\Delta\phi_{1/2}$ in the along- and cross-track directions. For an antenna at an altitude h and incidence angle θ, the cross-track swathwidth X_S in the surface plane is

$$X_S = \Delta\theta_{1/2} R_0 / \cos\theta = \Delta\theta_{1/2} h / \cos^2\theta = \lambda h / w\cos^2\theta \qquad (13.1)$$

where the distance from the radar to the surface is $R_0 = h/\cos\theta$. Derivation of (13.1) depends on the assumption that $\Delta\theta_{1/2} \ll \theta$, while the additional $\cos\theta$ term converts the beamwidth that is normal to the boresight direction into the surface swathwidth. A typical value of X_S is 100 km. Similarly, the along-track swathwidth Y_S is approximately given by

$$Y_S = \Delta\phi_{1/2} R_0 = \Delta\phi_{1/2} h / \cos\theta = \lambda h / l \cos\theta \qquad (13.2)$$

so that Y_S is inversely proportional to the antenna length l. For RADARSAT, a characteristic value of Y_S is about 3 km, so that the footprint has a very narrow aspect ratio.

For both SAR and SLR and following Section 10.4, the pulse length determines the cross-track resolution. The two imagers differ in their determination of the along-track resolution where, because this resolution is a function of azimuth angle, it is also called the azimuthal resolution. Because the SLR relies only on range binning, its azimuthal resolution corresponds to Y_S in Equation (13.2) and thus improves with increasing l and decreasing range. If the RADARSAT antenna could be operated as a SLR, it would be incapable of distinguishing two objects unless their cross-track separation was greater than about 3 km.

In contrast, the SAR can achieve an optimum azimuthal resolution equal to $l/2$ or half the antenna length. The SAR achieves this resolution through the following procedure. Physically, the SAR divides into two parts, the antenna and its associated transmitter/receiver, and its memory or *echo store*. Consider the point A in Figure 13.1. In the radar coordinate system, the point enters the swath to the left and exits on the right. The point takes about 0.5 s to pass through the RADARSAT swath, during which time it is illuminated with about 10^3 pulses. For the echo from each pulse, the SLR only records the amplitude time history. In contrast, the SAR records the time history of both amplitude and phase, creating what is called a *coherent* radar. Within this stored data, every spatial point that passes through the illuminated footprint has a unique history in terms of time, range and Doppler shift. If, during the illumination period, the relative positions of the surface elements do not change, then a computationally intensive analysis of the pulse sequence produces a high resolution surface image in both range and azimuth. This computation is approximately equivalent to the synthesis of an antenna aperture equal in length to the swathwidth, or for RADARSAT to about 3 km. In actuality, the SAR works even better than a long antenna, because as Section 13.4.1 shows, the combination of range and Doppler processing produces an azimuthal resolution that is independent of range.

13.2.2 Resolution and pixel size

There are two definitions of resolution used in remote sensing. The first is the FOV diameter; the second is the minimum separation between two objects at which they can be distinguished. First, for the VIR imagers and as Section 1.6.5 describes, the resolution is defined as the nadir FOV diameter and equals the pixel size; for the microwave conical scanner, it equals the FOV diameter. To illustrate the first case, Figure 13.2a shows a series of FOVs with two objects represented by vertical bars and separated by the FOV diameter Δx. Because the image represents these objects as adjacent dark pixels, they cannot be separately resolved. For radars, because the resolution is defined as the minimum separation between two objects at which they can be discriminated, the pixel size must equal half the resolution distance (Raney, 1998, pp. 12–14). Figure 13.2b shows that two objects separated by this distance can be resolved. Finally, Figure 13.2c

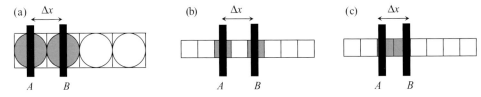

Figure 13.2. Example of the two definitions of resolution, where A, B are two objects and Δx is the constant resolution distance. (a) Resolution in the VIR, where the resolution is defined as the FOV diameter and equals the pixel size, and the two objects are separated by Δx. Because the objects generate dark adjacent pixels, they cannot be separately distinguished in the image. (b) Radar resolution, where Δx is the minimum separation at which the two objects can be discriminated. On the figure, the object separation equals the resolution, the pixel size is half the resolution, and the two objects are separately visible in the image. (c) The two objects separated by less than the radar resolution distance, showing that they cannot be discriminated. See text for further description.

shows that, for an object separation less than the resolution distance, the objects cannot be distinguished. Another way to demonstrate the difference between pixel size and resolution comes from the Nyquist criterion (Jenkins and Watts, 1968). This states that to resolve a spatial variation of wavelength λ_S the signal must be sampled at a minimum spacing of $\lambda_S/2$, so that the pixel spacing is again half the resolution distance.

13.2.3 Polarization

SAR pulses are polarized, generally in the horizontal (H) or vertical (V) plane. Antennas that broadcast and receive in both H or V are called HH or VV antennas. As Section 10.2.2 describes, SARs that use all four polarization modes (HH, HV, VV, VH) are called polarimetric SARs; when all four polarizations are measured, the radar is in a *quad-pol* mode. In quad-pol, the SAR first transmits a V-pulse, and measures the V and H returns (VV, VH). The SAR then transmits an H-pulse and measures the H and V returns (HH, HV). The reason for alternating between transmission of H and V pulses instead of transmitting both pulses simultaneously is that ambiguities occur in distinguishing between VV and HV, and HH and VH. The polarimetric SARs have the advantage that they provide more information about the surface; their disadvantage is that they are much more data-intensive than single polarization SARs. The applications of polarimetric SAR to the open ocean and ice are topics of current research. Boerner *et al.* (1998) discuss its applications to oceanography and in particular its potential use in estimation of the thickness of oil slicks; Drinkwater *et al.* (1992) describe its applications to sea ice research.

13.2.4 Interferometric radars

SAR and SLR interferometry consists of taking data from the same area from either two different locations or at two different times, then using the combined data to determine variations in surface displacement or velocity (Madsen and Zebker, 1998; Rosen *et al.*, 2000).

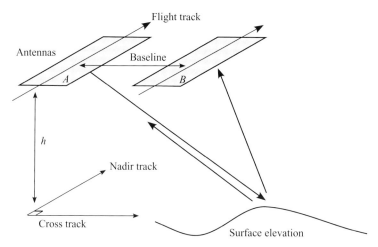

Figure 13.3. The geometry of a cross-track interferometer. The two antennas are at a specific altitude in a parallel track, and are separated by a precisely determined baseline. The antennas make simultaneous observations of the same surface area from two different locations.

From Massonnet and Feigl (1998), interferometry is widely employed in land mapping and in studies of earthquake and ice sheet deformation. Its use is just beginning in oceanography, where it has been proposed for measurement of surface height and ocean currents. The two kinds proposed for oceanography are called cross-track and along-track interferometry.

Cross-track interferometry consists of two antennas at the same along-track, but different cross-track locations taking simultaneous radar images of the same surface area (Figure 13.3). The antennas are separated in the cross-track direction by a carefully measured and maintained *baseline* distance or separation, typically of order meters. The interferometer geometry is determined by the antenna size, the baseline separation and the instrument altitude. There are at least two possible antenna operating configurations. The first is that one antenna transmits and receives (*A* in the above figure); while the *B* antenna only receives, so that the reflection from the transmitted signal is received at two antennas. An alternative configuration is the so-called *ping-pong* mode, where antenna *A* transmits with *A* and *B* receiving, followed by *B* transmitting with *A* and *B* receiving, and so forth. For each pulse and each surface pixel, combination of the returns yields the phase difference between the signals, where this phase difference is proportional to the difference in path length from the antennas to each pixel. Given precise knowledge of the geometry, the sea surface height at each pixel can then be calculated. As Chapter 14 describes, two different satellite cross-track interferometers are being developed for this purpose: the European SAR Interferometric Radar Altimeter (SIRAL), which will be flown on CryoSat and will investigate ice sheet and pack ice properties, and an SLR Wide Swath Ocean Altimeter (WSOA), which will be flown on the OSTM successor to JASON-1 and will make wide-swath measurements of sea surface height.

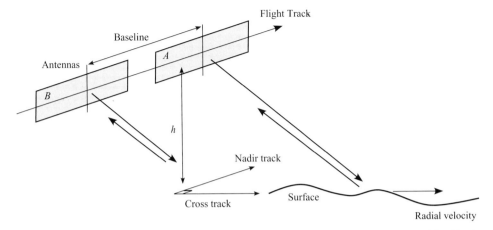

Figure 13.4. The geometry of an along-track interferometer. The two antennas are at a specific altitude in the same track, and are separated by a known baseline. The specific surface area is first illuminated by antenna A, then at a slightly later time, by antenna B.

   Along-track interferometry consists of the use of two antennas on the same trajectory to take two images of the same surface area from the same orbit location but at different times (Figure 13.4). For this case, the phase differences observed at each pixel can be analyzed to yield radial displacement. Goldstein *et al.* (1994) describe its use in the measurement of ocean surface velocities.

13.2.5 Past, present and proposed satellite SAR missions

Table 13.1 lists some of the past, present and proposed civilian satellite SAR missions. The table gives the satellite names, the responsible organization or country, the SAR frequency and polarizations, the launch year and approximate lifetime, and the presence or absence of onboard data storage. SARs without onboard data storage were restricted to operate only within the receiving masks of their ground stations. The first civilian SAR was the NASA SEASAT SAR launched in 1978. This was also the last US civilian satellite SAR; since that time, US civilian SARs have only flown on short duration Space Shuttle missions. SEASAT was followed by the Russian ALMAZ (diamond) SAR, and the ERS-1 and ERS-2 SARs. During 1995, ERS-1 and ERS-2 operated in the same orbit and made along-track interferometric measurements. The Japanese SARs consist of the JERS-1 low frequency L-band SAR and the proposed L-band PALSAR. The Canadian RADARSAT SAR, which is the first operational wide swath satellite SAR, has a polarimetric successor, RADARSAT-2. The ERS SAR successor is the ENVISAT ASAR that operates in both VV and HH. Following ERS-2, all SAR missions have onboard data storage. Raney (1998) and Johannessen (2000) give more information about these missions; Section 13.7 discusses ASAR, RADARSAT-2 and PALSAR.

Table 13.1. *Civilian satellite SAR missions*

Satellite	Agency or country	f (GHz), polarization	Launch date/lifetime	Onboard storage?
SEASAT	NASA	1.3, HH	1978/3 mo	N
ALMAZ	USSR	3, HH	1991/1.5 yr	Y
ERS-1	ESA	5.3, VV	1991/3 yr	N
JERS-1	Japan	1.3, HH	1992/2 yr	Y
ERS-2	ESA	5.3, VV	1995/5 yr	N
RADARSAT-1	Canada	5.3, HH	1996/5 yr	Y
ENVISAT (ASAR)	ESA	5.3, VV/HH	2002/5 yr (est.)	Y
RADARSAT-2	Canada	5.3, Quad-pol	2004/7 yr (est.)	Y
ALOS (PALSAR)	Japan	1.3, Quad-pol	2005/—	Y

JERS is the Japanese Earth Remote Sensing satellite; ASAR is the ESA Advanced SAR, ALOS is the Advanced Land Observing Satellite, PALSAR is the Phased Array L-Band SAR. In the last column, N means no, Y means yes.
Pre-1999 data adapted from Raney, 1998, Tables 2-4, 2-5 and 2-6

13.3 SLR resolution

This section derives the resolution of the SLR, where in the cross-track direction the SLR and SAR resolution are determined using the same technique. As Figure 13.1 shows, the SLR looks off to one side of the satellite track, sends out short pulses of energy, then receives the backscattered energy and bins it by range. Because the spacecraft velocity is much less than the speed of light, as the SLR moves along its flight track the image is built up line by line as the echo from each pulse is received and binned. A SLR flew on the USSR KOSMOS series of satellites, on the subsequent Ukraine/Russia OKEAN satellites, and is proposed for flight on the Ukraine/Russia OKEAN-O (Operational) successor to the OKEAN series. Mitnik and Kalmykov (1992) describe the operational SLR on the KOSMOS satellites and give examples of imagery; in Russia, the SLR data were received on the equivalent of a fax machine. The KOSMOS SLR has a ground resolution of 1–3 km and can be thought of as an all-weather AVHRR. An SLR was also a separate operating mode of the USSR ALMAZ SAR, and as Chapter 14 describes, OSTM will carry the WSOA interferometric SLR.

The SLR resolution is a function of range in both the along-track and azimuthal directions. In the azimuthal direction, if the separation of two targets at the same range is so small that both targets lie within the Y_S from Equation (13.2), then energy from the same pulse is simultaneously reflected from both targets so that they cannot be distinguished from one another. Therefore, the optimum SLR along-track resolution Δy_{SLR} equals the swathwidth, so that

$$\Delta y_{SLR} = Y_S = \Delta\phi_{1/2} R_0 = R_0 \lambda / l \qquad (13.3)$$

Equation (13.3) shows that the along-track resolution decreases linearly with R_0 or with distance from the satellite.

In the cross-track direction and from Section 10.4, the SLR cross-track resolution Δx_{SLR} equals half the projection of the pulse length onto the surface or in terms of the pulse duration τ,

$$\Delta x_{SLR} = c\tau/2 \sin \theta \qquad (13.4)$$

Equations (13.3) and (13.4) show that as θ increases from nadir, Δx_{SLR} decreases and Δy_{SLR} increases. Two limiting cases occur for $\theta \to 0$ and $\theta \to \pi/2$. First, as $\theta \to 0$ or near nadir, the energy is reflected back simultaneously from the surface, so that $\Delta x_{SLR} \to \infty$, $\Delta y_{SLR} = \Delta\theta_{1/2} h$, and the SLR is unusable.

Second, as $\theta \to \pi/2$ or for a horizontal look angle, $\Delta x_{SLR} \to c\tau/2$ and achieves its minimum, and $\Delta y_{SLR} \to \infty$, so that the SLR is again unusable. Between these two extremes, the resolutions vary with θ, so that the SLR resolution in both dimensions depends on θ. The cross-track resolution can be improved through use of shorter pulse length, but the along-track resolution can only be improved through use of a longer antenna and a smaller $\Delta\phi_{1/2}$.

13.4 How the SAR achieves its resolution

This section describes how the SAR achieves its resolution and discusses some of the constraints on its operation. Specifically, Section 13.4.1 derives the SAR azimuthal resolution; Section 13.4.2 discusses the constraints imposed on its resolution by the PRF. Section 13.4.3 describes the constraints imposed by instrument and environmental noise; Section 13.4.4 describes *speckle*, which is the noise created by random surface backscatter. Finally, Sections 13.4.5 and 13.4.6 discuss image problems that occur with SARs. These are the need for radiometric balancing of the image and the problem of *range walk*, which is the image distortion generated by relative motion within the surface footprint.

13.4.1 SAR resolution derived from Doppler beam sharpening

The optimum azimuthal resolution of a SAR antenna equals half the antenna length, or $l/2$, where this result is independent of cross-track range and frequency. Ulaby *et al.* (1982) derive this result in several ways; of these, this section presents the one called *Doppler beam sharpening*. This analysis involves Doppler tracking of an individual target across the surface footprint and yields a derivation of the azimuthal resolution.

For a non-rotating Earth and an antenna looking at right angles to the spacecraft trajectory, Figure 13.5 shows several characteristic isodops and lines of constant range within the surface footprint. If the surface position relative to the spacecraft is defined by x and y, then within the footprint y is much less than x, where $\delta = y/x$ is defined as the azimuth angle relative to the cross-track direction. Because the technique involves Doppler shifts,

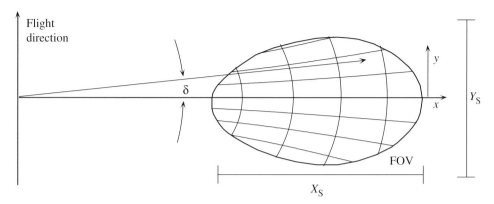

Figure 13.5. The SAR surface footprint, the lines of constant range and the orthogonal surface isodops, which lie approximately at right angles to the flight direction. The y-dimension is greatly exaggerated relative to x.

f_0 and λ_0 are chosen as the center frequency and wavelength of the incident radiation. For these definitions, Equation (10.32) shows that the dependence of the Doppler shift Δf on incidence angle is

$$\Delta f = 2U_0\delta \sin\theta/\lambda_0 \tag{13.5}$$

Substitution of δ and $x = R_0 \sin\theta$ from Figure 13.1 into (13.5) gives

$$\Delta f = 2U_0 y/\lambda_0 R_0 \tag{13.6}$$

From Equation (13.2), the y-location of the leading footprint edge $y_{max} = Y_S/2$ can be written

$$y_{max}/R_0 = \Delta\phi_{1/2}/2 = \lambda_0/2l \tag{13.7}$$

with an equal but opposite relation for the trailing edge. Substitution of Equation (13.7) into (13.6) gives the frequency shift Δf_{SAR} at the leading and trailing edge of the surface footprint as

$$\Delta f_{SAR} = \pm U_0/l = \pm 1/\tau_s \tag{13.8}$$

In (13.8), τ_s is the time it takes for the satellite to travel one antenna length. For RADARSAT, $l = 15$ m and $U_0 = 6.5$ km s^{-1} so that $\Delta f_{SAR} = \pm 430$ Hz.

Figure 13.5 shows that, relative to the spacecraft, as a fixed target crosses the footprint its range decreases until its location is described by $y = 0$, then increases, while for the same period the Doppler frequency decreases nearly linearly. Since the target range varies across the footprint, the variable-range tracking produces what is called a focused SAR. For comparison, an unfocused SAR assumes that the target is at a constant range.

For the focused SAR and from Equation (13.5), Figure 13.6 shows the target position as a function of time and Doppler shift. Relative to the spacecraft, the target enters the footprint

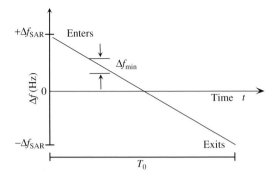

Figure 13.6. Tracking an object in frequency space across the SAR footprint, where the target enters at upper left and exits at lower right.

with a Doppler shift of $+\Delta f_{SAR}$ and exits with $-\Delta f_{SAR}$. Suppose the target is tracked using a Doppler filter, where Δf is the center frequency of the filter, which decreases with time, and where the carrier frequency f_0 is removed from the return. If Δf_{min} is the minimum frequency interval to which Δf can be determined, then from reorganization of Equation (13.6), the minimum along-track resolution Δy_{min} can be written as

$$\Delta y_{min} = \Delta f_{min}\, \lambda_0 R_0 / 2U_0 \qquad (13.9)$$

Equation (13.9) shows that given Δf_{min}, Δy_{min} is easily calculated.

The frequency resolution Δf_{min} is determined from the illumination time T_0, which is the time for a point on the surface to traverse the swath, or equivalently the time in which the satellite travels a distance equal to the along-track swathwidth, so that from (13.7),

$$T_0 = R_0 \lambda_0 / l U_0 \qquad (13.10)$$

From basic time series constraints (Jenkins and Watts, 1968),

$$\Delta f_{min} = 1/T_0 = l U_0 / R_0 \lambda_0 \qquad (13.11)$$

For RADARSAT, T_0 is about 0.5 s, so that $\Delta f_{min} = 1.2$ Hz. Substitution of Δf_{min} into (13.9) shows that the minimum along-track resolution equals half the antenna length, or that

$$\Delta y_{min} = (l U_0 \lambda_0 R_0)/(\lambda_0 R_0 2U_0) = l/2 \qquad (13.12)$$

Paraphrasing Elachi (1987, 204–205), this is an unusual result, in that $l/2$ is independent of frequency and range and the shorter the antenna, the better the resolution. The reason for the range independence is that when the surface point in question is further from the actual antenna, the footprint is wider, so that the synthetic antenna is longer. This increase in synthetic length exactly compensates for the resolution decrease caused by the greater distance. Second, the improvement in resolution with decreasing antenna length occurs because a shorter antenna yields a wider footprint and a longer synthetic aperture, thereby producing a finer surface resolution. This does not mean that a very small antenna can be used to obtain a very fine resolution; as the next section shows, the constraints imposed by the PRF mean that the antenna area cannot be smaller than a PRF- and frequency-dependent minimum.

13.4.2 Constraints on the PRF

For (13.12) to be valid, the PRF must satisfy the two constraints, one setting a floor on the PRF, the other a ceiling. The floor depends on the antenna length, the ceiling on the antenna width. The combination of these constraints sets a minimum antenna area.

The PRF floor is determined from the antenna resolution. For Equation (13.12) to be valid and from the Nyquist criterion, the PRF must equal at least twice the largest Doppler shift that occurs in the sample, so that

$$\text{PRF} \geq 2\Delta f_{\text{SAR}} \tag{13.13}$$

To obtain a SAR resolution of $l/2$, from (13.8) the PRF must satisfy

$$\text{PRF} \geq 2U_0/l \tag{13.14}$$

Equation (13.14) shows that for the azimuthal resolution to equal $l/2$, the PRF must equal at least two pulses per antenna length. Given this constraint and for the 15-m long RADARSAT antenna with $U_0 = 6.5 \text{ km s}^{-1}$, the PRF must be greater than about 900 Hz. Equation (13.14) sets a PRF floor, and means that for a very short antenna to yield a resolution of $l/2$, the PRF must be very large. If, however, the PRF is less than the $2U_0/l$ lower limit in Equation (13.14), the antenna continues to work, but with $\Delta y_{\text{min}} > l/2$ (Anthony Freeman, private communication, 1999).

The maximum or ceiling PRF is set by the constraint that the return from each pulse must be unambiguously identified without confusion from earlier or later pulses. From Section 10.4.2, this means that the PRF must satisfy Equation (10.24). To derive the maximum possible PRF for the narrow cross-track SAR shown in Figure 13.7, the cross-track beamwidth $\Delta\theta_{1/2}$ is assumed to be much less than θ, where at the mean incidence angle θ_{m}, R_0 is the distance to the surface. Dropping the subscript on θ_{m} and λ_0 and after some trigonometry, the distance d_{p} between consecutive pulses must satisfy

$$d_{\text{p}} = c\tau_{\text{p}} > 2 \tan\theta\, R_0\lambda/w \tag{13.15}$$

Because $\text{PRF} = \tau_{\text{p}}^{-1}$, Equation (13.15) can be written

$$\text{PRF} < cw/2R_0\lambda \tan\theta \tag{13.16}$$

For the RADARSAT antenna with $\theta = 45°$, $\lambda = 5.6 \text{ cm}$ and $R_0 = 1100 \text{ km}$, Equation (13.16) shows that the PRF must be less than 3600 Hz. From (13.1) a broad cross-track beamwidth corresponds to a small w, so that as (13.16) shows, a narrow antenna with its corresponding broad swath requires a small PRF. This decrease in PRF with increasing swathwidth is the reason that broad-swath SARs have a poor resolution and is also why most high resolution SARs have a relatively narrow cross-track swathwidth.

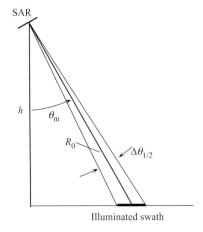

SAR

h

θ_m

R_0

$\Delta\theta_{1/2}$

Illuminated swath

Figure 13.7. The SAR geometry used in discussion of the relation between the cross-track swathwidth and PRF for a narrow-swath SAR.

Combination of the inequalities in (13.14) and (13.16) yields

$$2U_0/l < \mathrm{PRF} < cw/2R_0\lambda\tan\theta \tag{13.17}$$

Reorganization of (13.17) shows that the antenna area must satisfy the following:

$$lw > 4U_0\lambda R_0\tan\theta/c \tag{13.18}$$

From (13.18) and at a look angle of $\theta = 45°$, an X-band (10 GHz) antenna requires that $lw > 2.8\,\mathrm{m}^2$; an L-band (1.3 GHz) antenna, $lw > 21.8\,\mathrm{m}^2$. Compared with the actual RADARSAT area of 22.5 m^2, at 5.3 GHz and 45°, its minimum required antenna area is 5.3 m^2. For the special case of the SeaWinds parabolic antenna described in Chapter 11, with $f = 13.4\,\mathrm{GHz}$ and $\theta = 50°$, $lw > 2.9\,\mathrm{m}^2$. This means for SeaWinds to operate as a SAR would require an antenna diameter of about 2 m, which is about twice its actual size.

13.4.3 Signal-to-noise constraints

Section 10.2.3 shows that the received power is the sum of the attenuated backscatter return, the instrument noise and the environmental blackbody radiation. The smallest signal that can be distinguished from the noise must be greater than the instrument noise floor, described in terms of the noise-equivalent-σ_0 ($NE\sigma_0$). Because of the additional noise contributions from environmental radiation, the actual noise floor is greater. For the radar, the ability to make the signal-to-noise ratio as large as possible requires a large power output, which is in turn restricted by the size of the solar panels and the capacity of the batteries. For SEASAT, ERS-1 and -2 and RADARSAT, a typical value of $NE\sigma_0$ is -24 dB (Raney, 1998).

13.4.4 Speckle

As Ulaby *et al.* (1982) and Rees (2001) discuss, as well as the instrument noise and the environmental blackbody emission terms, there is an additional noise source generated by backscatter from a uniform surface. For this case, even though adjacent surface elements have the same σ_0, the interaction within each element of the fine-scale structure with the incident beam creates a statistical uncertainty in σ_0 from one pixel to the next, generating variations in image brightness called *speckle*. Speckle is reduced by averaging of adjacent pixels, where the number averaged in this way is termed the *number of looks*. This averaging, which is generally presented as the number of azimuth-averaged samples times the range-averaged, as in 2×4, reduces the statistical variance of the image and improves the spatial resolution.

13.4.5 Radiometric balancing

Another characteristic of SAR images is caused by the backscatter dependence on θ. For a constant oceanic wind speed, Figure 10.16 shows that σ_0 decreases with increasing θ. This backscatter dependence on θ means that the SAR brightness decreases with distance across the swath, so that the image is brighter on the near side and darker in the far. In the processing this decrease is sometimes reduced by removal of a linear trend in brightness across the image, where this correction is called *radiometric balancing*. Several examples of unbalanced images are shown below.

13.4.6 Range walk

Within the surface footprint, the SAR ability to produce a realistic image depends on there being no relative motion of the surface features. Because Doppler processing responds to surface motion, the SAR imagery is distorted by its response to any moving object, current or ocean wave with a cross-track velocity component. For example, consider the 1.3 GHz SEASAT SAR incident on the ocean surface at $\theta = 22°$. Assume that, within the antenna footprint, a ship moves in the cross-track direction toward the SAR, with a velocity of 10 knots or 5 m s^{-1}. As Equation (10.29) shows with the satellite velocity U_0 replaced by 5 ms^{-1}, instead of observing a zero Doppler shift, in the cross-track direction the SAR observes $\Delta f = 20$ Hz. From Equation (13.6), this corresponds to a Δy of approximately 0.3 km, so that on the image, the ship is displaced in the positive Doppler direction by this distance from its wake and its actual position. Similarly, a ship traveling away from the SAR in the cross-track direction is displaced in the negative Doppler direction. This velocity-induced apparent shift in position is called *range walk*.

From a SEASAT SAR image of the Caribbean, Figure 13.8 shows two examples of range walk. The white circles show the locations of two ships and their wakes; the ships are bright from specular reflection, and the wakes are dark, from either oil discharges or currents in the wake suppressing the Bragg scatterers. The dark area in the center of the image is probably

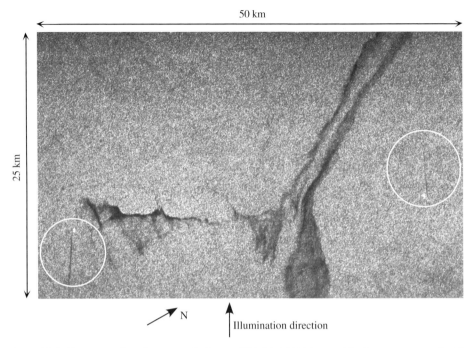

50 km

25 km

N

Illumination direction

Figure 13.8. Two examples of range walk from a SEASAT SAR image. The image was acquired in the Caribbean on October 3, 1978. The white circles show the locations of two ships and their wakes; the ships are bright from specular reflection; their wakes are dark. (From Fu and Holt, 1982, used with permission, courtesy Ben Holt).

pollutants displaced by winds and currents. The wake locations relative to the ships show that the ships are moving in opposite directions. The left-hand ship is moving away from the SAR, so that the ship is displaced to the right, or in the direction of positive Doppler shift. The right-hand ship is moving toward the SAR, and is displaced in the opposite direction. Such Doppler shifts associated with moving objects are the cause of such image problems as a locomotive displaced from its tracks or cars displaced from a highway. Range walk is also caused by irregular satellite motion, such as spacecraft yaw and tilt, or by orbit-changing maneuvers.

13.5 RADARSAT SAR

To illustrate the SAR imaging modes and operational constraints, this section describes the Canadian RADARSAT SAR. RADARSAT was launched on November 4, 1995 by a NASA spacecraft. In exchange, Canada provided NASA with a portion of the SAR data and with the two periods of Antarctic RADARSAT coverage described below. Figure 13.9 shows a line drawing of RADARSAT and Table 13.2 lists some of its properties. The figure shows

Table 13.2. *The RADARSAT SAR characteristics*

SAR property	Value
Altitude	800 km
Frequency/wavelength	5.3 GHz/5.6 cm (C-band)
Swathwidth	10–500 km
Look angle	20–49°
Polarization	HH
PRF	1270–1390 Hz
Pulse lengths (compressed)	33, 57, 86 ns
Noise floor ($NE\sigma_0$)	−23 dB
Range resolution	10–100 m
Antenna length, width	15, 1.5 m

Ahmed, *et al.*, 1990; Raney, 1998

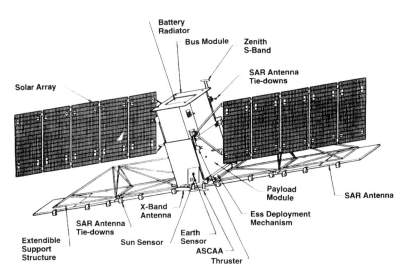

Figure 13.9. The configuration of the RADARSAT spacecraft; the antenna measures 15 m by 1.5 m. (Figure 1 from Moore *et al.* 1993, © 1993 Canadian Aeronautics and Space Institute, used with permission).

the SAR antenna, the X-band downlink and the solar array, which has an area of 27 m² and produces 3 kilowatts of peak power. The instrument has a 5-year design life.

The SAR antenna consists of 32 rows of waveguide radiators that transmit and receive in H-pol (Luscombe *et al.*, 1993). Its 15-m length, which is about 50% longer than the ERS SARs, has a fixed azimuthal beamwidth of about 0.2°. In the cross-track direction, however, and because of the separate waveguide radiators, the antenna has a flexible beam-shaping capability. The shape and incidence angles of the beams generated by these radiators are

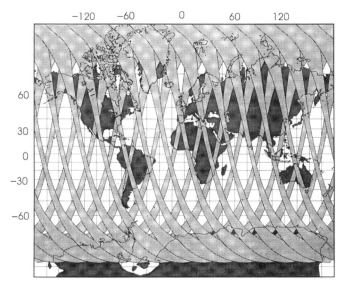

Figure 13.10. One-day RADARSAT coverage of the Earth. The swathwidth is 500 km (Figure 2 from Parashar, Langham, McNally and Ahmed, 1993, © 1993 Canadian Aeronautics and Space Institute, used with permission).

electronically controlled, so that to switch from one beam to another takes less than 15 μs. As shown below, this rapid beam switching makes the ScanSAR mode possible.

As Raney, *et al.* (1991) and Raney (1998) describe, RADARSAT is in a dawn–dusk sun-synchronous orbit, specifically chosen to maximize the exposure of the solar panels to sunlight. Except over the South Pole, the sun fully powers the satellite, which reduces the need for batteries. Under normal operations, the antenna looks to the right or to the north of the sub-satellite track, so that there is near daily coverage above 70° N and no coverage below 79° S. For the 500-km wide ScanSAR swath, Figure 13.10 shows the one-day coverage that was designed for Canada and its adjacent waters. Because of this excellent coverage at high northern latitudes, RADARSAT is used for fisheries monitoring and enforcement in high latitude oceans. During periods in 1997 and 2000 and under the Antarctic Mapping Mission (AMM), the satellite was rotated about its nadir axis by180°, which reversed the hemispheric bias and allowed RADARSAT to acquire complete composite images of Antarctica.

13.5.1 Image modes

RADARSAT uses its cross-track electronic beam shaping capability to generate a variety of image modes. As Figure 13.11 and Table 13.3 show, these include Standard, Wide Swath, Fine Resolution, Extended and ScanSAR (Raney, 1998; Moore *et al.*, 1993). For these modes, RADARSAT operates at incidence angles between 20° and 50°, and uses the

Table 13.3. *Imaging modes of the RADARSAT SAR*

Beam mode	Nominal swathwidth (km)	Incidence angle (deg)	Number of looks	Resolution (m × m)
Standard	100	20–50	1 × 4	25 × 28
Wide	150	20–45	1 × 4	25 × 28
Fine	50	37–48	1 × 1	10 × 9
ScanSAR narrow	300	20–46	2 × 2	50 × 50
ScanSAR wide	500	20–50	2 × 4	100 × 100
Ultra-fine wide	20	30–40	1	3 × 3
Ultra-fine narrow	10	30–40	1	3 × 3

In the last two columns, the number of looks and the resolution are given as range times azimuth.

Luscombe *et al.* 1993

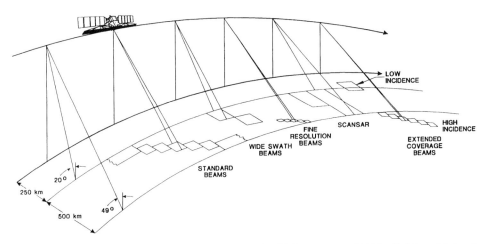

Figure 13.11. The different imaging modes for RADARSAT; see text and Table 13.3 for additional description (Figure 1 from Luscombe *et al.*, 1993, © 1993 Canadian Aeronautics and Space Institute, used with permission).

three different pulse lengths shown in Table 13.2 combined with different PRFs to obtain flexibility in surface range resolution. For oceanographic purposes, the most commonly used modes are Standard and ScanSAR-Wide. The Standard mode produces a 100×100 km^2 image, with a 25-m resolution or a 12.5-m pixel size. The ScanSAR wide mode has a 500-km width with a characteristic resolution of 100 m.

ScanSAR operates very differently from the Standard mode. For this case, the antenna beam is electronically switched among a number of parallel sub-swaths at a fast enough rate that a synthetic aperture is formed within each sub-swath, allowing the synthesis of a 300–500 km wide image. Specifically, the imaged area is divided into a series of sub-swaths

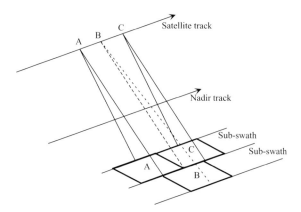

Figure 13.12. The sequence of beam positions used in the generation of a two-beam ScanSAR image. The letters on the flight track show the mean position of the SAR during acquisition of the corresponding surface image. The along-track widths of the sub-frames are greatly exaggerated (Adapted from Figure 8, Raney *et al.*, 1991).

and sub-frames, through which the antenna cycles sufficiently rapidly that the sub-frames are contiguous (Raney, *et al.*, 1991). The 300-km wide ScanSAR mode involves two sub-swaths; the 500-km mode involves four sub-swaths. Figure 13.12 shows the simplest case of two sub-swaths. In this example, ScanSAR begins with the inner swath and samples frame A, switches to the outer swath and samples frame B, then switches back to the inner swath and samples frame C. For ScanSAR to work, frames A and B and frames B and C must overlap within their respective azimuthal beamwidths. ScanSAR permits the generation of a wide swath image, but because the PRF condition applies to the entire swath, the resolution is worse than for the Standard swath.

13.5.2 Data storage and data rates

The early SARs had no onboard data storage, but only operated within the receiving mask of a ground station. Newer SARs carry tape recorders (RADARSAT) or solid state storage devices (ENVISAT) that record out-of-mask data for later downloading at one or more receiving stations. There are two constraints on the number of out-of-mask images, one imposed by the amount of onboard storage, the other by the downlink rate. The RADARSAT tape recorder has a 70 Gb storage capacity that can operate for 14 minutes at 85 megabits-per-second (Mbps). Because a 100 km × 100 km Standard image typically requires about 1 Gb of storage, the recorder has room for about 70 images.

The necessity of downloading this recorded data also restricts the number of out-of-mask images. Given that the satellites are only over their receiving stations for short periods of time and the bandwidth constraints imposed on the data transfer by other spectrum users, there is a limit as to how much data can be downloaded in a single pass. For this purpose, RADARSAT contains two X-band downlinks, one for direct broadcast of imagery gathered within the receiving station mask at 105 Mbps; the other for transfer of tape recorded data at

85 Mbps. With the condition that the satellite be at least 5° above the horizon, the maximum downlink period over the ground station is about 12 minutes, which also limits the number of images taken per orbit.

The planned and proposed polarimetric SARs have larger storage and data rate requirements. These are being met in part by replacement of onboard tape recorders with high density solid state recorders. For example, the RADARSAT-2 solid state recorder will have a 200 Gb capacity with a 400 Mbps data transfer rate from the SAR to the recorder. In the more distant future, multifrequency and multipolarization SARs will require downlink rates as large as 300 Mbps and onboard storage capacities greater than 300 Gb. To deal with the downloading of these large amounts of data, there are plans to use optical downlinks, where the data would be downloaded via a broadband communications laser to cloud-free sites such as Mauna Loa on Hawaii (Anthony Freeman, private communication, 1999).

13.6 Applications and examples

SAR is used in open ocean and polar pack ice studies. For the open ocean, Section 13.6.1 shows that SAR can view surface and internal waves, ocean fronts and eddies, and monitor fishing fleets and oil spills. For the polar pack ice, Section 13.6.2 shows that SAR can identify different ice types, track ice floes and icebergs, and through the Alaska SAR Facility (ASF) provide sequential maps of the Arctic ice cover.

13.6.1 Open ocean

Because SARs generally operate at look angles greater than 20°, the open ocean radar return is dominated by Bragg scatter from short ocean waves. Exceptions include Bragg scatter from raindrop splashes and backscatter from bright specular reflectors such as offshore structures, ships and icebergs. The SAR response to Bragg scattering means that the instrument can view any large scale ocean or atmospheric feature that generates, damps or modulates these waves. The oceanic features include surface slicks, ocean currents, long period surface waves and internal waves; the atmospheric features include rain, wind bursts and weather fronts.

The presence of nonuniform ocean currents and bottom topography also affects the short waves (Phillips, 1977). An adverse current steepens the wave slopes and gives rise to parasitic capillaries; a current in the wave direction reduces the slopes. Such currents are generated by local winds, long-period surface waves, internal waves and by large scale systems such as the Gulf Stream. Long waves that propagate over bottom topography are also steepened, yielding short wave growth and making the bottom topography visible in the imagery. Fu and Holt (1982) describe these modulation mechanisms in detail and illustrate them with an extensive collection of SEASAT SAR images; Mouchot and Garello (1998) also describe the application of SAR to oceanography and show many of the images in Fu and Holt. The following discusses three general examples: ocean swell, surface slicks and oil discharges, and internal waves.

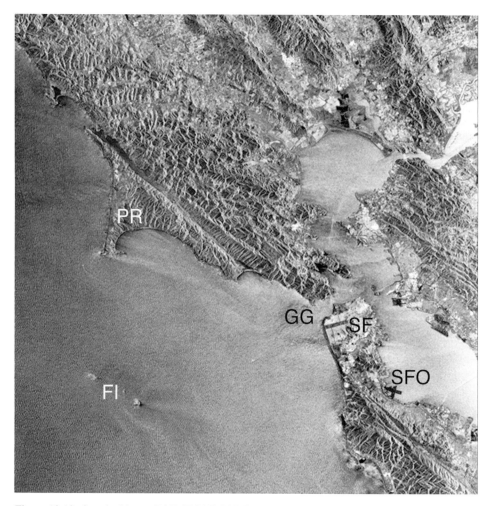

Figure 13.13. Standard beam RADARSAT SAR image of the San Francisco Bay area acquired on November 22, 2001 at 1424 UTC during a descending orbit. The image measures 100 km by 100 km. On the image, PR is Point Reyes, SF is the city of San Francisco, SFO is the airport, FI is the Faralon Islands, and GG is the Golden Gate. The illumination direction is from the right, the image is oriented approximately north/south (RADARSAT data © Canadian Space Agency/Agence Spatiale Canadienne 2001. Processed and distributed by RADARSAT International, courtesy Ben Holt).

Ocean swell   As an example of the SAR ability to observe ocean swell, Figure 13.13 shows a Standard beam RADARSAT SAR image of the San Francisco Bay region and the adjacent Pacific Ocean on November 22, 2001. For the same scene, Figure 13.14 shows an enlarged view of the area around Point Reyes. Both images have a 25-m resolution, a 12.5-m pixel size and are illuminated from the right. At the image acquisition time, the wind is from the west with speeds of 4–8 m s^{-1} (from http://sfports.wr.usgs.gov/wind/). Both images are radiometrically unbalanced, with enhanced brightness to the right. In the open Pacific, a

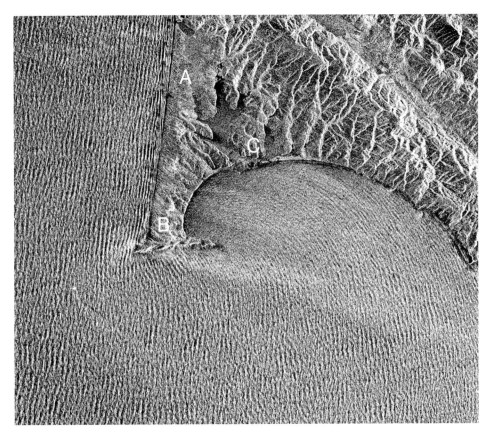

Figure 13.14. Enlarged view of the Point Reyes peninsula from Figure 13.13; the image measures approximately 25 km by 25 km (RADARSAT data © Canadian Space Agency/Agence Spatiale Canadienne 2001. Processed and distributed by RADARSAT International, courtesy Ben Holt).

long period ocean swell is visible as a linear pattern of bright and dark bands propagating toward the coast; in contrast, San Francisco Bay is characterized by an absence of swell, but with patches of brightness associated with wind-generated Bragg scatterers. Also visible on this image is the long linear San Andreas fault, running just inland of Point Reyes, then south through San Francisco.

The reason swell is visible is that the capillary waves associated with Bragg scatter form preferentially on and just ahead of the crests, in part because of the curvature and in part because the troughs are sheltered from the winds while the crests are exposed. This variation in capillary wave amplitude creates the observed bright/dark pattern. Because the waves are propagating, they are slightly distorted by range walk. From the SAR image, the deep water wavelength of this swell is 350 m, corresponding to a 15-s wave period. Coincident TOPEX data acquired off the California coast shows that the swell has an $H_{1/3}$ of 5–6 m (from the TOPEX quicklook archive at http://podaac.jpl.nasa.gov/topex/www/ql_archive. html). This means that in deep water, the swell has a small wave slope.

Figure 13.14 shows the details of the wave diffraction around Point Reyes, and the decrease in wavelength that occurs as the waves propagate into shallow water. The waves incident on the exposed coast at A become shorter and brighter as they approach the coast, indicating an amplitude increase as they move into shallow water. At the tip of Point Reyes marked by B, there is a bright region of wave breaking. As the waves move past Point Reyes, they are diffracted by the topography so that, as the waves move into shallow water, their crests rotate to become parallel to the coast. There is also a wave shadow in the embayment at C. The image illustrates wave diffraction around an obstacle and shows the usefulness of SAR in studies of the interaction of ocean swell with coasts and harbors. In the larger image, the Faralon Islands provide another example of wave breaking and diffraction.

Slicks SARs can also observe the location and extent of surface slicks. As Section 2.2.5 describes, the surface slicks associated with human-induced oil or chemical spills and with naturally occurring petroleum or biological slicks damp out waves with lengths less than about 0.3 m, which greatly reduces the Bragg scatter. Because ships and offshore structures are specular reflectors and appear bright in the imagery while oil slicks damp out the Bragg scatterers and appear dark, SAR provides a technique for monitoring off-shore oil wells, shipping and fisheries.

Figure 13.15 a shows a portion of a ScanSAR image of fishing vessels and factory ships off Cape Navarin in the Russian Bering Sea during the walleye pollock season. On the image,

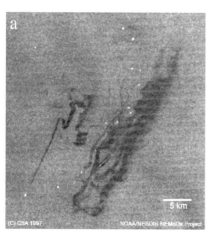

Figure 13.15. Comparison of ScanSAR and surface imagery of fishing operations. (a) A portion of a RADARSAT ScanSAR image of fishing and factory ships off Cape Navarin in the Russian Bering Sea during the walleye pollock fishing season. The image was acquired on August 24, 1997 at 1817 UTC at the Alaska SAR facility, the horizontal banding was induced by the processing (Figure 5a from Pichel and Clemente-Colón, 2000, reprinted from *Johns Hopkins APL Technical Digest* with permission, © 2000 The Johns Hopkins University Applied Physics Laboratory; RADARSAT data © Canadian Space Agency/Agence Spatiale Canadienne 1997. Processed and distributed by RADARSAT International, courtesy Pablo Clemente-Colón.). (b) 1991 surface photograph of pollock fisheries waste being discharged from a factory ship in the Bering Sea (Photograph © Robert Visser/Greenpeace, used with permission).

← ———————— 20 km ———————— →

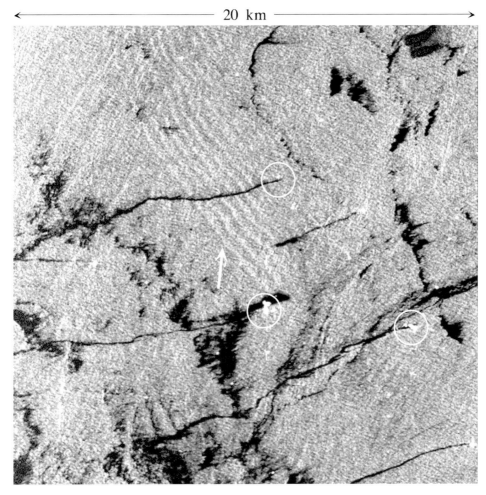

Figure 13.16. An offshore drilling field about 150 km west of Bombay, India, in the Arabian Sea. The image is centered at 19.25° N and 71.34° E and measures 20 km by 20 km. The image was acquired by the three-frequency Spaceborne Imaging Radar-C/X SAR (SIR-C/X SAR) from the Space Shuttle Endeavour on October 9, 1994 (Courtesy of NASA/JPL/Caltech, used with permission).

the bright dots show the fleet, where the factory ships have lengths up to 90 m. The dark areas are either organic waste products released from the factory ships or oil released by bilge pumping. Figure 13.15b shows a surface photograph of organic waste being discharged from a factory ship similar to those in the SAR image, with many birds feeding off the discharge. The figures demonstrate the ability of ScanSAR to monitor fishing operations and their waste discharges. Because of the value of such images to real-time monitoring of coastal waters, they are extensively used in the NOAA CoastWatch program (Pichel and Clemente-Colón, 2000).

Figure 13.16 shows an offshore oil field about 150 km west of Bombay, India, in the Arabian Sea. The image was acquired by the three-frequency Spaceborne Imaging

Radar-C/X SAR (SIR-C/X SAR) from the Space Shuttle Endeavour on October 9, 1994. SIR-C/X was a joint venture between the US (NASA/JPL), Germany (Deutsche Forschungsanstalt für Luft- und Raumfahrt e.V.) and Italy (Agenzia Spaziale Italiana). On the image, which shows the C- and L-band channels only, the drilling platforms are specular reflectors and appear as bright white spots. The long, thin dark streaks extending from the three circled platforms are probably oil discharges spreading downwind. The larger dark patches have the feathered appearance of dispersed slicks that may have been discharged earlier when the wind was from a different direction. The white arrow marks the surface signature of an internal wave packet, discussed further below. The less distinct crests at right angles to the internal waves are long period ocean swell. (Description courtesy of the JPL SIR-C/X SAR website, http://www.jpl.nasa.gov/radar/sircxsar.)

Internal waves Because as Gasparovic, Apel and Kasischke (1988) describe, internal waves generate surface regions of convergent and divergent currents, SAR can also observe patterns of internal waves (Figure 13.17). Specifically, when the wind velocity is in the same direction as the induced current, the capillary wave amplitudes are reduced; when the current opposes the wind, the amplitudes are enhanced. Figure 13.18 shows two examples of internal waves. For the continental slope off New Jersey, Figure 13.18a shows the propagation of internal waves in about 35 m of water, where these waves are generated by the interaction of the semi-diurnal tide with the shelf slope (Li, Clemente-Colón and Friedman, 2000). Along the white line in Figure 13.18a, two packets of internal waves are visible with average wavelengths of about 700 m and with about eight wave crests per packet. Li *et al.* (2000) use such images to estimate the wavelength and phase speed of the observed waves, from which they can infer the water column stratification.

For the South China Sea, Figure 13.18b shows internal waves in a portion of a ScanSAR image analyzed by Hsu and Liu (2000). The image shows the westward propagation of

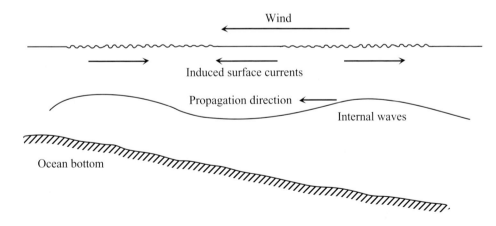

Figure 13.17. The generation of surface roughness by the surface currents induced by long internal waves (Adapted from Figure 2 of Hsu and Liu, 2000).

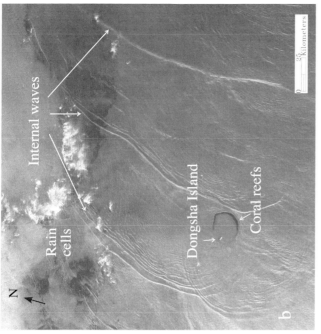

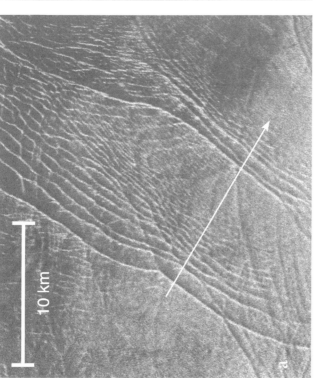

Figure 13.18. SAR observations of internal waves. (a) Propagation of internal waves on the continental slope off New Jersey taken from a Standard beam RADARSAT image taken on 2240 UTC on 31 July, 1996. The wave crests are approximately parallel to the isobaths. The white line and arrow show the direction of wave propagation (Figure 2 from Li *et al.*, 2000, reprinted from *Johns Hopkins APL Technical Digest* with permission, figure © 2000 The Johns Hopkins University Applied Physics Laboratory, RADARSAT data © Canadian Space Agency/Agence Spatiale Canadienne 1996. Processed and distributed by RADARSAT International, courtesy Pablo Clemente-Colón). (b) RADARSAT ScanSAR-Wide image taken on April 26, 1998 of the westward propagation of internal waves in the South China Sea and in the vicinity of Dongsha Island and its surrounding coral reefs. The image measures about 240 km by 240 km. Pixel size in original image is 100 m. See text for further description (Adapted from Figure 5 of Hsu and Liu, 2000, figure © 2000 Canadian Aeronautics and Space Institute, RADARSAT data © Canadian Space Agency/Agence Spatiale Canadienne 1996. Processed and distributed by RADARSAT International, courtesy Antony Liu).

357

internal waves and, at the lower left, Dongsha Island and its surrounding coral reefs. The waves are generated from the interaction of the Kuroshio current with the shallow topography in Luzon Strait, which lies outside the image to the east. The internal waves propagate toward Dongsha Island, where at the reefs they divide into two packets that interact with each other west of the island. Hsu and Liu (2000) use such observations to verify theoretical models of internal wave interactions. In the upper part of the image, the bright highly reflective areas that resemble clouds are Bragg scatterers generated by rain cells.

13.6.2 Sea ice

Onstott (1992) describes radar backscatter from different classes of sea ice. As the sea ice thickness increases from open water, to thin, young, first-year and multiyear ice, the surface roughness and backscatter generally increase, so that SAR allows for discrimination of some ice types. Exceptions to this general increase include open water, which because its brightness depends on wind speed, can be either brighter or darker than the adjacent ice. Because the pancake ice shown in Figure 2.9b has small diameters, raised rims and a quasiperiodic distribution of floes, Bragg scatter also makes it appear very bright (Wadhams and Holt, 1991). A phenomenon called 'frost flowers' that forms on the surface of new ice also induces a bright transient return from Bragg scatter (Nghiem *et al.*, 1997).

With this as background, this section discusses five examples of sea ice imagery. The first three show the Arctic pack ice at three different scales, basin-wide, 500 km and 10 km. Of these, the basin-wide image is an example of the Arctic snapshot, the medium scale shows the pack ice within the snapshot, and the small scale image gives a sequence of pack ice images analyzed with the RGPS. The last two are a multifrequency image ScanSAR image of the southern ocean ice edge and a combined SAR and AVHRR image of an open water region in the Bering Sea pack ice.

First, for November 2–5, 1997, Figure 13.19 shows the entire Arctic 500-km wide ScanSAR swath coverage, or Arctic snapshot, as processed at the Alaska SAR facility. The swaths are radiometrically uncorrected, as is especially shown in the swath marked 'Chukchi Sea'. The coasts are outlined in white; the Chukchi Sea, Alaska and Russia are labeled. The most prominent feature is the open water in the Chukchi Sea, which is maintained by the warm water flux through the Bering Strait. Because the snapshot consists of both descending and ascending passes made over a 3-day period and under different wind and temperature conditions, the swath brightnesses differ from one another. These snapshots are repeated at 3 to 6-day intervals, and are used in the analysis of the pack ice motion and deformation.

Figure 13.20 shows an image of the Beaufort Sea pack ice measuring about 500 km square and taken from the Arctic snapshot. There are two classes of ice in the image. To the left, the dark ice is thin first-year ice that formed adjacent to the coast during the fall freeze-up. To

Figure 13.19. The Arctic snapshot, or the 3-day overlay of RADARSAT ScanSAR imagery of ice and open water in the Arctic Ocean taken within the Alaska SAR Facility receiving mask. The snapshot includes swaths from days 306 to 30° or November 2–5, 1997; the ScanSAR was processed at a 300-m resolution. (RADARSAT data © Canadian Space Agency/Agence Spatiale Canadienne 1997. Processed and distributed by RADARSAT International, figure courtesy Nettie LaBelle-Hamer and the Alaska SAR Facility, used with permission).

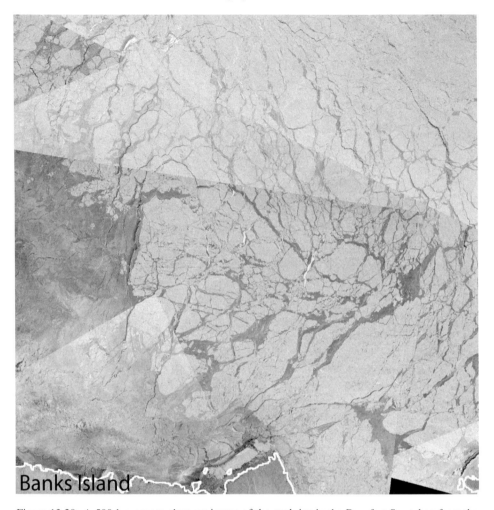

Banks Island

Figure 13.20. A 500-km square close-up image of the pack ice in the Beaufort Sea taken from the Arctic snapshot in Figure 13.19. See text for further description (RADARSAT data © Canadian Space Agency/Agence Spatiale Canadienne 1997. Processed and distributed by RADARSAT International, figure courtesy Nettie LaBelle-Hamer and the Alaska SAR Facility, used with permission).

the right, the figure shows the large multiyear ice floes characteristic of the central Arctic pack with their bright backscatter, separated from each other by darker leads covered with thin ice. The large floes have characteristic scales of 25–75 km. The very bright linear features are caused by either specular scattering from pressure ridges, or Bragg scattering from leads covered with frost flowers.

The snapshots are processed through the RGPS to derive ice statistics and motion. As an example of this processing and for the 1996 autumn, Figure 13.21 shows a series of nine pack ice images taken over a 41-day period (Kwok *et al.*, 1999). Because older

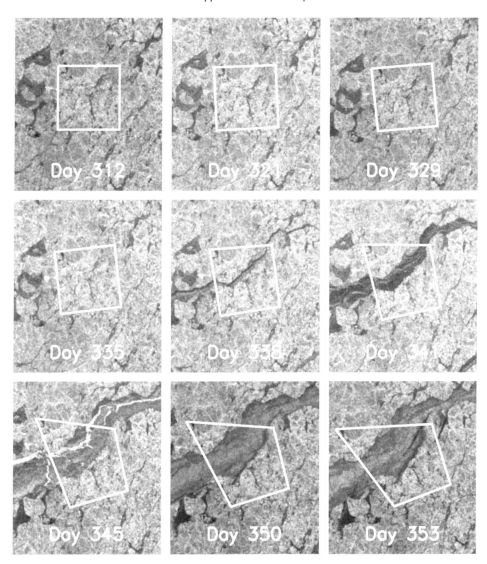

Figure 13.21. Time series of RADARSAT observations of sea ice at various intervals in the Beaufort Sea during 1996, showing the deformation of an initial 10-km square box over a 41-day period from Day 312 (November 8) to 353 (December 19). The white outlined square on Day 312 is the 10-km box; the successive images show its deformation (Figure 1 in Kwok *et al.* 1999, © 1999 American Geophysical Union, reproduced/modified by permission of AGU; RADARSAT data © Canadian Space Agency/Agence Spatiale Canadienne 1996. Processed and distributed by RADARSAT International, courtesy Ron Kwok).

ice generally has a greater backscatter than young ice, old ice is white, young ice is black. For each day, the geographic area is not fixed, rather the RGPS uses correlation methods to track common ice features, so that the same ice features appear in each image. In the Day 312 image, the white square or cell measures 10 km square; its distortion in the later images illustrates the shear and divergence associated with the ice motion. The cell area remains nearly constant until Day 338, when a lead containing open water and thin ice opens within the cell. The lead continues to open between Days 341 and 345, adding to the area of new ice and distorting the original cell. At the end of the 41-day period, thin new ice occupies more than 50% of the cell. Such imagery is used on a large scale to determine the ice statistics used in the verification of numerical models.

All of the previous SAR images have been presented as single channel, gray scale images. Multiple observing frequencies are also used to generate false color images. As an example from the Weddell Sea ice edge, Plate 20 shows a radar image obtained on October 5, 1994 from the SIR-C/X SAR described above. The image is oriented approximately east–west; the image dimensions are 240 km by 350 km. The colors correspond to the following: red is C-band VV; green is L-band HV; and blue is L-band VV. Historically, this was the first ScanSAR image. The image shows the boundary between pack ice and open ocean in the Weddell Sea and also shows two large clockwise or cyclonic eddies in the ice. The open ocean to the north is a uniform blue, due to the generation of Bragg scatterers by strong winds. The dark green ice at the lower right labeled A is first-year pack ice, with typical thicknesses of 0.5 m. The large black region to the center right labeled B is an area of grease ice as discussed in Chapter 2 and shown in Figure 2.9a. Grease ice is a slurry of small ice crystals, with characteristic crystal scales of about 1 mm, which collects on the ocean surface and damps out the Bragg scatterers, so that it is dark and non-reflective. The figure also shows the bright, white or light blue ice that is advected by the ocean eddies labeled C; this is probably pancake ice. Although multifrequency SARs have only been flown experimentally on the Space Shuttle, they may be flown operationally in the future. (Description courtesy of the JPL SIR-C/X SAR website, http://www.jpl.nasa.gov/radar/sircxsar.)

Finally, Plate 21 shows separate and combined ScanSAR and AVHRR images of the open water or polynya region south of St. Lawrence Island in the Bering Sea. Such large persistent openings in the ice cover are regions of strong atmospheric heat flux and large ice and brine generation (Martin, 2001). The smaller images show near simultaneous AVHRR and ScanSAR images, where the AVHRR pixel size is 1 km, and the ScanSAR pixel size is 200 m. The images are oriented so that north is toward the upper right-hand corner. The wind velocity is approximately northerly at 20 m s^{-1}, the air temperature over the polynya is about $-15°$ C.

Because of the northerly winds, the area south of the island is swept clear of pack ice. Within this region, the combination of wind and waves generates frazil ice in the water, where a Langmuir circulation herds the ice into the long linear streaks that are parallel to the wind and visible in the image. The AVHRR image is presented in terms of ice surface temperature, calculated using the split-window algorithm of Key, *et al.* (1997). The temperatures show that the frazil ice region south of the island is relatively warm and the thick pack ice to the north is cold. The large lower image shows a superposition of the AVHRR and SAR images; the AVHRR temperature provides the color, SAR provides the texture. The combined image shows that the region of Langmuir streaks is relatively warm while the surrounding thicker pack ice is colder. The image illustrates the importance of polynyas in the Arctic heat balance.

13.7 Satellite SARs 2002–2005

The three SARs launched or scheduled for launch between 2002 and 2005 include the European ENVISAT ASAR (Advanced SAR), launched in March 2002, the pending 2004 Canadian RADARSAT-2 and the 2005 Japanese ALOS PALSAR (Advanced Land Observing Satellite Phased Array L-band SAR). These SARs offer a variety of polarimetric modes that will improve the ability to study and model sea ice and to monitor oil spills.

13.7.1 ASAR

The European ENVISAT was launched in March 2002 and carries an Advanced Synthetic Aperture Radar (ASAR). ASAR operates at C-band (5.6 GHz) and is an advanced version of the AMI on ERS-1 and ERS-2 (Desnos *et al.*, 2000). ASAR looks to the right of the flight path, and of the three AMI modes described in Chapter 11, includes the SAR and wave mode, but not the wind mode. The ENVISAT solid state recorder is designed to hold 10 minutes of data at 100 Mbps. The SAR antenna is made up of 320 transmit/receive modules that are used for beam shaping. The SAR features include enhanced capability in terms of coverage, range of incidence angles, polarization, and modes of operation, including a 400-km wide ScanSAR mode. Although ASAR does not have a quad-pol mode, it has an alternating polarization mode similar in operation to ScanSAR. In this mode, instead of carrying out ScanSAR imaging of two adjacent swaths at the same polarization, ASAR images the same surface swath at two polarizations. These include HH and VV, or HH and HV, or VV and VH, where the cross-track width of this mode can be as large as 100 km. The ScanSAR mode, however, is only available in HH or VV (from http://envisat.estec.esa.nl/instruments/asar/descr/concept.html).

13.7.2 RADARSAT-2

The Canadian RADARSAT-2 is scheduled for launch in 2004 and has a design life of 7 years (Thompson, *et al.*, 2001). RADARSAT-2 will be at a different position along the same orbit as RADARSAT-1. The satellite carries a solid state recorder with a 200-Gb capacity and a data transfer rate from the antenna of 400 Mbps. For ground communications and data transfer, RADARSAT-2 has a high power X-band downlink, so that the ground stations can operate with a 3-m diameter antenna. The downlink transfer rate remains at about 100 Mbps. The satellite orbital position will be determined from a combination of

GPS and a numerical orbit model. RADARSAT-2 supports all the existing RADARSAT-1 beam modes and also operates at 5.6 GHz. It offers two quad-pol modes (VV, VH, HH, HV), both with a 25-km swathwidth, but with differing resolutions. For the wider swath Standard and ScanSAR modes, the quad-pol mode is not available, instead the satellite transmits in H or V, and receives either in H or V, or alternatively in H and V. Because the RADARSAT-2 SAR is mounted directly beneath the satellite, it can acquire images both to the left and right of the satellite nadir track, which will double the accessibility swath (from http://radarsat.mda.ca).

13.7.3 ALOS PALSAR

The Japanese Advanced Land Observing Satellite (ALOS) is a sun-synchronous mission planned for launch in about 2005 that will carry PALSAR, a joint project between NASDA and the Japan Resources Observation System Organization (JAROS). PALSAR offers the Standard and 250–350 km wide ScanSAR modes, as well as an experimental quad-pol SAR. To utilize this data, ALOS has a high speed and large capacity data handling technology. PALSAR provides a 240-Mbps downlink via the geosynchronous Japanese Data Relay Technology Satellite and a 120-Mbps direct broadcast downlink to ground stations (http://alos.nasda.go.jp/index-e.html).

13.8 Further reading and acknowledgements

The collection of papers in Henderson and Lewis (1998) provide an excellent introduction to SAR; Elachi (1987) and Ulaby *et al.* (1981, 1982, 1986) describe the SAR principles of operation. This chapter draws on an unpublished lecture by Anthony Freeman called "The Future of SAR Systems", which was presented as a short course at the International Geoscience and Remote Sensing Symposium (IGARSS) meeting in Hamburg, Germany on June 27, 1999. Other references include the collection of papers in Carsey (1992) and in Carsey, McNutt, and Rothrock (1994). The author particularly acknowledges the contribution of Ben Holt to the discussion of the San Francisco SAR images in Figures 13.13 and 13.14, and thanks Pablo Clemente-Colón, Ron Kwok, Nettie LaBelle-Hamer, Antony Liu and Dale Weinbrenner for their contributions. Leonid Mitnik provided the information on the Ukrainian and Russian space program. I also thank the Canadian Space Agency for their help in obtaining the imagery used in this chapter.

14

Future oceanographic satellite systems: 2004 to 2019

14.1 Introduction

There are a large number of satellite oceanography missions approved or proposed for the next two decades, where "approved" means that the mission is funded and has a launch date and "proposed" means it is under consideration. Previous chapters discuss some of these missions; they include the Gravity Field and Steady-State Ocean Circulation Explorer (GOCE) in Chapter 12, the ESA Advanced Scatterometer (ASCAT) in Chapter 11 and a variety of SARs in Chapter 13. The following sections describe five additional categories of future missions and instruments, their purpose and approximate timing.

First, Section 14.2 describes the European METOP contribution to a new US/Europe POES system, and discusses the transition from the POES and DMSP operational systems to the new National Polar-orbiting Operational Environmental Satellite System (NPOESS). As the successor to POES and DMSP, NPOESS will also continue some of the EOS TERRA and AQUA observations. As Section 14.3 describes, NPOESS will carry two new oceanographic instruments, the Conical-scanning Microwave Imager/Sounder (CMIS) and the Visible/Infrared Imager/Radiometer Suite (VIIRS).

Second, there are three new altimetry missions, one launched, one approved and one proposed. Section 14.4 describes the first two missions, the NASA Geoscience Laser Altimeter System (GLAS) on the Ice, Cloud and land Elevation Satellite (ICESat) launched in January 2003 and the ESA SAR Interferometric Radar Altimeter (SIRAL) on CryoSat scheduled for 2004. Both will determine the surface topography of polar ice caps and sea ice to a better resolution than the traditional altimeter. Section 14.5 describes the third, the Wide Swath Ocean Altimeter (WSOA) mission, which is the JASON-2/Ocean Surface Topography Mission (OSTM) successor to JASON-1. WSOA combines the JASON altimeter with an interferometric real aperture radar and will make 200-km wide measurements of sea surface height along the TOPEX ground track.

Third, Section 14.6 describes the ESA Soil Moisture and Ocean Salinity (SMOS) and NASA Aquarius satellites. These will retrieve sea surface salinity (SSS), a physical property that is critical to understanding the ocean circulation and global water cycle and is not presently measured from space. SMOS retrieves SSS using passive microwave radiometry and is approved; Aquarius uses a mixture of passive radiometry and radar and is funded for future study. Fourth, Section 14.7 describes the Argo program that will survey the upper 2 km of the ocean with a global array of about 3000 profiling buoys reporting their observations by satellite. These buoys are the oceanic equivalent of weather balloons in meteorology; their data will contribute to numerical oceanic forecast models.

Fifth, Section 14.8 discusses the Japanese Global Change Observation Missions (GCOM) that are the successors to ADEOS-2. GCOM divides into the GCOM-A atmospheric observation satellites and the GCOM-B oceanographic satellites. GCOM-B1 includes an ocean color instrument, a passive microwave conical scanner and the scatterometer successor to SeaWinds. Section 14.9 summarizes the approved and proposed missions for the period 2000–2019, with the missions divided into the categories of vector wind retrieval, ocean and ice topography, ocean color and SST. Section 14.10 concludes with some final thoughts.

14.2 POES to NPOESS transition

The purpose of NPOESS is to merge the US Department of Defense DMSP meteorological satellites, the Department of Commerce POES satellites and certain aspects of the NASA TERRA and AQUA satellites into a single operational system. NPOESS will replace the current four US operational satellites with one European and two US. The system is planned in part as a cost-saving measure, in part to expand the POES and DMSP capabilities to include ocean color retrieval and passive microwave retrieval of vector winds, and in part to continue certain of the EOS observations that began with TERRA and AQUA.

A 1994 US presidential directive initiated NPOESS, which is administered by the Integrated Project Office (IPO) with representatives from NASA, Commerce and Defense. Figure 14.1 shows the timeline for this transition. As the following sections discuss in detail, the timeline begins in 1999 with the current POES system and the launch of TERRA, then continues in 2002 with the launch of AQUA and the NOAA-17 replacement for NOAA-15. In 2003, WindSat was launched; in about 2005, METOP-1 is scheduled to replace the current NOAA-17 and the NPOESS Preparatory Project (NPP) satellite will replace TERRA. In 2009, the first NPOESS satellite will replace the early morning DMSP; in 2011, the second NPOESS will replace NOAA-N or NOAA-N′; in 2013, NPOESS 'lite' will be launched. Post 2014, the three NPOESS orbits have equatorial crossing times of approximately 0530 descending, 0930 descending and 1330 ascending. For any particular geographic area, the ascending and descending NPOESS day and night overpasses will occur at approximately 4-hour intervals. The first launch in the preparation for NPOESS occurred in 2002 when the mid-morning 1000 descending NOAA-17 replaced the early morning 0730 descending NOAA-15. The purpose of this orbit change was to satisfy the NPOESS observing requirement for a mid-morning satellite; the NOAA-15 early morning observations were taken over by the 0730 DMSP satellite.

The transition from POES to NPOESS takes place in three stages. The first is under a NOAA/EUMETSAT program called the Initial Joint Polar-orbiting operational satellite System (IJPS) that replaces the mid-morning POES satellite with the European METOP

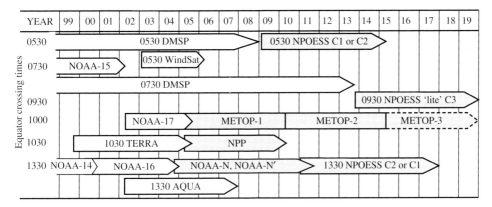

Figure 14.1. The various satellites, their launch years and duration for the transition from POES to NPOESS. The horizontal axis is in years, the vertical axis gives the equator crossing times. All of these satellites are sun-synchronous. A white background means that in 2003 the satellite is in orbit; gray means the mission is approved; a dashed outline means the mission is under consideration. See text for further description (Figure design courtesy Michael Van Woert, National Ice Center and William Patzert, JPL).

satellite. The second is the test of certain NPOESS concepts and instruments on the Coriolis and NPP satellites. The third is the transition from the POES, DMSP and METOP satellites into the single NPOESS system.

14.2.1 IJPS and the European contribution to POES

In about 2005 and under IJPS, when the current mid-morning NOAA-17 satellite approaches the end of its lifetime, METOP-1 will replace it in the same orbit (Figure 14.2). Then after an additional 5 years, METOP-2 will replace METOP-1, also in the same orbit. Three METOP spacecraft (METOP-1, -2, -3) will be built, each with a 5-year lifetime and with 2005 as the earliest possible launch date for METOP-1. Under IJPS, NOAA has provided EUMETSAT with a variety of meteorological instruments for the flights of METOP-1 and METOP-2 including two AVHRR/3. METOP will also carry specific EUMETSAT instruments such as the ASCAT wind scatterometer described in Chapter 11.

The current POES system, with METOP-1 replacing the mid-morning NOAA satellite, will continue through about 2008. Then, as Section 14.2.3 describes in more detail, the first NPOESS satellite will replace the early morning 0530 DMSP satellite. The next replacements occur for the early afternoon 1330 orbit that is presently occupied by AQUA and NOAA-16. When NOAA-16 approaches the end of its life, it will be replaced by NOAA-N, which at the end of its life will be replaced by either NOAA-N′ or, in about 2011, by an NPOESS satellite. METOP-3 is tentatively scheduled for launch in 2015 and does not come under the IJPS; in 2003 its payload remains under discussion. Under IJPS, NOAA has reserved an additional AVHRR/3 as a spare for NOAA-N and -N′ and METOP-1

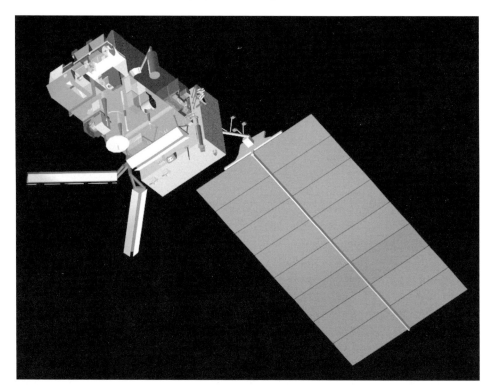

Figure 14.2. Artist's conception of METOP-1. The antennas for the Advanced Scatterometer (ASCAT) are visible beneath the spacecraft (Courtesy ESA, used with permission).

and -2. Because the NPOESS VIIRS is physically too large and has data rates that cannot be accommodated by METOP-3, EUMETSAT wants to fly the spare AVHRR/3 on METOP-3. Consequently, the transition to NPOESS may not be complete until METOP-3 comes to the end of its lifetime, or until 2019.

14.2.2 Preparation for NPOESS

Independent of METOP, the transition to NPOESS involves two intermediate satellites. The first was the January 2003 launch of the WindSat polarimetric microwave radiometer on the Coriolis satellite that is testing the concept of passive microwave vector wind retrieval. The second is the NPP satellite scheduled for launch in 2005. NPP will carry the most complicated and data-intensive NPOESS instruments, including VIIRS, CMIS and the non-oceanographic Cross-track Infrared Sounder (CrIS), which will not be discussed further. VIIRS will continue the AVHRR observations and certain of the MODIS bands at a better resolution; CMIS will continue the passive microwave observations represented by SSM/I, TMI, AMSR-E and WindSat.

The NPP satellite will be in a 1030 sun-synchronous descending orbit at an altitude of 833 km. This orbit matches the TERRA crossing time and the NPOESS altitude, where the TERRA altitude is 720 km. NPP serves at least three purposes. First, except for fluorescence measurements, VIIRS will continue the TERRA MODIS ocean color observations. Second, during the period when the POES and DMSP satellites and their associated ground systems remain intact, NPP will provide users with an opportunity to gain experience with the much greater NPOESS data rates and the more complex instruments. Third, CMIS will continue the radiometer observations of vector winds that began with WindSat. During at least part of the 5-year NPP operational period and until the launch of the first NPOESS satellite, the system will consist of five satellites: DMSP satellites at 0530 and 0730, METOP-1 at 1000, NPP at 1030 and NOAA-N or -N′ at 1330.

A major concern of the NPOESS transition involves the increased rates of data transfer to ground stations. NPOESS will broadcast at a high and low data rate. The High Rate Data (HRD) will transmit all information at 20 Mbps over an X-band downlink. The Low Rate Data (LRD) will transmit the CMIS microwave data and certain of the VIIRS image channels in real time at 3.5 Mbps over the POES 1.7 GHz downlinks. The LRD format will be approximately compatible with the Advanced High Resolution Picture Transmission (AHRPT) proposed for METOP. For comparison, POES has characteristic data rates of 0.6 Mbps that take place over a 1.7 GHz downlink; DMSP has rates of 1.0 Mbps at 2.2 GHz. The EOS satellites have downlink rates of 15 Mbps and, as described in Chapter 13, RADARSAT has downlink rates of 100 Mbps. The NPOESS HRD rates are comparable to the EOS rates and represent an increase of 20:1 over POES; the LRD rates are six times larger than POES (Nelson and Cunningham, 2002).

14.2.3 NPOESS

Five NPOESS satellites will be built with two additional payloads for EUMETSAT spacecraft, where each NPOESS satellite has an estimated 7-year lifetime (Figure 14.3). The tentative schedule for replacement of the POES and DMSP satellites with NPOESS satellites is as follows. In about 2008, the 0530 DMSP satellite will be replaced by the first NPOESS C1 or C2 satellite that will also carry a radar altimeter. Because this will not occur until the DMSP satellite begins to fail, the exact replacement date is unknown.

In about 2011, the second NPOESS satellite will replace the early afternoon 1330 POES satellite. This crossing time is presently occupied by AQUA satellite and NOAA-16. NOAA-16 with its replacements NOAA-N and N′ should continue through 2011 and AQUA is designed to last until about 2007, while the earliest availability of the second NPOESS satellite is 2009. This suggests that, with a possible two-year gap, NPOESS will continue most of the AQUA ocean color observations.

The second 0730 DMSP satellite will be supported through 2013, at which time it will be replaced by the NPOESS 'lite' C3 satellite in an 0930 descending orbit. NPOESS 'lite' will carry a reduced set of NPOESS instruments, primarily VIIRS and CMIS. At this time, the NPOESS three-satellite system, with crossing times of 0530, 0930 and 1330 will be in

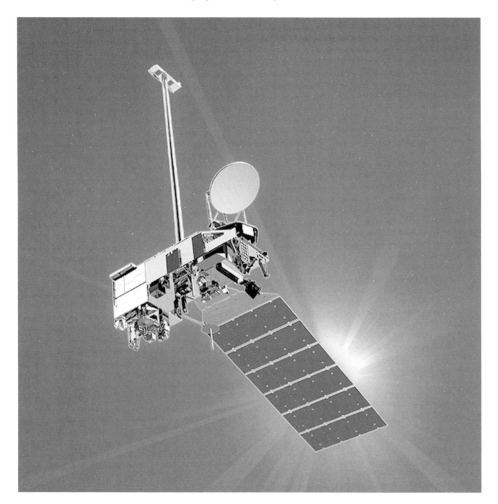

Figure 14.3. Artist's conception of the NPOESS spacecraft; the solar panel is to the rear, the 2.2-m diameter CMIS radiometer antenna is on top (Figure © Northrop Grumman Corporation, 2000. All rights reserved. Republished by kind permission of Northrop Grumman Corporation).

place, although with NPOESS 'lite' in the mid-morning orbit. The transition to a full US/ European three-satellite system should occur in about 2019, when a fully NPOESS-equipped European satellite is scheduled to replace NPOESS 'lite' in the mid-morning orbit.

14.3 NPOESS oceanographic instruments

The first NPOESS instrument is the WindSat precursor to CMIS that was launched on the Coriolis satellite in 2003. The next set of NPOESS oceanographic instruments are the VIIRS and CMIS scheduled for NPP and NPOESS, where VIIRS replaces the AVHRR, OLS, SeaWiFS and MODIS imagers, and CMIS replaces SSM/I and SSMI/S on DMSP, the WindSat radiometer, and certain of the functions of TMI on TRMM and of AMSR-E on AQUA. The next sections respectively discuss VIIRS, WindSat and CMIS.

14.3.1 Visible/Infrared Imager/Radiometer Suite (VIIRS)

VIIRS is a hybrid cross-track instrument (Section 1.6.4) with 22 bands at wavelengths between 0.3 and 14 μm (Hommel *et al.*, 2002). The smaller number of VIIRS bands compared with the 36 MODIS bands reduces the VIIRS complexity, cost and weight. The VIIRS instrument consists of a rotating telescope combined with the MODIS elliptical FOV and linear arrays of along-track sensors. This should eliminate the problems that occur with the MODIS two-sided rotating mirror. VIIRS has a cross-track view angle of ±56° and a 3000-km swathwidth, which is 30% greater than the MODIS swathwidth.

At nadir, the VIIRS FOV extends 11.87 km in the along-track direction and 750 m in the along-scan direction. Within the instrument, the FOV radiances are focused onto linear detector arrays. Each array consists of either 16 or 32 detectors in the along-track direction and 3 detectors in the along-scan direction, where for reasons discussed below, the along-scan detectors improve the off-nadir resolution. At nadir, the arrays of 16 along-track detectors yield an along-track and along-scan resolution of 750 m, called a *Moderate* resolution; the arrays of 32 along-track detectors yield a resolution of 375 m called *Imaging*.

Table 14.1 compares the spatial resolution and wavelengths of the VIRSS bands with selected AVHRR, SeaWiFS and MODIS bands, where the VIIRS bands are listed as M for Moderate and I for Imaging. In the table, VIIRS bands that either match or lie in the wavelength range of the AVHRR/3 bands are underlined; the boldface bands match the MODIS ocean color bands. Examination of the underlined bands shows that with an improved spatial resolution, VIIRS almost exactly matches AVHRR bands 3B, 4 and 5, so that the AVHRR SST algorithms should work with the VIIRS bands. Although AVHRR bands 1 and 2 do not exactly equal their VIIRS counterparts, comparable VIIRS bands with a 375-m resolution nest within the corresponding AVHRR bands, so that the AVHRR cloud algorithms should also apply. The table also shows at 3 and 10 μm, VIIRS has similar capabilities to MODIS, while at 8 μm, the MODIS bands are reduced to a single VIIRS band.

Examination of the ocean color bands shows that similar to MODIS, VIIRS has aerosol correction bands at 746 nm and 865 nm, a sediment band at 410 nm, and approximately matches the 442-nm, 488-nm and 551-nm bands used in the OC3M algorithm described in Section 6.6.3. The VIIRS bands have a characteristic width of 20 nm, compared with the MODIS 10-nm width. VIIRS also has no equivalent for MODIS band 11 at 530 nm or for SeaWiFS band 4 at 510 nm. A major difference between VIIRS and MODIS is that because VIIRS lacks the equivalent of the triplet of MODIS bands 13, 14 and 15 described in Section 6.5.2, it cannot measure fluorescence line height (FLH). Instead, VIIRS combines MODIS bands 13 and 14 into a single band at 662–682 nm that extends across the fluorescence peak.

An advantage of VIIRS is that it partially corrects for the bowtie effect or FOV distortion at large look angles. As Chapter 1 describes, one problem with cross-track scanners is that as the scan angle increases from nadir to large oblique angles, the FOV also increases. This

Table 14.1. *Comparison of VIIRS ocean color and thermal bands with SeaWiFS, MODIS and AVHRR/3*

VIIRS			SeaWiFS	MODIS		AVHRR/3	
Band	Wavelength (nm)	Nadir resolution (m)	Band	Band	Wavelength (nm)	Band	Wavelength (μm)
DNB	500–900						
M1	**402–422**	**750**	**1**	**8**	**405–420**		—
M2	**436–454**	**750**	**2**	**9**	**438–448**		—
M3	**478–498**	**750**	**3**	**10**	**483–493**		—
M4	**545–565**	**750**	**5**	**12**	**546–556**		—
I1	600–680	375		1	620–670	1	0.58–0.68
M5	662–682	750	6	13	662–672		—
				14	673–683		
M6	**739–754**	**750**	**7**	**15**	**743–753**		
I2	**846–885**	**375**		**2**	**841–876**	2	0.725–1.00
M7	**846–885**	**750**	**8**	**16**	**862–877**		
	(μm)				(μm)		
M8	1.23–1.25	750		5	1.23–1.25		
M9	1.37–1.39	750		26	1.36–1.39		
I3	1.58–1.64	375		—		3A	1.58–1.64
M10	1.58–1.64	750		—		3A	1.58–1.64
M11	2.22–2.28	750		—			
I4	3.55–3.93	375		—		3B	3.55–3.93
M12	3.66–3.84	750		20	3.66–3.84	3B	3.55–3.93
M13	3.97–4.13	750		22	3.93–4.00		
				23	4.02–4.08		
M14	8.40–8.70	750		29	8.40–8.70		
M15	10.26–11.26	750		31	10.78–11.28	4	10.3–11.3
I5	10.50–12.40	375					
M16	11.54–12.50	750		32	11.77–12.27	5	11.5–12.5

Ocean color bands are in bold face; AVHRR bands are underlined. The SeaWiFS wavelengths are not shown, but are approximately given by the MODIS wavelengths. M stands for Moderate resolution, I stands for Imaging resolution, DNB is the day/night band exclusive to OLS.
From Hommel *et al.*, 2002

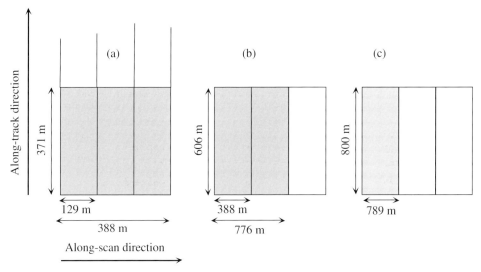

Figure 14.4. For the VIIRS Imaging resolution bands, the along-scan configuration of the number of detectors used to determine the FOV as a function of zenith angle. The gray sensors are used in the retrieval of the surface radiance; the adjacent dimensions give the size of the surface FOV. (a) Nadir view; (b) $\theta_Z = 32°$; (c) $\theta_Z = 45°$.

increase is due to geometric effects and the Earth's curvature, and is much greater in the along-scan than in the along-track direction. As the following shows, VIIRS compensates for this increase by having the number of along-scan sensors decrease with increasing zenith angle.

Figure 14.4 shows the configuration of the along-scan sensors, and, for specific values of the zenith angle θ_Z, the FOV dimensions. For $0° < \theta_Z < 32°$, three sensors determine the FOV, where, as Figure 14.4a shows for the nadir case, the FOV generated by the sensors is nearly square and measures 371×388 m^2. For $32° < \theta_Z < 45°$, the number of sensors determining the FOV decreases from three to two, yielding at $32°$ an FOV of 606×776 m^2, so that it remains approximately square. For angles greater than $45°$, the number of sensors decreases from two to one, yielding at $45°$ a square FOV measuring about 800×800 m^2. From this dependence of the number of viewing sensors on zenith angle, the linear FOV dimensions increase by a factor of 2; without it, the along-scan dimension would increase by a factor of 6. The Moderate resolution scans use an identical technique to maintain the FOV size.

In summary, VIIRS includes all of the AVHRR bands at a finer resolution and provides sufficient visible and NIR bands for application of a cloud algorithm. The VIIRS thermal IR bands provide for SST retrieval at both 12 and 4 μm. Although VIIRS cannot retrieve FLH, VIIRS has sufficient ocean color bands for aerosol correction and for application of both a semianalytic algorithm and a maximum band ratio empirical algorithm similar to OC3M. From VIIRS, both the mid-morning and early afternoon NPOESS satellites will provide ocean color and SST.

14.3.2 WindSat/Coriolis

The WindSat instrument on the Coriolis satellite launched in 2003 is a polarimetric microwave radiometer designed to test the concept of vector wind retrievals from multi-frequency V- and H-pol measurements as well as from measurements of all four Stokes parameters (St. Germain, Poe and Gaiser, 1998; St. Germain and Gaiser, 2000). As the primary instrument on Coriolis, WindSat is funded jointly by NPOESS and the US Navy. WindSat operates at an altitude of 830 km in a sun-synchronous 0600 dawn–dusk descending orbit designed to minimize solar reflectance (Gaiser, 1999). WindSat consists of a conically scanned radiometer mounted on top of the Coriolis satellite with a 1.83-m diameter antenna that observes the ocean surface at look angles of 50–55° over almost 360° of rotation (Figure 14.5). The instrument has a 1000-km swathwidth and a surface resolution of about 20 km.

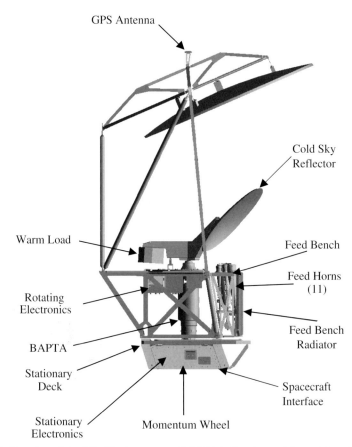

Figure 14.5. Drawing of the WindSat instrument. BAPTA is Bearing and Power Transfer Assembly. The instrument can view the sea surface over almost 360° of rotation (Figure 2 from St. Germain and Gaiser, 2000, © 2000 IEEE, used with permission).

Table 14.2. *WindSat frequencies, polarizations and Stokes parameters*

Frequency (GHz)	V-pol	H-pol	Third Stokes	Fourth Stokes
6.6	X	X	—	—
10.6	X	X	X	X
18.7	X	X	X	X
23.8	X	X	—	—
37	X	X	X	X

For each frequency, the Xs mean that the parameter is acquired
St. Germain and Gaiser, 2000

For the different observational frequencies, Table 14.2 lists the polarizations and Stokes parameters observed by WindSat. As Sections 9.4.4 and 9.4.5 describe, WindSat is designed to retrieve vector winds in two ways: first, following Wentz (1992), from two HH and VV looks at the same surface area from ahead and behind the satellite; second, from a single look at all four Stokes parameters. The single-look wind retrieval exploits the dependence of different Stokes components on the azimuth angle relative to the wind direction shown in Figures 9.11 and 9.12. Given the additional cost and complexity of the two-look retrieval, the hope is that WindSat will validate the concept of a single look, vector wind retrieval.

The WindSat observations are calibrated with independent measurements of surface winds and atmospheric transmissivity made at the surface and by other satellites. These include winds from the SSM/I and SSMI/S radiometers on DMSP, the AMSR-E radiometer on AQUA, and the NDBC surface wind observations. Combined with WindSat observations, these should resolve the questions raised in Section 9.4.5 about the accuracy of the passive microwave vector wind retrieval and its dependence on wind magnitude.

14.3.3 Conical-scanning Microwave Imager/Sounder (CMIS)

The polarimetric CMIS microwave radiometer is scheduled for flight on the NPP and NPOESS satellites (Gasster and Flaming, 1998). Figure 14.6 compares a model of CMIS with TMI and SSM/I, and illustrates the much greater size of its 2.2-m diameter antenna. This antenna is a conically scanned rotating dish that reflects the received radiances into microwave feedhorns. On each rotation, CMIS also observes a cold space reflector and a warm calibration source. The instrument has a swathwidth of 1700 km and approximately 70 channels between 6 and 183 GHz, most of which are used as atmospheric sounders. Exclusive of the sounder bands, the CMIS bands include 6.6, 10.6, 18.7, 23.8, 36.5 and 89 GHz, so that CMIS will continue the passive microwave retrievals of SST, water vapor, liquid water, rain and sea ice described in Chapter 9. Because CMIS will be on multiple platforms, it should provide daily near global observations of SST. For the non-sounding

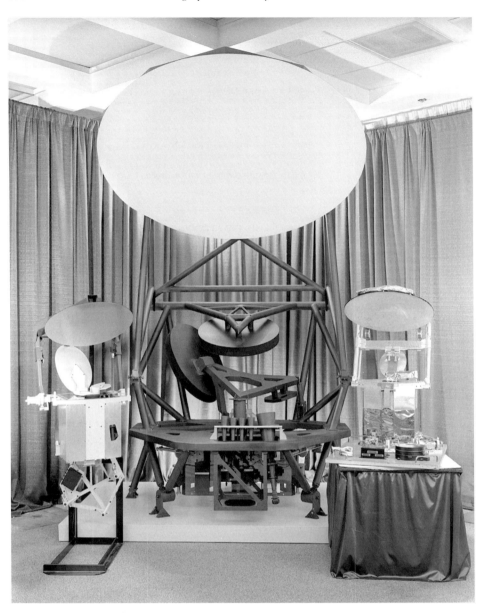

Figure 14.6. Comparison of a model of the CMIS polarimetric microwave imager (center) with the TMI (left) and SSM/I (right) (Figure © 2002 Boeing Satellite Systems, used with permission, courtesy James Jewell, NPOESS).

Table 14.3. *CMIS frequencies and their polarizations, Stokes parameters and footprint dimensions*

Frequency (GHz)	V-pol	H-pol	Third Stokes	Fourth Stokes	Footprint (km × km)
6.6	X	X	—	—	68 × 39
10.6	X	X	X	X	46 × 25
18.7	X	X	X	X	24 × 16
23.8	X	X	—	—	24 × 16
36.5	X	X	X	X	17 × 10
89	X	X	—	—	—

For each frequency, the Xs mean that the parameter is acquired.
Adapted from a presentation by Wentz and Meissner at the IGARSS NPOESS Workshop, June 23, 2002

channels, Table 14.3 lists the CMIS frequencies, polarizations, Stokes parameters and size of its surface footprints. CMIS retrieves all four Stokes parameters at 10.6, 18.7 and 36.5 GHz, so that it should be able to continue the WindSat single and two-look retrievals of vector wind speed.

14.4 GLAS and SIRAL altimeters

This and the following section describe three new altimeters. This section describes the GLAS laser altimeter on ICESat, launched in 2003, and the SIRAL interferometric radar altimeter on CryoSat, which is scheduled for launch in 2004. Both satellites are designed to investigate the topography of land and sea ice. Because GLAS has a spatial resolution of about 100 m but can be obscured by clouds, while SIRAL has a lower but cloud-independent resolution, the two complement each other. Section 14.5 describes the open ocean Wide Swath Ocean Altimeter (WSOA) successor to JASON-1 that is scheduled for launch in 2006.

The GLAS and SIRAL land ice investigations will determine the topography of the large ice sheets and the smaller but equally important mountain glaciers; the sea ice investigations will determine the topography of Arctic and Antarctic pack ice. The land ice consists of the Greenland and Antarctic ice caps and the glaciers located on the west coast mountain ranges of North and South America, the north coast of Europe and Asia, and the Tibetan plateau. The oceanographic importance of the land ice involves its potential melting and contribution to sea level rise. The pack ice investigation concerns determination of its extent and thickness and particularly of any changes in its thickness. These measurements are motivated by the recent submarine observations of Rothrock, Yu and Maykut (1999), which suggest that the Arctic pack is decreasing in thickness. Since the thinning or disappearance of the pack would yield a very different Arctic heat balance, these observations are critical to both oceanography and climatology.

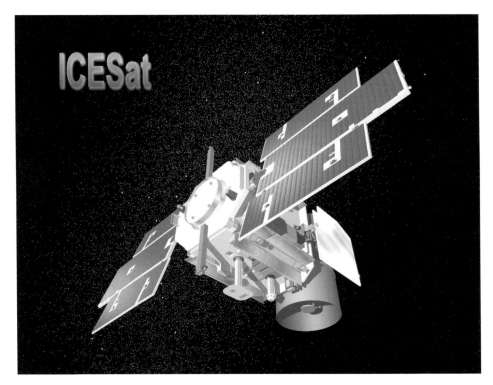

Figure 14.7. The GLAS instrument on ICESat. The figure shows the solar panels facing outward, the lasers pointing downward, and at the rear of the satellite, the 1-m diameter telescope that collects the reflected radiance (Figure courtesy Ball Aerospace & Technologies Corporation, used with permission).

14.4.1 ICESat and GLAS

ICESat operates in a near pole-to-pole, non sun-synchronous orbit at an altitude of 600 km (Figure 14.7). Over a 15-year period, the program assumes that three such satellites, each with a 5-year lifetime, will fly in sequence. The ICESat ground track has a 183-day repeat cycle with a 15-km equatorial separation, which will provide dense spatial coverage of the polar regions. The satellite carries three identical lasers, where only one operates at any time (Schutz, 1998).

   The satellite was designed on the assumption that two such lasers should provide for five years of operation, while the third was a spare. Unfortunately, due to problems with the photo-diodes, the first laser failed after just 36 days of operation. After evaluation of the failure, the second laser was turned on with the plan that it would operate for 45 days, following which it would be turned off for six months. At this writing, the long-term status of the satellite is uncertain.

   GLAS operates at two wavelengths. The first, in the NIR at 1.064 μm, was chosen to enhance the reflection of the laser radiance from the snow surface. The second, in the green at 532 nm, measures aerosols and other atmospheric characteristics. The NIR laser pulses have a length of 5 ns, a surface spot diameter of 70 m and a PRF of 40 Hz, so that the surface separation between pulses is 175 m. A 1-m diameter telescope collects the reflected radiance. In calculation of the ice topography, a combination of laser retroreflectors and GPS determines the ICESat orbital position; a star camera and gyroscopes determine the laser orientation.

The surface height of the illuminated spot is determined from the time delay between the pulse transmission and reception, the satellite position and the laser pointing angle. The advantage of the NIR laser is that it provides high resolution surface topography. Its disadvantage is that even though the laser is powerful enough to penetrate thin clouds and haze, the Arctic cloudiness varies from 60% in winter to 80–90% during June through October (Beesley and Moritz, 1999). Inevitably, periods will occur when it is impossible to view the surface.

14.4.2 CryoSat

CryoSat is a 3-year ice sheet and sea ice radar altimetry mission that will operate in a nearly circular, pole-to-pole orbit at an altitude of 720 km and an inclination of 92° (Figure 14.8). The orbit has a repeat period of 369 days and an equatorial track spacing of 7.5 km. The DORIS system and a laser retroreflector array will be used to measure the CryoSat orbit (Section 12.4.4; Phalippou *et al.*, 2001).

The principal instrument on CryoSat is the SAR Interferometric Radar Altimeter (SIRAL). SIRAL consists of two Cassegrain elliptical antennas mounted side-by-side to form a cross-track interferometer. The antennas measure 1.15 m by 1.4 m with their long axes parallel to the satellite trajectory and are separated by a baseline distance of 1.15 m. The purpose of the elliptical design is to accommodate both the launch vehicle fairing and the different beamwidth requirements in the along- and cross-track directions.

SIRAL operates at a single Ku-band frequency and in three modes (Francis, 2001). The first is a low-resolution mode where SIRAL acts as a classic single-frequency altimeter, using a single antenna to transmit and receive pulses, where the DORIS system will provide the ionospheric correction. This conventional pulse-limited mode will be used over the ocean and in the pack ice interior where the roughness is less than at the ice edge. For this case, a characteristic FOV diameter is about 15 km. The second is a SAR mode, where, to obtain better resolution in the along-track direction, the instrument uses a single antenna to transmit and receive with a PRF that is roughly ten times the low resolution mode. SAR/Doppler processing is used to divide the footprint in the along-track direction into 64 sub-bins. Each bin measures about 250 m in the along-track direction and, depending on surface roughness, up to 15 km in the cross-track direction. This mode will be primarily used over the rougher ice that occurs at the pack ice edge.

The third is a SAR-interferometric mode designed to measure ice sheet height over regions of sloping ice sheet topography. In this mode, at a PRF that is twice that of the

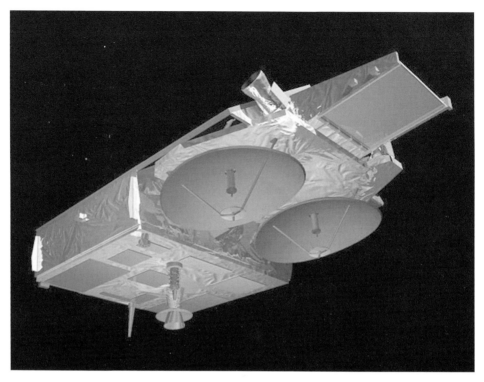

Figure 14.8. Artist's conception of CryoSat. The solar array is mounted in a shed-roof configuration on top of the satellite, the two SIRAL elliptical radar altimeter antennas are visible beneath the spacecraft. See text for further description (Figure courtesy Mark Drinkwater and the European Space Agency, © European Space Agency, used with permission).

SAR mode, the instrument transmits from one antenna and receives from both. Because the presence of a sloping surface means that the first return can be from a point other than nadir, the purpose of the interferometer is to determine this off-nadir angle and its associated range. This mode does not produce images, rather, at along-track intervals of about 250 m, it produces a single range measurement. Depending on the slope of the ice sheet, the surface location of each measurement wanders from side to side of the nadir track. Thus the combination of SIRAL and GLAS should provide complementary measurements of the pack and glacial ice topography.

14.5 JASON-2/Ocean Surface Topography Mission (OSTM)

As Chapter 12 shows, a single TOPEX altimeter can resolve sea surface topographic features with spatial wavelengths as small as 300 km and time scales of 20 days. This means that TOPEX can observe long period Rossby and Kelvin waves and interannual phenomena such as El Niño, but cannot observe shorter time and length scale phenomena such as upwelling, internal waves or short period Rossby waves. Although the launch of JASON-1 and the adjustment of the TOPEX ground track to lie midway between two adjacent JASON-1 tracks improves this resolution, the need remains for higher resolution measurements.

One goal of the OSTM replacement for JASON-1 is to determine sea surface topography at shorter spatial and time scales. The report of the High-Resolution Ocean Topography Science Working Group (HOTSWG), edited by Chelton (2001), summarizes the scientific requirements for future ocean topography measurements at higher spatial and temporal resolutions and describes three new altimeter technologies. These include multiple, inexpensive altimeter satellites that occupy adjacent orbits with ground tracks separated longitudinally from each other, the Wide Swath Ocean Altimeter (WSOA) satellite described below, and the use of reflected GPS signals to provide high resolution surface topography. The HOTSWG report proposes the following: (1) that planning begin immediately to build and launch a series of three to five altimetric satellites into orbits offset from the TOPEX/JASON-1 orbit; (2) that WSOA be incorporated into OSTM; (3) that testing continue on the GPS concept.

WSOA consists of a POSEIDON dual-frequency nadir-viewing altimeter, a nadir-viewing microwave radiometer, LRA, DORIS and GPS receivers for precision orbit determination, and a cross-track real aperture radar interferometer with a 7-m baseline between the two antennas (Figure 14.9). The central altimeter measures sea surface height (SSH) along the nadir ground track; the interferometer measures SSH in 100-km wide swaths on each side of this track. The ionospheric and tropospheric delays measured by the POSEIDON altimeter will be used to calibrate the interferometer return.

Within the swaths and consistent with the discussion of the SLR resolution in Section 13.3, the interferometric radars have an along-track resolution of about 13.5 km and a cross-track resolution in the near range of 670 m, improving in the far range to 100 m. The returns will be binned into 15-km cells. Because the equatorial separation of successive TOPEX/JASON ground tracks is about 300 km, over much of the globe the successive WSOA tracks will either overlap or be less than 100 km apart. The presence of the interferometer means that, between latitudes of $\pm66°$, any surface area will be imaged at least twice during the 10-day orbit repeat cycle and often more frequently (Rodriguez and Pollard, 2001). Within the swaths, WSOA will produce a two-dimensional sea surface topography, yielding direct measurements of geostrophic vector velocities.

14.6 Sea surface salinity

Sea surface salinity (SSS) is currently the most important oceanic variable not retrieved by satellite. Salinity is critical for determining the global water balance, for understanding ocean current systems and for estimating evaporation rates. As Dickson *et al.* (1988) describe in their study of the North Atlantic "great salinity anomaly", variations in SSS can be harbingers of climate change. As Chapter 9 explains, one reason for the past neglect of SSS is that, because it must be measured at 1.4 GHz, determination of SSS to a 50-km resolution at an altitude of 860 km requires at least a 4-m diameter antenna.

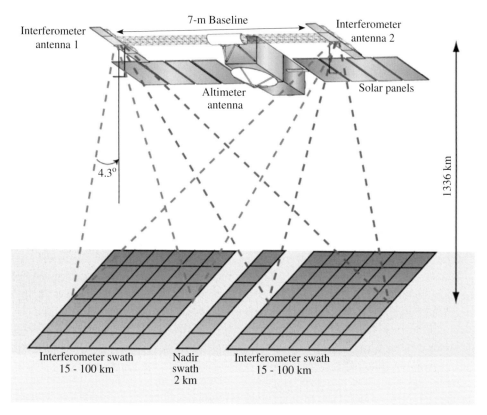

Figure 14.9. A drawing of the WSOA satellite, which combines the central JASON altimeter with a pair of interferometric real aperture radars. See text for further description (Figure 1 from Rodriguez and Pollard, 2001, used with permission).

There are now two instruments designed to provide SSS at this resolution; the European Soil Moisture and Ocean Salinity (SMOS) satellite (Figure 14.10), which is approved, and the US Aquarius satellite (Figure 14.11), which is funded but awaiting further study before flight approval. SMOS uses a fixed two-dimensional interferometric antenna; Aquarius uses a large non-rotating dish. Both satellites will retrieve salinity with an accuracy of 0.1–0.2 precision salinity units (psu) at a resolution of about 50 km. Because both instruments operate over a range of incidence angles, they differ from the passive microwave imagers described in Chapter 8 that operate at fixed incidence angles. This section first describes issues common to both missions, then discusses the specific instruments.

There are four concerns regarding SSS retrieval: (1) the magnitude of the reflected solar and galactic radiances; (2) the additional dependence of the emitted radiance on SST and surface roughness; (3) the need for a large antenna; (4) the requirement for a salinity accuracy of 0.1–0.2 psu. Addressing these concerns in this order, first, Table 9.1 shows that the solar brightness temperature at 1.4 GHz is about 2×10^5 K, so that to minimize solar reflection into the antenna feedhorns, both satellites will operate in a sun-synchronous dawn–dusk orbit (Section 9.3). For this orbit, the near grazing solar incidence angle will minimize solar

Figure 14.10. Artist's conception of the SMOS satellite (Courtesy Alcatel Space Industry and Yann Kerr; image courtesy MIRA Productions, used with permission).

reflection to the antenna. Aquarius further minimizes sun glint by having its antenna point to the night side of the orbit.

Second, Section 9.5 shows that the 1.4 GHz surface brightness temperature depends not only on SSS, but also on SST and surface roughness. For both instruments, observations from other satellites will provide SST, and for SMOS, will also provide surface roughness. In contrast, Aquarius carries an active radar that will retrieve near simultaneous measurements of surface roughness. Third, to fit the required large antenna into the launch vehicle, the SMOS antenna is a folding three-armed phased array; the Aquarius antenna is a folding clamshell. Finally, the salinity accuracy is constrained by the small dynamic range of brightness temperature associated with the global range of surface temperatures and salinities (Figure 9.14). This range corresponds to about 5 K, so that a SSS accuracy of 0.2 psu requires a brightness temperature accuracy of 0.1 K, which in turn requires an accurate, low noise radiometer.

14.6.1 SMOS

SMOS is approved for flight in 2006 and will measure both soil moisture and ocean salinity. SMOS employs a three-armed Y-shaped antenna, where each arm measures 4.5 m in length,

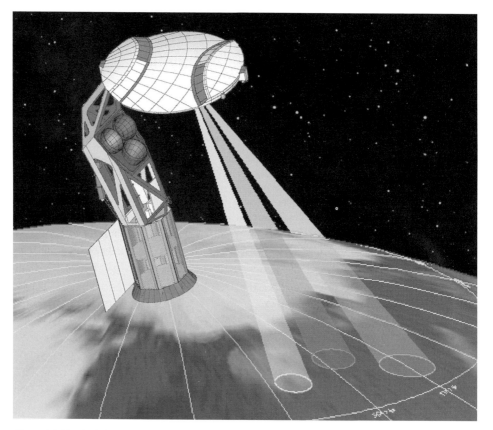

Figure 14.11. Artist's conception of the Aquarius satellite. The design details are tentative and subject to change (Courtesy C. Koblinsky, NASA GSFC).

contains 24 radiometers and folds for launch. The arms do not rotate but are fixed to the satellite; the polarized signals collected by the radiometers are cross-correlated to construct maps of brightness temperatures over a 1000-km swathwidth. The swath is divided into pixels with a spatial resolution of 35 km at nadir, increasing to about 50 km at the swath edge. The satellite will operate in a sun-synchronous 0600 ascending orbit at an altitude of 760 km. The choice of orbit and swathwidth means that the salinity measurements have a 3-day repeat cycle. The minimum mission duration is 3 years, the planned is 5 years.

14.6.2 *Aquarius*

Aquarius is an active/passive pushbroom instrument scheduled for launch in 2006 or 2007. The passive instrument consists of a 3-m diameter folding clamshell antenna with three offset feedhorns that operate at 1.41 GHz and receive V-pol, H-pol and the third Stokes parameter. Using the same feedhorns and antenna, Aquarius makes near simultaneous 1.2 GHz scatterometer measurements of the surface roughness. To simplify the satellite operation and reduce cost, the antenna does not rotate, instead the feedhorns receive radiances from three FOVs offset to the night side of the orbit (Le Vine *et al.*, 2001). The FOV diameters are 60, 75 and 90 km; the swathwidth is 300 km. The satellite will operate in a dawn–dusk sun-synchronous orbit at a 600-km altitude, with an 8-day repeat cycle. The instrument is calibrated by internal reference sources, rotation of the entire spacecraft to view the cold sky and viewing of instrumented surface sites. Its SSS retrieval accuracy is estimated at 0.2 psu.

14.7 Argo: a global array of ocean interior profiling floats

Because satellites cannot view the ocean depths, much less is known about the ocean interior than the surface properties. As a remedy and during the 2001–2010 decade, the Argo project is deploying approximately 3000 profiling buoys across the global ocean at north–south and east–west intervals of approximately 300 km. The buoys operate as follows. Beginning at the surface, the buoys will take profiles down to depths of about 2 km, drift at depth for about 10 days, then return to the surface and report position and profile data via satellite (Figure 14.12). The buoys are designed to operate for about 4 years; they are deployed from ships of opportunity, aircraft and research vessels. They provide the oceanographic analog of atmospheric radiosonde balloon soundings. These data combined with the satellite data sets such as SST, sea surface height, and surface winds will allow oceanic numerical forecast models to be processed in a manner similar to the present day numerical weather prediction models.

14.8 Global Change Observation Mission-B (GCOM-B)

The Japanese successor to ADEOS-2 is called Global Change Observation Mission-B1 (GCOM-B1), where the GCOM-B series of satellites will carry oceanographic instruments and the GCOM-A will carry atmospheric instruments. GCOM-B1 will be in a sun-synchronous 1000 descending orbit, and will carry the Second-generation Global Imager (SGLI), which is a visible/infrared imager with ocean color channels similar to MODIS and VIIRS, and a tilt capability similar to SeaWiFS. GCOM-B1 also carries the Advanced Microwave Scanning Radiometer Follow-On (AMSR F/O) and the Ocean Vector Wind Mission (OVWM) instrument that is also called AlphaScat, where OVWM is the SeaWinds successor being developed at JPL. GCOM and NPOESS are the two major satellite programs planned for the next two decades.

14.9 Future missions

For the period 1999–2019, this section discusses the timelines of the following kinds of oceanographic missions: vector winds, ocean and ice surface topography, ocean color and SST. Because satellite launch dates are subject to delays, the timelines are tentative.

Figure 14.12. Photograph of a profiling ARGO float held by Dana Swift. The satellite antenna and oceanographic sensors are at the top of the buoy; the buoyancy control is at the base. (Courtesy Stephen Riser, used with permission).

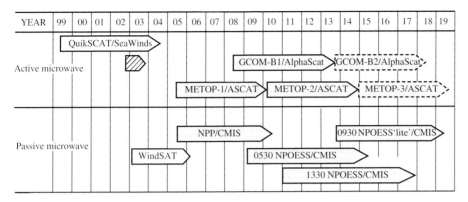

Figure 14.13. The vector wind retrieval instruments and satellites for 1999–2019. The horizontal axis is in years, the horizontal arrows show the mission duration. The missions divide into active scatterometers and passive polarimetric radiometers. A white background means that the satellite is in orbit; gray, that the mission is approved; a dashed outline, that the mission is under consideration. The short arrow with diagonal hatching represents the SeaWinds sensor on ADEOS-2 (Figure design courtesy Michael Van Woert, National Ice Center and William Patzert, JPL).

14.9.1 Vector wind retrieval

Figure 14.13 shows the timelines of the satellite microwave instruments designed specifically for vector wind retrieval. These divide into the active microwave scatterometers and the passive polarimetric radiometers. For the scatterometers, the figure shows that the 2002 successor to SeaWinds on QuikSCAT consisted of the identical instrument on ADEOS-2. Because ADEOS-2 failed in 2003, for the next two years, the availability of scatterometer winds is in doubt. In about 2005 and as described in Chapter 11, vector wind measurements from the ESA Advanced Scatterometer (ASCAT) on METOP will become available. In about 2007, the Japanese GCOM-B1 will carry the OVWM AlphaScat successor to Sea-Winds, where the possibility exists for a similar instrument on GCOM-B2. After about 2005, the combination of METOP and the GCOM-B1 mission means that at least two scatterometers will be in orbit.

The polarimetric wind sensors consist of WindSat and CMIS, and after about 2010 two NPOESS CMIS should be in orbit. Although it is unclear at this time as to which concept, the passive or active vector wind retrieval, will be more successful, the WindSat observations should resolve this question. If the polarimetric radiometer wind retrieval is successful, then either group of satellites could provide the daily global wind coverage required by the numerical weather prediction models.

14.9.2 Ocean and ice topography

Figure 14.14 shows the four mission categories involved in the retrieval of sea and ice surface topography. The first three employ altimetry, the fourth employs gravity measurements. The

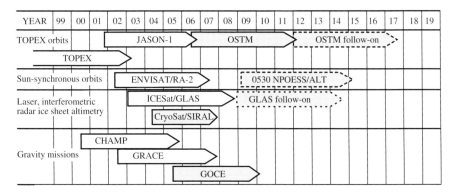

Figure 14.14. The satellite instruments used in the retrieval of sea and ice surface topography. As the vertical axis shows, the missions divide into altimeters in the TOPEX and JASON-1 orbits, altimeters in sun-synchronous orbits, the specialized ice sheet laser and interferometric radar altimeters and the related gravity missions. See caption of Figure 14.13 and text for more information.

first category consists of those altimeters that are located in the original or adjacent non sun-synchronous TOPEX orbits, namely TOPEX, JASON-1 and OSTM. These provide altimeter coverage between ±66° of latitude and have the great advantage that they do not alias the diurnal tide. Their disadvantage is that the TOPEX orbit does not extend to sufficiently high latitudes for glacial or pack ice studies.

The second category consists of the altimeters in sun-synchronous orbits, such as the dual-frequency ENVISAT Radar Altimeter-2 (RA-2) and the altimeter proposed for the 0530 NPOESS satellite. These will supplement the OSTM observations and because their orbits extend to high latitudes, they will also provide coverage of the pack and glacier ice. The third consists of the GLAS laser altimeter and the CryoSat interferometric altimeter that are designed specifically for polar ice. Although they can also be used for ocean studies, their orbits are most appropriate to the slowly varying ice sheet topography. The fourth category is the related gravity missions, CHAMP, GRACE and GOCE. These will produce an improved geoid and may also make direct measurements of changes in the deep ocean currents, both of which will contribute to the understanding of the ocean circulation. The figure also illustrates the growing complexity of the altimeter missions; in the 2001–2010 decade, the basic altimeter concept will expand to include a wide swath interferometric SLR altimeter, a laser altimeter and a SAR/interferometric ice sheet altimeter.

14.9.3 Ocean color

Figure 14.15 shows some of the existing, proposed and planned ocean color missions, where all of the satellites shown are in sun-synchronous orbits. Each arrow contains the

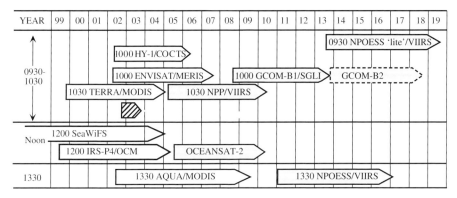

Figure 14.15. The satellites and instruments used for ocean color retrieval, listed by their equator crossing times as shown within the arrows and on the left-hand scale. The short arrow with diagonal hatching represents the GLI sensor on ADEOS-2. See caption of Figure 14.13 and text for more information.

satellite and instrument name, and when necessary, the equator crossing time. The majority have crossing times between 0930 and 1030; this is preferable to later times because of the relatively high sun angle and smaller cloud extent. The SeaWiFS and Indian satellites have local noon crossings, the early afternoon orbit is occupied only by MODIS and VIIRS. Given the large number of countries that maintain ocean color satellites, including China, ESA, India, Japan, the Republic of Korea, the Republic of China and the US, their combined observations should provide a multidecadal time series of ocean colors and their biological and related products.

14.9.4 SST

As Figure 14.16 shows, the SST instruments operate in either the visible/infrared or microwave. Although the infrared instruments are high resolution, they are hampered by clouds and in some cases by the necessity for continuous surface calibration. The microwave instruments are lower resolution, but require only an initial calibration and operate under all weather conditions except heavy rain. As the long experience with AVHRR and the shorter experience with TMI demonstrate, both kinds of SST retrievals are successful.

In the infrared, the NPOESS VIIRS will replace the current AVHRR and MODIS observations, with additional potential contributions from a variety of other sensors. In the passive microwave, TMI presently provides tropical TMI SST observations, and AMSR-E on AQUA provides near global observations at 2-day intervals. These will be continued with the WindSat and CMIS observations. This suggests that in the future there will be two separate and redundant daily SST global data sets; one from the IR, the other from passive microwave.

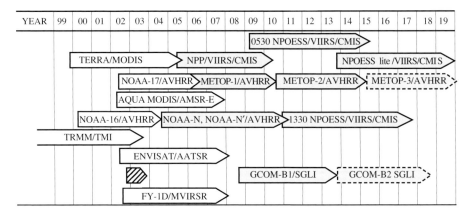

Figure 14.16. The satellites and the infrared and passive microwave instruments used for SST retrieval. Because many satellites carry both kinds of instruments, the satellites are not divided by type of observation. The short arrow with diagonal hatching represents the GLI and AMSR sensors on ADEOS-2. See caption of Figure 14.13 and text for more information.

14.10 Final thoughts

This and the preceding chapters show how a combination of scientific knowledge, engineering skills, computer and programming resources and a relatively small number of satellites, whose size and weight range from that of a small car to a large bus, have revolutionized our view of the planet. During the past two decades, satellite observations of such features as ocean color and altimetry have been transformed from experimental procedures to routine observations. Quantities that were previously unmeasurable, such as the global ocean carbon uptake and its dependence on the El Niño cycle, can now be calculated. Furthermore, the global availability of environmental variables such as pack ice concentration and extent, surface wind speed, SST, mid-ocean tides, the propagation of planetary waves and their relation to climate change are now available on public websites. In this period of concern about global warming and climate variability, and the effect of population pressure on marine resources, it is appropriate and necessary that the tools and data sets for interpreting these changes be publicly and readily available. In the next two decades, it is the task of researchers and policy makers to understand these changes and relate them to national and international societal concerns.

14.11 Further reading and acknowledgements

The NPOESS and NPP projects are described in downloadable presentations at http://npoesslib.ipo.noaa.gov. GLAS and ICESat are described at http://www.csr.utexas.edu/glas/; SMOS at http://www.esa.int; CMIS at http://npoesslib.ipo.noaa.gov/S_cmis.htm. The International Ocean Colour Coordinating Group website at http://www.ioccg.org describes present and future ocean color instruments. The Chelton *et al.* (2001) HOTSWG report is at http://www.oce.orst.edu/po/research/hotswg/. Schutz (1998) is at http://www.csr.utexas.edu/glas/Publications. I thank Stan Wilson and Michael Mignogno for help with IJPS, NPOESS, POES and the METOP payloads, Peter Minnett and Wayne Esaias for help with VIIRS, Mark Drinkwater and Yann Kerr for help with CryoSat, Gary Lagerloef and Charles Koblinsky for help with Aquarius, Karen St. Germain and James Jolley for help with WindSat and CMIS, and Meric Srokosz for help with SMOS.

Appendix

Table A.1. *The letter designations used for frequency bands in the microwave*

Letter designation	Frequency range (GHz)	Wavelength range (cm)
L	0.39–1.55	76.9–19.3
S	1.55–5.20	19.3–5.77
C	3.90–6.20	7.69–4.84
X	5.20–10.90	5.77–2.75
K	10.90–36.00	2.75–0.834
K_u	10.9–18.0	2.75–1.67
K_l	15.35–24.50	1.74–1.22
K_a	33.00–36.00	0.909–0.834

Adapted from Kramer, 1994

Table A.2. *MODIS technical specifications and applications*

Primary use	Band[a]	Bandwidth (nm)	Saturation reflectance	Required SNR[b]
250-m resolution	**1**[c]	620–670	1.6	128
	2	841–876	1.05	201
500-m resolution	3	459–479	1.07	243
Bands 1–7 are used for	4	545–565	1.01	228
discrimination of land, clouds	5	1230–1250	0.84	74
and aerosols	6	1628–1652	1.03	275
	7	2105–2155	0.33	110
1-km resolution	8	405–420	0.37	880
Ocean color	9	438–448	0.26	838
	10	483–493	0.19	802
	11	526–536	0.16	754
	12	546–556	0.12	750
	13	662–672	0.08	910
	14	673–683	0.06	1087
	15	743–753	0.07	586
	16	862–877	0.06	516
Water vapor	17	890–920	0.75	167
Cirrus	**18**	931–941	1.14	57
	19	915–965	0.85	250
High cirrus	**26**	1360–1390	0.89	150

[a] Bands 1–19 and 26 are in nm; Bands 20–25; 27–36 are in μm   (*cont.*)
[b] SNR: Signal-to-noise ratio
[c] Bold face bands are used for cloud identification

Appendix

Table A.2. *(cont.)*

Primary use	Band	Bandwidth (μm)	Saturation brightness temperature (K)	Required $NE\Delta T^d$ (K)
SST	**20**	3.660–3.840	333	0.05
Forest fires	21	3.929–3.989	429	2.00
SST	**22**	3.929–3.989	329	0.07
SST	**23**	4.020–4.080	329	0.07
Atmospheric	24	4.433–4.498	318	0.25
temperature	25	4.482–4.549	314	0.25
Cloud properties	**27**	6.535–6.895	323	0.25
	28	7.175–7.475	320	0.25
	29	8.400–8.700	330	0.05
Ozone	30	9.580–9.880	364	0.25
SST	**31**	10.780–11.280	399	0.05
	32	11.770–12.270	391	0.05
Cloud top altitude	33	13.185–13.485	335	0.25
	34	13.485–13.785	341	0.25
	35	13.785–14.085	339	0.25
	36	14.085–14.385	371	0.35

[d] $NE\Delta T$: Noise-equivalent delta-temperature

Data courtesy NASA Goddard MODIS project.

References

Ackerman, S., Strabala, K., Menzel, P., Frey, R., Moeller, C., Gumley, L., Baum, B., Schaaf, C., & Riggs, G. (1997). *Discriminating Clear Sky from Cloud with MODIS, Version 3.2.* MODIS Algorithm Theoretical Basis Document, ATBD-MOD-06. Greenbelt, MD: NASA Goddard Space Flight Center.

Ahmed, S., Warren, H. R., Symonds, M. D., & Cox, R. P. (1990). The Radarsat System. *IEEE Trans. Geosci. Remote Sens.*, **28**, 598–602.

Aiken, J., Moore, G. F., Trees, C. C., Hooker, S. B., & Clark, D. K. (1995). *The SeaWiFS CZCS-Type Pigment Algorithm.* SeaWiFS Technical Report Series, Vol. 29, ed. S. B. Hooker & E. R. Firestone, NASA Tech. Memo. 104566. Greenbelt, MD: NASA Goddard Space Flight Center.

Anderson, G. P., Kneizys, F. X., Chetwynd, J. H., Wang, J., Hoke, M. L., Rothman, L. S., Kimball, L. M., & McClatchey, R. A. (1995). FASCODE/MODTRAN/LOWTRAN: Past/Present/Future. *18th Annual Review Conference on Atmospheric Transmission Models*, 6–8 June 1995. Lexington, MA: Phillips Laboratories.

Anon. (ed.) (1994). TOPEX/POSEIDON: Geophysical Evaluation. Special section, *J. Geophys. Res.*, **99**, 24 369–25 062.

Asrar, G., & Dozier, J. (1994). *EOS: Science Strategy for the Earth Observing System.* Woodbury, NY: AIP Press.

Atlas, R., Bloom, S. C., Hoffman, R. N., Brin, E., Ardizzone, J., Terry, J., Bungato, D., & Jusem, J. C. (1999). Geophysical validation of NSCAT winds using atmospheric data and analyses. *J. Geophys. Res.*, **104**, 11 405–11 424.

Atlas, R., Hoffman, R. N., Leidner, S. M., Sienkiewicz, T.-W., Bloom, S. C., Brin, E., Ardizzone, J., Terry, J., Bungato, D., & Jusem, J. C. (2001). The effects of marine winds from scatterometer data on weather analysis and forecasting. *Bull. Amer. Meteor. Soc.*, **82**, 1965–1990.

Attema, E. P. W. (1991). The active microwave instrument on-board the ERS-1 satellite. *Proc. IEEE*, **79**, 791–799.

Baker, M. E. (1997). Cloud microphysics and climate. *Science*, **276**, 1072–1078.

Balanis, C. A. (1982). *Antenna Theory: Analysis and Design.* New York: John Wiley.

Balch, W. M., & Byrne, C. F. (1994). Factors affecting the estimate of primary production from space. *J. Geophys. Res.*, **99**, 7555–7570.

Balch, W. M., Drapeau, D. T., Cucci, T. L., Vaillancourt, R. D., Kilpatrick, K. A., & Fritz, J. J. (1999). Optical backscattering by calcifying algae – separating the contribution by particulate inorganic and organic carbon fractions. *J. Geophys. Res.*, **104**, 1541–1558.

Baldy, S. (1993). A generation-dispersion model of ambient and transient bubbles in the close vicinity of breaking waves. *J. Geophys. Res.*, **98**, 18 277–18 293.

Barnes, W. L., & Salomonson, V. V. (1993). MODIS: A global imaging spectroradiometer for the Earth Observing System. *Crit. Rev. Opt. Sci. Technol.*, **CR47**, 285–307.

Barnes, W. L., Pagano, T. S., & Salomonson, V. V. (1998). Prelaunch characteristics of the Moderate Imaging Spectroradiometer (MODIS) on EOS-AM1. *IEEE Trans. Geosci. Remote Sens.*, **36**, 1088–1100.

Barrick, D. E., & Swift, C. T. (1980). The Seasat microwave instruments in historical perspective. *IEEE J. Oceanic Eng.*, **OE-5**, 74–80.

Barton, I. J. (1995). Satellite-derived sea surface temperatures: current status. *J. Geophys. Res.*, **100**, 8777–8790.

Beesley, J. A., & Moritz, R. E. (1999). Toward an explanation of the annual cycle of cloudiness over the Arctic Ocean. *J. Climate*, **12**, 395–415.

Behrenfeld, M. J., & Falkowski, P. G. (1997). Photosynthetic rates derived from satellite-based chlorophyll concentration. *Limnol. Oceanogr.*, **42**, 1–20.

Bernstein, R. L., & Chelton, D. B. (1985). Large-scale sea surface temperature variability from satellite and shipboard measurements. *J. Geophys. Res.*, **90**, 11 619–11 630.

Bertiger, W. I., Bar-Sever, Y. E., Christensen, E. J., Davis, E. S., Guinn, J. R., Haines, B. J., Ibanez-Meier, R. W., Jee, J. R., Lichten, S. M., Melbourne, W. G., Muellerschoen, R. J., Munson, T. N., Vigue, Y., Wu, S. C., & Yunck, T. P. (1994). GPS precise tracking of TOPEX/POSEIDON: results and implications. *J. Geophys. Res.*, **99**, 24 449–24 464.

Bindschadler, R. A., Born, G., Chase, R. R. P., Fu, L.-L., Mouginis-Mark, P., Parsons, C., & Tapley, B. (1987). Altimetric System. Vol. IIh, *Earth Observing System Reports*. Washington, DC: NASA.

Bird, J. (1998). *The Upper Atmosphere: Threshold of Space*. NASA Publication NP-105. Washington, DC: NASA.

Birks, A. R., Delwart, S., Mutlow, C. T., & Llewellyn-Jones, D. (1999). The ENVISAT-1 Advanced Along Track Scanning Radiometer processor and data products. *Proc. IEEE Trans. Geosci. Remote Sens. Symposium, 1999 (IGARSS 1999)*, **3**, 1703–1705.

Boerner, W.-M., Mott, H., Lüneburg, E., Livingstone, C., Brisco, B., Brown, R. J., & Paterson, J. S. (1998). Polarimetry in radar remote sensing: basic and applied concepts. In *Manual of Remote Sensing*, 3rd edn., ed.-in-chief, R. A. Ryerson, Vol. 2, *Principles and Applications of Imaging Radar*, ed. F. M. Henderson & A. J. Lewis, pp. 271–357. New York: Wiley & Sons.

Born, M., & Wolf, E. (1999). *Principles of Optics*, 7th edn. Cambridge: Cambridge University Press.

Bourassa, M. A., Zamudio, L., & O'Brien, J. J. (1999). Non-inertial flow in NSCAT observations of Tehuantepec winds. *J. Geophys. Res.*, **104**, 11 311–11 320.

Brown, O. B., & Minnett, P. J. (1999). *MODIS Infrared Sea Surface Temperature Algorithm, Version 2.0*. MODIS Algorithm Theoretical Basis Document, ATBD-MOD-25. Washington, DC: NASA.

Brown, O. B., Evans, R. H., Minnett, P. J., Kearns, E. J., & Kilpatrick, K. (2002). Sea surface temperature measured by the MODerate resolution Imaging Spectroradiometer (MODIS). Presentation at MODIS Internal Working Group (IWG) meeting, November 18, 2002. Greenbelt, MD: NASA Goddard Space Flight Center.

Brown, R. A. (2000). On satellite scatterometer model functions. *J. Geophys. Res.*, **105**, 29 195–29 205.

Cannizzaro, J. P., Carder, K. L., Chen, F. R., Heil, C. A., & Vargo, G. A. (2004). A novel technique for detection of the toxic dinoflagellate, *Karenia brevis*, in the Gulf of Mexico from remotely sensed ocean color data. *Continental Shelf Res.* (submitted June 2003).

Carder, K. L. (2002). Performance of MODIS semi-analytic ocean color algorithms: chlorophyll a, absorption coefficients, and absorbed radiation by phytoplankton. Presentation at MODIS Science Team Meeting, July 22–24, 2002. Greenbelt, MD: NASA Goddard Space Flight Center.

Carder, K. L., Chen, F. R., Lee, Z. P., Hawes, S., & Kamykowski, D. (1999). Semi-analytic MODIS algorithms for chlorophyll a and absorption with bio-optical domains based on nitrate-depletion temperatures. *J. Geophys. Res.*, **104**, 5403–5421.

Carsey, F. D. (ed.) (1992). *Microwave Remote Sensing of Sea Ice*, Geophysical Monograph 68. Washington, DC: American Geophysical Union.

Carsey, F. D., McNutt, L., & Rothrock, D. A. (ed.) (1994). The Alaska Synthetic Aperture Radar Facility. Special section. *J. Geophys. Res.*, **99**, 22 389–22 490.

Casey, K. S., & Cornillon, P. (1999). A comparison of satellite and in situ based sea surface temperature climatologies. *J. Climate*, **12**, 1848–1863.

Cavalieri, D. J. (1994). A passive microwave technique for mapping new and young sea ice in seasonal sea ice zones. *J. Geophys. Res.*, **99**, 12 561–12 572.

Cavalieri, D. J., Gloersen, P., & Campbell, W. J. (1984). Determination of sea ice parameters with the Nimbus 7 SMMR. *J. Geophys. Res.*, **89**, 5355–5369.

Cavalieri, D. J., Parkinson, C., Gloerson, P., & Zwally, H. J. (1999, updated 2002). *Sea Ice Concentrations from Nimbus-7 SMMR and DMSP SSM/I Passive Microwave Data* (CD-ROM). Boulder, CO: National Snow and Ice Data Center.

Chahine, M. T., McCleese, D. J., Rosenkranz, P. W., & Staelin, D. H. (1983). Interaction mechanisms within the atmosphere. In *Manual of Remote Sensing*, 2nd edn., ed.-in-chief, R. N. Colwell, Vol. 1, *Theory, Instruments and Techniques*, ed. D. S. Simonett, pp. 165–230. Falls Church, VA: American Society of Photogrammetry.

Charman, W. N. (1995). Optics of the eye. In *Handbook of Optics*, 2nd edn., Vol. 1, ed. M. Bass, pp. 24.3–24.54. New York: McGraw-Hill.

Chavez, F. P., Strutton, P. G., Friederick, G. E., Feely, R. A., Feldman, G. C., Foley, D. G., & McPhaden, M. J. (1999). Biological and chemical response of the equatorial Pacific Ocean to the 1997–98 El Niño. *Science*, **286**, 2126–2131.

Chelton, D. B. (1998). *WOCE/NASA Altimeter Algorithm Workshop*. US WOCE Tech. Rep. No. 2. College Station, TX: US Planning Office for WOCE.

Chelton, D. B. (ed.) (2001). *Report of the High-Resolution Ocean Topography Science Working Group Meeting*. Reference 2001–4. Corvallis, OR: College of Oceanic and Atmospheric Sciences, Oregon State University.

Chelton, D. B., & Schlax, M. G. (1996). Global observations of oceanic Rossby waves. *Science*, **272**, 234–238.

Chelton, D. B., Esbensen, S. K., Schlax, M. G., Thum, N., Freilich, M. H., Wentz, F. J., Gentemann, C. L., McPhaden, M. J., & Schoff, P. S. (2001a). Observations of coupling between surface wind stress and sea surface temperature in the eastern tropical Pacific. *J. Climate*, **14**, 1479–1498.

Chelton, D. B., Ries, J. C., Haines, B. J., Fu, L.-L., & Callahan, P. S. (2001b). Satellite altimetry. In *Satellite Altimetry and Earth Sciences*, ed. L.-L. Fu & A. Cazenave, pp. 1–131. San Diego: Academic Press.

Cheney, R. (ed.) (1995). TOPEX/POSEIDON: Scientific Results. Special section, *J. Geophys. Res.*, **100**, 24 893–25 382.

Cheney, R., Miller, L., Agreen, R., Doyle, N., & Lillibridge, J. (1994). TOPEX/POSEI-DON: The 2-cm solution. *J. Geophys. Res.*, **99**, 24 555–24 563.

Christensen, E. J., Haines, B. J., Keihm, S. J., Morris, C. S., Norman, R. A., Purcell, G. H., Williams, B. G., Wilson, B. D., Born, G. H., Parke, M. E., Gill, S. K., Shum, C. K., Tapley, B. D., Kolenkiewicz, R., & Nerem, R. S. (1994). Calibration of TOPEX/POSEIDON at Platform Harvest. *J. Geophys. Res.*, **99**, 24 465–24 485.

Clemente-Colón, P., & Yan, X.-H. (2000). Low-backscatter ocean features in synthetic aperture radar imagery. *Johns Hopkins APL Technical Digest*, **21**(1), 116–121.

Colton, M. C., & Poe, G. A. (1999). Intersensor calibration of DMSP SSM/I's: F-8 to F-14, 1987–1997. *IEEE Trans. Geosci. Remote Sens.*, **37**, 418–439.

Comiso, J. C., Cavalieri, D. J., Parkinson, C. P., & Gloersen, P. (1997). Passive microwave algorithms for sea ice concentration – a comparison of two techniques. *Remote Sens. Environ.*, **60**, 357–384.

Cox, C., & Munk, W. (1954). Statistics of the sea surface derived from sun glitter. *J. Marine Res.*, **13**, 198–227.

Cracknell, A. P. (1997). *The Advanced Very High Resolution Radiometer*. London: Taylor & Francis.

Cushman-Roisin, B. (1994). *Introduction to Geophysical Fluid Dynamics*. Englewood Cliffs, NJ: Prentice Hall.

Desnos, Y.-L., Buck, C., Guijarro, J., Levrini, G., Suchail, J.-L., Torres, R., Laur, H., Closa, J., & Rosich, B. (2000). The ENVISAT advanced synthetic aperture radar system. *Proc. IEEE Trans. Geosci. Remote Sens. Symposium, 2000 (IGARSS 2000)*, **3**, 1171–1173.

Dickson, R. R., Meincke, J., Malmberg, S. A., & Lee, A. J. (1988). The "great salinity anomaly" in the North Atlantic 1968–1982. *Prog. Oceanog.*, **20**, 103–151.

DiGiacomo, P. M., & Holt, B. (2001). Satellite observations of small coastal eddies in the Southern California Bight. *J. Geophys. Res.*, **106**, 22 521–22 543.

Donelan, M. A., & Pierson, W. J. (1987). Radar scattering and equilibrium ranges in wind-generated waves with application to scatterometry. *J. Geophys. Res.*, **92**, 4971–5029.

Donlon, C. J., Minnett, P. J., Gentemann, C., Nightingale, T. J., Barton, I. J., Ward, B., & Murray, J. (2002). Toward improved validation of satellite sea surface skin temperature measurements for climate research. *J. Climate,* **15**, 353–369.

Drinkwater, M. R., Kwok, R., Rignot, E., Israelsson, H., Onstott, R. G., & Winebrenner, D. P. (1992). Potential applications of polarimetry to the classification of sea ice. In *Microwave Remote Sensing of Sea Ice*, ed. F. D. Carsey, Geophysical Monograph 68, pp. 419–430. Washington, DC: American Geophysical Union.

Duck, K. I., & King, J. C. (1983). Orbital mechanics for remote sensing. In *Manual of Remote Sensing*, 2nd edn., ed.-in-chief, R. N. Colwell, Vol. 1, *Theory, Instruments and Techniques*, ed. D. S. Simonett, pp. 699–717. Falls Church, VA: American Society of Photogrammetry.

Elachi, C. (1987). *Introduction to the Physics and Techniques of Remote Sensing*. New York: Wiley-Interscience.

Emery, W. J., Yu, Y., Wick, G. A., Schluessel, P., & Reynolds, R. W. (1994). Correcting infrared satellite estimates of sea surface temperature for atmospheric water vapor contamination. *J. Geophys. Res.*, **99**, 5219–5236.

Eos Science Steering Committee (1989). *From Pattern to Process: the Strategy of the Earth Observing System*. Eos Science Steering Committee Report, Vol. II. Washington, DC: NASA.

Eplee, R. E., & Barnes, R. A. (2000). Lunar data analysis for SeaWiFS calibration. In *SeaWiFS Postlaunch Technical Report Series*, Vol. 9, ed. S. B. Hooker & E. R. Firestone, NASA/TM-1999-206892, pp. 17–27. Greenbelt, MD: NASA Goddard Space Flight Center.

Esaias, W. E., Abbott, M. R., Barton, I., Brown, O. B., Campbell, J. W., Carder, K. L., Clark, D. K., Evans, R. J., Hoge, F. E., Gordon, H. R., Balch, W. M., Letelier, R., & Minnett, P. J. (1998). An overview of MODIS capabilities for ocean science observations. *IEEE Trans. Geosci. Remote Sens.*, **36**, 1250–1265.

Evans, R. H., & Gordon, H. R. (1994). Coastal zone color scanner "system calibration": a retrospective examination. *J. Geophys. Res.*, **99**, 7293–7307.

Ezraty, R., & Cavanié, A. (1999). Intercomparison of backscatter maps over Arctic sea ice from NSCAT and the ERS scatterometer. *J. Geophys. Res.*, **104**, 11 471–11 483.

Francis, C. R. (2001). *CryoSat Mission and Data Description*. ESA Document Number CS-RP-ESA-SY-0059. Noordwijk: ESTEC.

Freilich, M. H. (2000). *SeaWinds Algorithm Theoretical Basis Document*. ATBD-SWS-01. Greenbelt, MD: NASA Goddard Space Flight Center.

Freilich, M. H., & Dunbar, R. S. (1999). The accuracy of the NSCAT 1 vector winds: comparison with National Data Center buoys. *J. Geophys. Res.*, **104**, 11 231–11 246.

Frouin, R., Schwindling, M., & Deschamps, P.-Y. (1996). Spectral reflectance of sea foam in the visible and near-infrared: In situ measurements and remote sensing implications. *J. Geophys. Res.*, **101**, 14 361–14 371.

Fu, L.-L., & Cazenave, A. (ed.) (2001). *Satellite Altimetry and Earth Sciences*. San Diego: Academic Press.

Fu, L.-L., & Holt, B. (1982). *Seasat Views Oceans and Sea Ice with Synthetic-Aperture Radar*. JPL Publication 81–120. Pasadena: NASA Jet Propulsion Laboratory, California Institute of Technology.

Fu, L.-L., Christensen, E. J., Yamarone, C. A., Lefebvre, M., Menard, Y., Dorrer, M., & Escudier, P. (1994). TOPEX/POSEIDON mission overview. *J. Geophys. Res.*, **99**, 24 369–24 381.

Gaiser, P. W. (1999). Windsat-satellite-based polarimetric microwave radiometer. *IEEE MTT-S Digest*, **1**, 403–406.

Gao, B.-C., Goetz, A. F. H., & Wiscombe, W. J. (1993). Cirrus cloud detection from airborne imaging spectrometer data using the 1.38 micron water vapor band. *Geophys. Res. Lett.*, **20**, 301–304.

Gasparovic, R. F., Apel, J. R., & Kasischke, E. S. (1988). An overview of the SAR internal wave signature experiment. *J. Geophys. Res.*, **93**, 12 304–12 316.

Gasster, S., & Flaming, G. M. (1998). Overview of the Conical Microwave Imager/Sounder development for the NPOESS program. *Proc. IEEE Trans. Geosci. Remote Sens. Symposium, 1998 (IGARSS 1998)*, **1**, 268–270.

Gill, A. E. (1982). *Atmosphere-Ocean Dynamics*. London: Academic Press.

Gloersen, P., & Barath, F. T. (1977). A Scanning Multichannel Microwave Radiometer for Nimbus-G and SeaSat-A. *IEEE J. Oceanic Eng.*, **OE-2**, 172–178.

Gloersen, P., & Cavalieri, D. J. (1986). Reduction of weather effects in the calculation of sea ice concentration from microwave radiances. *J. Geophys. Res.*, **91**, 3913–3919.

Gloersen, P., Campbell, W. J., Cavalieri, D. J., Comiso, J. C., Parkinson, C. L., & Zwally, H. J. (1992). *Arctic and Antarctic Sea Ice, 1978–1987: Satellite Passive-Microwave Observations and Analysis*. NASA Spec. Publ. 511, Washington, DC: NASA.

Goldstein, R. M., Li, F., Smith, J., Pinkel, R., & Barnett, T. P. (1994). Remote sensing of surface waves: The Surface Wave Process Program experiment. *J. Geophys. Res.*, **99**, 7945–7950.

Gonzales, A. E., & Long, D. G. (1999). An assessment of NSCAT ambiguity removal. *J. Geophys. Res.*, **104**, 11 449–11 457.

Gordon, H. R., & Balch, W. M. (1999). *MODIS Detached Coccolith Concentration Algorithm Theoretical Basis Document, Version 4*. NASA Algorithm Theoretical Basis Document ATBD-MOD-23. Greenbelt, MD: NASA Goddard Space Flight Center.

Gordon, H. R., & Castaño, D. J. (1987). Coastal Zone Scanner atmospheric correction algorithm: multiple scattering effects. *Appl. Opt.*, **26**, 2111–2122.

Gordon, H. R., & Clark, D. K. (1981). Clear water radiances for atmospheric correction of coastal zone color scanner imagery. *Appl. Opt.*, **20**, 4175–4180.

Gordon, H. R., & Morel, A. (1983). *Remote Assessment of Ocean Color for Interpretation of Satellite Visible Imagery, a Review*. New York: Springer-Verlag.

Gordon, H. R., & Voss, K. J. (1999). *MODIS normalized water-leaving radiance*. MODIS Algorithm Theoretical Basis Bocument ATBD MOD-17, April 30, 1999. Greenbelt, MD: NASA Goddard Space Flight Center.

Gordon, H. R., & Wang, M. (1992). Surface-roughness considerations for atmospheric correction of ocean color sensors. II. Error in the retrieved water-leaving radiance. *Appl. Opt.*, **31**, 4261–4267.

  (1994a). Retrieval of water-leaving radiance and aerosol optical thickness over the oceans with SeaWiFS: a preliminary algorithm. *Appl. Opt.*, **33**, 443–452.

  (1994b). Influence of oceanic whitecaps on atmospheric correction of SeaWiFS. *Appl. Opt.*, **33**, 7754–7763.

Gordon, H. R., Brown, O. B., & Jacobs, M. M. (1975). Computed relationships between the inherent and apparent optical properties of a flat homogeneous ocean. *Appl. Opt.*, **14**, 417–427.

Gordon, H. R., Clark, D. K., Brown, J. W., Brown, O. B., Evans, R. H., & Broenkow, W. W. (1983). Phytoplankton pigment concentrations in the Middle Atlantic Bight: comparison of ship determinations and CZCS estimates. *Appl. Opt.*, **22**, 20–36.

Haines, B., Bertiger, W., Desai, S., Kuang, D., Munson, T., Young L., & Willis, P. (2002). Initial orbit determination results for Jason-1: towards a 1-cm orbit. *Proc. Inst. Navigation GPS 2002 Conference*, 2011–2021.

Hamilton, G. D. (1986). National Data Buoy Center programs. *Bull. Amer. Meteor. Soc.*, **67**, 411–415.

Hansen, J. E., & Travis, L. D. (1974). Light scattering in planetary atmospheres. *Space Sci. Rev.*, **16**, 527–610.

Heimbach, P., & Hasselmann, K. (2000). Development and application of satellite retrievals of ocean wave spectra. Chapter 2 in *Satellites, Oceanography and Society*, ed. D. Halpern, pp. 5–33. Amsterdam: Elsevier.

Henderson, F. M., & Lewis, A. J. (ed.) (1998). *Principles and Applications of Imaging Radar*. Vol. 2 of *Manual of Remote Sensing*, 3rd edn., ed.-in-chief, R. A. Ryerson. New York: John Wiley.

Hoepffner, N., & Sathyendranath, S. (1993). Determination of the major groups of phytoplankton pigments from the absorption spectra of total particulate matter. *J. Geophys. Res.*, **98**, 22 789–22 803.

Holligan, P. M., Fernandez, E., Aiken, J., Balch, W. B., Boyd, P., Burkill, P. H., Finch, M., Groom, S. B., Malin, G., Muller, K., Purdie, D. A., Robinson, C., Trees, C. C.,

Turner, S. M., & van der Wal, P. (1993). A biogeochemical study of the coccolithophore, *Emiliania huxleyi*, in the North Atlantic. *Global Biogeochem. Cycles*, **7**, 879–900.

Hollinger, J. P., Peirce, J. L., & Poe, G. A. (1990). SSM/I instrument evaluation. *IEEE Trans. Geosci. Remote Sens.*, **28**, 781–790.

Hommel, D., Carter, C., Liu, Q., & Carder, K. (2002). *Ocean Color/Chlorophyll, Visible/Infrared Imager/Radiometer Suite*. Algorithm Theoretical Basis Document, Version 5, SRBS Document # Y2408. Lanham, MD: Raytheon Systems Company.

Hooker, S. B., & McClain, C. R. (2000). The calibration and validation of SeaWiFS data. *Progress in Oceanography*, **45**, 427–465.

Hooker, S. B., Esaias, W. E., Feldman, G. C., Gregg, W. W., & McClain, C. R. (1992). *An Overview of SeaWiFS and Ocean Color*. In SeaWiFS Technical Report Series, Vol. 1, ed. S. B. Hooker & E. R. Firestone, NASA Tech. Memo. 104566. Greenbelt, MD: NASA Goddard Space Flight Center.

Hsu, M.-K., & Liu, A. K. (2000). Nonlinear internal waves in the South China Sea. *Can. J. Remote Sens.*, **26**, 72–81.

Huang, N. E., Tung, C.-C., & Long, S. R. (1990). Wave spectra. In *Ocean Engineering Science*, Vol. 9, Part B, *The Sea*, ed. B. Le Méhauté & D. M. Hayes, pp. 197–237. New York: John Wiley.

Hunt, G. E. (1973). Radiative properties of terrestrial clouds at visible and infrared thermal window wavelengths. *Q. J. R. Meteorol. Soc.*, **99**, 364–369.

IOCCG (1999). *Minimum Requirements for an Operational Ocean-colour Sensor for the Open Ocean*. Reports of the International Ocean-Colour Coordinating Group, No. 1, Bedford Institute of Oceanography, Dartmouth, Nova Scotia, Canada.

Irisov, V. G., Kuzmin, A. V., Pospelov, M. N., Trokhimovsky, J. G., & Etkin, V. S. (1991). The dependence of sea brightness temperature on surface wind direction and speed. Theory and experiment. *Proc. IEEE Geosci. Remote Sens. Symposium, 1991* (*IGARSS 1991*), 1297–1300.

Jackson, J. D. (1975). *Classical Electrodynamics,* 2nd edn. New York: John Wiley.

Janssen, M. A., Ruf, C. S., & Keihm, S. J. (1995). TOPEX/Poseidon microwave radiometer (TMR): II. Antenna pattern correction and brightness temperature algorithm. *IEEE Trans. Geosci. Remote Sens.*, **33**, 138–146.

Jeffrey, S. W., & Mantoura, R. F. C. (1997). Development of pigment methods for oceanography: SCOR-supported Working Groups and objectives. In *Phytoplankton Pigments in Oceanography: Guidelines to Modern Methods*, ed. S. W. Jeffrey, R. F. C. Mantoura & S. W. Wright, pp. 19–36. Paris: UNESCO Publishing.

Jeffrey, S. W., & Vesk, M. (1997). Introduction to marine phytoplankton and their pigment signatures. In *Phytoplankton Pigments in Oceanography: Guidelines to Modern Methods*, ed. S. W. Jeffrey, R. F. C. Mantoura & S. W. Wright, pp. 37–84. Paris: UNESCO Publishing.

Jenkins, G. M., & Watts, D. G. (1968). *Spectral Analysis and its Applications*. San Francisco, CA: Holden-Day.

Jessup, A. T., & Zappa, C. J. (1997). Defining and quantifying microscale wave breaking with infrared imagery. *J. Geophys. Res.*, **102**, 23 145–23 153.

Johannessen, J. A. (2000). Coastal observing systems: the role of Synthetic Aperture Radar. *Johns Hopkins APL Technical Digest*, **21**(1), 41–48.

Johnson, J. W., Williams, L. A., Bracalente, E. M., Beck, F. B., & Grantham, W. L. (1980). Seasat-A satellite scatterometer instrument evaluation. *IEEE J. Oceanic Eng.*, **OE-5**(2), 138–144.

Jones, W. L., & Schroeder, L. C. (1978). Radar backscatter from the ocean: dependence on surface friction velocity. *Boundary-Layer Meteorol.*, **13**, 133–149.

Jones, W. L., Schroeder, L. C., & Mitchell, J. L. (1977). Aircraft measurements of the microwave scattering signature of the ocean. *IEEE J. Oceanic Eng.*, **OE-2**(1), 52–61.

Jones, W. L., Wentz, F. J., & Schroeder, L. C. (1978). Algorithm for inferring wind stress from SEASAT-A. *J. Spacecr. Rockets*, **15**(6), 368–374.

Katsaros, K. B. (1980). The aqueous thermal boundary layer. *Boundary-Layer Meteorol.*, **18**, 107–127.

Katsaros, K. B., Forde, E. B., & Liu, W. T. (2001). QuikSCAT facilitates early identification of tropical depressions in 1999 hurricane season. *Geophys. Res. Lett.*, **28**, 1043–1046.

Kawai, S. (1979). Generation of initial wavelets by instability of a coupled shear flow and their evolution to wind waves. *J. Fluid Mech.*, **93**, 661–703.

Kelly, K. A., Dickinson, S., & Yu, Z. (1999). NSCAT tropical wind stress maps: implications for improving ocean modeling. *J. Geophys. Res.*, **104**, 11 291–11 310.

Key, J., Collins, J., Fowler, C., & Stone, R. (1997). High-latitude surface temperature estimates from thermal satellite data. *Remote Sens. Environ.*, **61**, 302–309.

Kidder, S. Q., & Vonder Haar, T. H. (1995). *Satellite Meteorology: An Introduction*. San Diego: Academic Press.

Killworth, P. D. (2001). Rossby waves. In *Encyclopedia of Ocean Sciences*, ed. J. H. Steele, K. K. Turekian & S. A. Thorpe, Vol. 4, pp. 2434–2443. London: Academic Press.

Kilpatrick, K. A., Podestá, G. P., & Evans, R. (2001). Overview of the NOAA/NASA advanced very high resolution radiometer Pathfinder algorithm for sea surface temperature and associated matchup database. *J. Geophys. Res.*, **106**, 9179–9197.

King, M. D., Kaufman, Y. J., Tanré, D., & Nakajima, T. (1999). Remote sensing of tropospheric aerosols from space: past, present, and future. *Bull. Am. Meteorol. Soc.*, **80**, 2229–2259.

Kinsman, B. (1984). *Wind Waves: Their Generation and Propagation on the Ocean Surface*. New York: Dover Publications.

Kirk, J. T. O. (1996). *Light and Photosynthesis in Aquatic Ecosystems*. Cambridge: Cambridge University Press.

Klein, L. A., & Swift, C. T. (1977). An improved model for the dielectric constant of sea water at microwave frequencies. *IEEE Trans. Antennas Propag.*, **AP-25**(1), 104–111.

Knauss, J. A. (1997). *Introduction to Physical Oceanography*, 2nd edn. Upper Saddle River, NJ: Prentice Hall.

Koepke, P. (1984). Effective reflectance of whitecaps. *Appl. Opt.*, **23**, 1816–1824.

Kopelevich, O. V. (1983). Small-parameter model of optical properties of sea water. Chapter 8 in *Ocean Optics, Vol. 1: Physical Ocean Optics*, ed. A. S. Monin. Moscow: Nauka Pub. (Russian).

Kramer, H. J. (1994). *Observation of the Earth and its Environment: Survey of Missions and Sensors*. Berlin: Springer-Verlag.

Kummerow, C., Barnes, W., Kozu, T., Shiue, J., & Simpson, J. (1998). The Tropical Rainfall Measuring Mission (TRMM) sensor package. *J. Atmos. Oceanic Technol.*, **15**, 809–817.

Kwok, R., Cunningham, G. F., LaBelle-Hamer, N., Holt, B., & Rothrock, D. (1999). Ice thickness derived from high-resolution radar imagery. *EOS, Trans. American Geophysical Union*, **80**(42), pp. 495, 497.

Kwok, R., Rignot, E., & Holt, B. (1992). Identification of sea ice types in spaceborne synthetic aperture radar data. *J. Geophys. Res.*, **97**, 2391–2402.

Lagerloef, G. (2000). Recent progress toward satellite measurements of the global sea surface salinity field. Chapter 18 in *Satellites, Oceanography and Society*, ed. D. Halpern, pp. 309–335. Amsterdam: Elsevier.

Lalli, C. M., & Parsons, T. R. (1993). *Biological Oceanography: An Introduction*. Oxford: Pergamon Press.

Lamarre, E., & Melville, W. K. (1996). Void-fraction measurements near the ocean surface. In *The Air-Sea Interface: Radio and Acoustic Sensing, Turbulence and Wave Dynamics*, ed. M. A. Donelan, W. H. Hui & W. J. Plant, pp. 693–698. Miami, FL: Rosenstiel School of Marine and Atmospheric Science, University of Miami.

Lamb, H. (1945). *Hydrodynamics*. New York: Dover Publications.

Le Provost, C. (2001). Ocean Tides. In *Satellite Altimetry and Earth Sciences,* ed. L.-L. Fu & A. Cazenave, pp. 267–303. San Diego: Academic Press.

Le Vine, D. M., & Abraham, S. (2001). Galactic noise and passive microwave remote sensing from space at L-band. *Proc. IEEE Trans. Geosci. Remote Sens. Symposium, 2001 (IGARSS 2001)*, **4**, 1581–1583.

Le Vine, D., Koblinsky, C., Pellerano, F., Lagerloef, G., Chao, Y., Yueh, S., & Wilson, W. (2001). The measurement of salinity from space: sensor concept. *Proc. IEEE Trans. Geosci. Remote Sens. Symposium, 2001 (IGARSS 2001)*, **3**, 1010–1012.

Lefèvre, J.-M., & Cotton, P. D. (2001). Ocean surface waves. In *Satellite Altimetry and Earth Sciences*, ed. L.-L. Fu & A. Cazenave, pp. 305–328. San Diego: Academic Press.

Letelier, R. M., & Abbott, M. R. (1996). An analysis of chlorophyll fluorescence algorithms for the Moderate Resolution Imaging Spectroradiometer (MODIS). *Remote Sens. Environ.*, **58**, 215–223.

Li, X., Clemente-Colón, P., & Friedman, K. S. (2000). Estimating oceanic mixed-layer depth from internal wave evolution observed from Radarsat-1 SAR. *Johns Hopkins APL Technical Digest*, **21**(*1*), 130–135.

Lighthill, J. (1980). *Waves in Fluids*. Cambridge: Cambridge University Press.

Liou, K.-N. (1980). *An Introduction to Atmospheric Radiation*. San Diego: Academic Press.

Liu, T. J. (1988). Moisture and latent heat flux variability in the tropical Pacific derived from satellite data. *J. Geophys. Res.*, **93**, 6749–6760.

Liu, W. T. (2002). Progress in scatterometer application. *J. Oceanogr.*, **58**, 121–136.

Liu, Y., Yan, X.-H., Liu, W. T., & Hwang, P. A. (1997). The probability density function of ocean surface slopes and its effects on radar backscatter. *J. Phys. Oceanogr.*, **27**, 782–797.

Long, D. G., & Drinkwater, M. R. (1999). Cryosphere applications of NSCAT data. *IEEE Trans. Geosci. Remote Sens.*, **37**, 1671–1684.

Luscombe, A. P., Ferguson, I., Sheperd, N., Zimcik, D. G., & Naraine, P. (1993). The RADARSAT synthetic aperture radar development. *Can. J. Remote Sens.*, **19**, 298–310.

Madsen, S. N., & Zebker, H. A. (1998). Imaging radar interferometry. In *Manual of Remote Sensing*, 3rd edn., ed.-in-chief, R. A. Ryerson, Vol. 2, *Principles and Applications of Imaging Radar*, ed. F. M. Henderson & A. J. Lewis, pp. 359–380. New York: Wiley and Sons.

Martin, S. (2001). Polynyas. In *Encyclopedia of Ocean Sciences*, ed. J. H. Steele, K. K. Turekian & S. A. Thorpe, Vol. 3, pp. 2241–2247. London: Academic Press.

Massom, R. (1991). *Satellite Remote Sensing of Polar Regions*. Boca Raton, FL: CRC Press.

Massonnet, D., & Feigl, K. L. (1998). Radar interferometry and its application to changes in the Earth's surface. *Rev. Geophys.*, **36**, 441–500.

Maul, G. A. (1985). *Introduction to Satellite Oceanography*. Dordrecht: Kluwer.

May, D. A., Parmeter, M. M., Olszewski, D. S., & McKenzie, B. D. (1998). Operational processing of satellite sea surface temperature retrievals at the Naval Oceanographic Office. *Bull. Am. Meteorol. Soc.*, **79**(3), 397–407.

McClain, C. R., Esaias, W. E., Barnes, W., Guenther, B., Endres, D., Hooker, S. B., Mitchell, B. G., & Barnes, R. (1992). In *SeaWiFS Calibration and Validation Plan*. SeaWiFS Technical Report Series, Vol. 3, ed. S. B. Hooker & E. R. Firestone, NASA Tech. Memo 104566. Greenbelt, MD: NASA Goddard Space Flight Center.

McClain, E. P., Pichel, W., & Walton, C. C. (1985). Comparative performance of AVHRR-based Multichannel Sea Surface Temperatures. *J. Geophys. Res.*, **90**, 11 587–11 601.

McPhaden, M. J. (1999). Genesis and evolution of the 1997–98 El Niño. *Science*, **283**, 950–954.

McPhaden, M. J., Busalacchi, A. J., Cheney, R., Donguy, J.-R., Gage, K. S., Halpern, D., Ji, M., Julian, P., Meyers, G., Mitchum, G. T., Niiler, P. P., Picaut, J., Reynolds, R. W., Smith, N., & Takeuchi, K. (1998). The Tropical Ocean-Global Atmosphere observing system: A decade of progress. *J. Geophys. Res.*, **103**, 14 169–14 240.

Meindl, E. A., & Hamilton, G. D. (1992). Programs of the National Data Buoy Center. *Bull. Amer. Meteorol. Soc.*, **73**, 985–993.

Meissner, T., & Wentz, F. J. (2002a). An updated analysis of the ocean surface wind direction signal in passive microwave brightness temperatures. *IEEE Trans. Geosci. Remote Sens.*, **40**, 1230–1240.

Meissner, T. & Wentz, F. J. (2002b). The ocean algorithm suite for the Conical-Scanning Microwave Imaging/Sounder (CMIS). *Proc. IEEE Trans. Geosci. Remote Sens. Symposium, 2002 (IGARSS 2002)*, **2**, 813–816.

Melville, W. K. (1996). The role of surface-wave breaking in air-sea interaction. *Ann. Rev. Fluid Mech.*, **28**, 279–321.

Minnett, P. J. (1995a). Sea surface temperature measurements from the Along-Track Scanning Radiometer on ERS-1. In *Oceanographic Applications of Remote Sensing*, ed. M. Ikeda & F. Dobson, pp. 131–143. Boca Raton, FL: CRC Press.

Minnett, P. J., (1995b). The Along-Track Scanning Radiometer: instrument details. In *Oceanographic Applications of Remote Sensing*, ed. M. Ikeda & F. Dobson, pp. 461–472. Boca Raton, FL: CRC Press.

Minnett, P. J., Evans, R. H., Kearns, E. J., & Brown, O. B. (2002). Sea-surface temperature measured by the Moderate Resolution Imaging Spectroradiometer (MODIS). *Proc. IEEE Trans. Geosci. Remote Sens. Symposium, 2002 (IGARSS 2002)*, **2**, 1177–1179.

Mitchell, B. G. (ed.) (1994). Ocean color from space: A coastal zone color scanner retrospective. Special section, *J. Geophys. Res.*, **99**, 7291–7570.

Mitnik, L. M., & Kalmykov, A. I. (1992). Structure and dynamics of the Sea of Okhotsk marginal ice zone from "Ocean" satellite sensing data. *J. Geophys. Res.*, **97**, 7429–7445.

Mobley, C. D. (1994). *Light and Water; Radiative Transfer in Natural Waters*. San Diego: Academic Press.

Mobley, C. D. (1995). The optical properties of water. In *Handbook of Optics*, 2nd edn., Vol. 1, ed. M. Bass, pp. 43.3–43.56. New York: McGraw-Hill.

Mobley, C. D. (1999). Estimation of the remote sensing reflectance from above-surface measurements. *Appl. Opt.*, **38**, 7442–7455.

Monahan, E. C., & O'Muircheartaigh, I. G. (1986). Whitecaps and the passive remote sensing of the ocean surface. *Int. J. Remote Sens.*, **7**, 627–642.

Moore, J. E., Sheppard, J., Pizacaroli, J., Lymna, P., & Warren, H. R. (1993). RADARSAT: the bus and solar array. *Can. J. Remote Sens.*, **19**, 289–297.

Moore, K. D., Voss, K. J., & Gordon, H. R. (2000). Spectral reflectance of whitecaps: their contribution to the water-leaving radiance. *J. Geophys. Res.*, **105**, 6493–6499.

Morel, A., & Prieur, L. (1977). Analysis of variations in ocean color. *Limnol. Oceanogr.*, **22**, 709–722.

Mouchot, M.-C., & Garello, R. (1998). SAR for oceanography. In *Manual of Remote Sensing*, 3rd edn., ed.-in-chief, R. A. Ryerson, Vol. 2, *Principles and Applications of Imaging Radar*, ed. F. M. Henderson & A. J. Lewis, pp. 631–675. New York: Wiley and Sons.

Naderi, F. M., Freilich, M. H., & Long, D. G. (1991). Spaceborne radar measurement of wind velocity over the ocean – an overview of the NSCAT Scatterometer System. *Proc. IEEE*, **79**(6), 850–866.

NASA Science Working Group (1984). *Passive Microwave Remote Sensing for Sea Ice Research*. Washington, DC: NASA Headquarters.

Nelson, C. S., & Cunningham, J. D. (2002). The National Polar-orbiting Operational Environmental Satellite System – Future U.S. Environmental Observing System. 6th Symposium on Integrated Observing Systems, Orlando, FL, 13–17 January 2002 (preprints). Boston MA: American Meteorological Society.

Nghiem, S. V., Martin, S., Perovich, D. K., Kwok, R., Drucker, R., & Gow, A. J. (1997). A laboratory study of the effect of frost flowers on C-Band radar backscatter from sea ice. *J. Geophys. Res.*, **102**, 3357–3370.

Njoku, E. G., & Bernstein, R. L. (ed.) (1985). Satellite sea surface temperature comparisons. *J. Geophys. Res*, **90**, 11 571–11 677.

Njoku, E. G., Christensen, E. J., & Cofield, R. E. (1980a). The Seasat Scanning Multichannel Microwave Radiometer (SMMR): antenna pattern corrections – development and implementation. *IEEE J. Oceanic Eng.*, **OE-5**(2), 125–137.

Njoku, E. G., Stacey, J. M., & Barath, F. T. (1980b). The SeaSat Scanning Multichannel Microwave Radiometer (SMMR): instrument description and performance. *IEEE J. Oceanic Eng.*, **OE-5**(2), 100–115.

O'Brien, J. J. (ed.) (1999). NSCAT Validation and Science. Special section, *J. Geophys. Res.*, **104**, 11 229–11 568.

Olsen, R. B., & Wahl, T. (2000). The role of wide swath SAR in high-latitude coastal management. *Johns Hopkins APL Technical Digest*, **21**, 136–140.

Onstott, R. G. (1992). SAR and scatterometer signatures of sea ice. In *Microwave Remote Sensing of Sea Ice*, ed. F. D. Carsey, pp. 73–104. Washington, DC: American Geophysical Union.

O'Reilly, J. E., Maritorena, S., Mitchell, B. G., Siegel, D. A., Carder, K. L., Garver, S. A., Kahru, N., & McClain, C. (1998). Ocean color chlorophyll algorithms for SeaWiFS. *J. Geophys. Res.*, **103**, 24 937–24 953.

Parashar, S., Langham, E., McNally, J., & Ahmed, S. (1993). RADARSAT mission requirements and concept. *Can. J. Remote Sens.*, **19**, 280–288.

Parkinson, C. L., & Greenstone, R. (2000). *EOS Data Products Handbook*, Vol. 2. Greenbelt, MD: NASA Goddard Space Flight Center.

Parkinson, C. L., Comiso, J. C., Zwally, H. J., Cavalieri, D. J., Gloersen, P., & Campbell, W. J. (1987). *Arctic Sea Ice, 1973–1976: Satellite Passive-Microwave Observations.* NASA Spec. Publ. 489. Washington, DC: NASA.

Pedlosky, J. (1987). *Geophysical Fluid Dynamics.* New York: Springer-Verlag.

Pegion, P. J., Bourrassa, M. A., Legler, D. M., & O'Brien, J. J. (2000). Objectively derived daily "winds" from satellite scatterometer data. *Mon. Wea. Rev.*, **128**, 3150–3167.

Perkowitz, S. (2000). *Universal Foam: from Cappuccino to the Cosmos.* New York: Walker.

Perry, K. L. (ed.) (2001). *QuikSCAT Science Data Product Users Manual, Version 2.2.* NASA Report No. D-18053. Pasadena, CA: Jet Propulsion Laboratory, California Institute of Technology.

Petzold, T. J. (1972). *Volume Scattering Functions for Selected Ocean Waters.* SIO Ref. 72–78. La Jolla, CA: Scripps Institution of Oceanography.

Phalippou, L., Rey, L., de Chateau-Thierry, P., Thouvenot, E., Steunou, N., Mavrocordatos, C., & Francis, R. (2001). Overview of the performances and tracking design of the SIRAL altimeter for the CryoSat mission. *Proc. IEEE Trans. Geosci. Remote Sens. Symposium (IGARSS 2001)*, **5**, 2025–2027.

Phillips, K. J. H. (1992). *Guide to the Sun.* Cambridge: Cambridge University Press.

Phillips, O. M. (1977). *The Dynamics of the Upper Ocean.* Cambridge: Cambridge University Press.

Pichel, W. G., & Clemente-Colón, P. (2000). NOAA CoastWatch SAR applications and demonstration. *Johns Hopkins APL Technical Digest*, **21**, 49–57.

Plant, W. J. (1990). Bragg scattering of electromagnetic waves from the air/sea interface. In *Surface Waves and Fluxes, II*, ed. G. L. Geernaert & W. J. Plant, pp. 41–108. Dordrecht: Kluwer.

Plant, W. J. (2001). Satellite remote sensing microwave scatterometers. In *Encyclopedia of Ocean Sciences*, ed. J. H. Steele, K. K. Turekian & S. A. Thorpe, Vol. 5, pp. 2539–2551. London: Academic Press.

Pond, S., & Pickard, G. L. (1986). *Introductory Dynamical Oceanography,* 2nd edn. New York: Pergamon Press.

Prabhakara, C. R., Fraser, R. S., Dalu, G., Wu, M. C., & Curran, R. J. (1988). Thin cirrus clouds: seasonal distribution over oceans deduced from Nimbus 4 IRIS. *J. Appl. Meteorol.*, **27**, 379–399.

Raman, C. V. (1922). On the molecular scattering of light in water and the colour of the sea. *Proc. R. Soc. Lond. A*, **101**, 64–80.

Raney, R. K. (1998). Radar fundamentals: technical perspective. In *Manual of Remote Sensing*, 3rd edn., ed.-in-chief, R. A. Ryerson, Vol. 2, *Principles and Applications of Imaging Radar*, ed. F. M. Henderson & A. J. Lewis, pp. 9–130. New York: Wiley & Sons.

Raney, R. K., Luscombe, A. P., Langham, E. J., & Ahmed, S. (1991). RADARSAT [SAR imaging]. *Proc. IEEE*, **79**, 839–849.

Rees, W. G. (2001). *Physical Principles of Remote Sensing,* 2nd edn. Cambridge: Cambridge University Press.

Reynolds, R. W., & Smith, T. M. (1994). Improved global sea surface temperature analysis using optimum interpolation. *J. Climate*, **7**, 929–948.

Robinson, I. S. (1985). *Satellite Oceanography.* Chichester, UK: Ellis Horwood.

Rodriguez, E., & Pollard, B. D. (2001). The measurement capabilities of wide-swath ocean altimeters. In *Report of the High-Resolution Ocean Topography Science Working Group Meeting*, ed. D. B. Chelton, pp. 190–215, Reference 2001–4. Corvallis, OR: College of Oceanic and Atmospheric Sciences, Oregon State University.

Roesler, C. S., & Perry, M. J. (1995). *In situ* phytoplankton absorption, fluorescence emission, and particulate backscattering spectra determined from reflectance. *J. Geophys. Res.*, **100**, 13 279–13 294.

Roesler, C. S., Perry, M. J., & Carder, K. L. (1989). Modeling *in situ* phytoplankton absorption from total absorption spectra in productive inland marine waters. *Limnol. Oceanogr.*, **34**, 1510–1523.

Rosen, P. A., Hensley, S., Joughin, I. R., Fuk, K. L., Madsen, S. A., Rodriguez, E., & Goldstein, R. M. (2000). Synthetic aperture radar interferometry. *Proc. IEEE*, **88**, 333–382.

Ross, D., & Jones, W. L. (1978). On the relationship of radar backscatter to wind speed and fetch. *Boundary-Layer Meteorol.*, **13**, 151–163.

Rossow, W. B. (1989). Measuring cloud properties from space. A review. *J. Climate*, **2**, 201–213.

Rostan, F. (2000). The calibration of the MetOp/Advanced Scatterometer (ASCAT). *Proc. IEEE Trans. Geosci. Remote Sens. Symposium, 2000 (IGARSS 2000)*, **5**, 2206–2208.

Rothrock, D. A., Yu, Y., & Maykut, G. A. (1999). Thinning of the Arctic sea-ice cover. *Geophys. Res. Lett.*, **26**, 3469–3472.

Ruf, C. S., & Giampaolo, J. C. (1998). Littoral deconvolution for a microwave radiometer. *Proc. IEEE Geosci. and Remote Sens. Symposium, 1998 (IGARSS 1998)*, **1**, 378–380.

Sabins, F. F. (1987). *Remote Sensing: Principles and Interpretation*, 2nd edn. New York: W. H. Freeman.

Saunders, R. W., & Kriebel, K. T. (1988). An improved method for detecting clear sky and cloudy radiances from AVHRR data. *Int. J. Remote Sens.*, **9**, 123–150.

Schutz, B. E. (1998). Spaceborne laser altimetry: 2001 and beyond. In *Book of Extended Abstracts, WEGENER-98*, ed. H. P. Plag. Honefuss, Norway: Norwegian Mapping Authority.

Segelstein, D. (1981). *The Complex Refractive Index of Water*. M.S. thesis, University of Missouri-Kansas City.

Shankaranarayanan, K., & Donelan, M. A. (2001). A probabilistic approach to scatterometer function verification. *J. Geophys. Res.*, **106**, 19 969–19 990.

Siegel, D. A., Wang, M., Maritorena, S., & Robinson, W. (2000). Atmospheric correction of satellite ocean color imagery: the black pixel assumption. *Appl. Opt.*, **39**, 3582–3591.

Smith, P. M. (1988). The emissivity of sea foam at 19 and 37 GHz. *IEEE Trans. Geosci. Remote Sens.*, **GE-26**, 541–547.

Smith, R. C., & Baker, K. S. (1981). Optical properties of the clearest natural waters. *Appl. Opt.*, **20**, 177–184.

Smith, W. H. F., & Sandwell, D. T. (1997). Global seafloor topography from satellite altimetry and ship depth soundings. *Science*, **277**, 1957–1962.

Spencer, M. W., Wu, C., & Long, D. G. (1997). Tradeoffs in the design of a spaceborne scanning pencil beam scatterometer: application to SeaWinds. *IEEE Trans. Geosci. Remote Sens.*, **35**, 115–126.

(2000). Improved resolution backscatter measurements with the SeaWinds pencil-beam scatterometer. *IEEE Trans. Geosci. Remote Sens.*, **38**, 89–104.

St. Germain, K. M., & Gaiser, P. W. (2000). Spaceborne polarimetric radiometry and the Coriolis WindSat system. *Aerospace Conf. Proc., 2000 IEEE*, **5**, 159–164.

St. Germain, K. M., Poe, G., & Gaiser, P. (1998). Modeling of the polarimetric microwave signal due to ocean surface wind vector. *Proc. IEEE Geosci. Remote Sens. Symposium, 1998 (IGARSS 1998)*, **5**, 2304–2306.

Stammer, D., & Wunsch, C. (1994). Preliminary assessment of the accuracy and precision of TOPEX/POSEIDON altimeter data with respect to the large-scale ocean circulation. *J. Geophys. Res.*, **99**, 24 584–24 604.

Stewart, R. H. (1985). *Methods of Satellite Oceanography*. Berkeley: University of California Press.

Stommel, H. (1966). *The Gulf Stream*, 2nd edn. Berkeley: University of California Press.

Stramski, D., & Kiefer, D. A. (1991). Light scattering by microorganisms in the open ocean. *Prog. Oceanogr.*, **28**, 343–383.

Swift, C. T., & McIntosh, R. E. (1983). Considerations for microwave remote sensing of ocean surface salinity. *IEEE Trans. Geosci. Rem. Sens.*, **GE-21**, 480–491.

Tapley, B. D., Ries, J. C., Davis, G. W., Eanes, R. J., Schutz, B. E., Sun, C. K., Watkins, M. M., Marshall, J. A., Nerem, R. S., Putney, B. H., Klosko, S. M., Luthcke, S. B., Williamson, R. G., & Zelensky, N. P. (1994). Precision orbit determination for TOPEX/POSEIDON. *J. Geophys. Res.*, **99**, 24 383–24 404.

Townsend, W. F., McGoogan, J. T., & Walsh, E. J. (1981). Satellite radar altimeters – present and future oceanographic capabilities. In *Oceanography from Space*, ed. J. F. R. Gower, pp. 625–636. New York: Plenum Press.

Thomas, A. (2001). Measuring the surface temperature of the ocean. *Backscatter*, **12**(2), 34–36.

Thomas, G. E., & Stamnes, K. (1999). *Radiative Transfer in the Atmosphere and Ocean*. Cambridge: Cambridge University Press.

Thompson, A. A., Luscombe, A. P., James, K., & Fox, P. (2001). New modes and techniques of the RADARSAT-2 SAR. *Proc. IEEE Geosci. Remote Sens. Symposium, 2001 (IGARSS 2001)*, **1**, 485–487.

Thompson, D. R., & Beal, R. C. (2000). Mapping high-resolution wind fields using synthetic aperture radar. *Johns Hopkins APL Technical Digest*, **21**(1), 58–67.

Trees, C. C., Clark, D. K., Bidigare, R. R., Ondrusek, M. E., & Mueller, J. L. (2000). Accessory pigments versus chlorophyll a concentrations within the euphotic zone: a ubiquitous relationship. *Limnol. Oceanogr.*, **45**(5), 1130–1143.

Ulaby, F. T., Moore, R. K., & Fung, A. K. (1981). *Microwave Remote Sensing: Active and Passive*, Vol. 1, *Microwave Remote Sensing: Fundamentals and Radiometry*. Boston: Addison-Wesley.

  (1982). *Microwave Remote Sensing: Active and Passive*, Vol. 2, *Radar Remote Sensing and Surface Scattering and Emission Theory*. Boston: Addison-Wesley.

  (1986). *Microwave Remote Sensing: Active and Passive*, Vol. 3, *From Theory to Application*. Boston: Addison-Wesley.

Valenzuela, G. R. (1978). Theories for the interaction of electromagnetic and oceanic waves – a review. *Boundary-Layer Meteorol.*, **13**, 61–85.

Vaughan, R. A. (2001). Satellite climatology – current and future European systems. In *Remote Sensing and Climate Change: the Role of Earth Observation*, ed. A. P. Cracknell, pp. 81–95. Chichester: Springer-Praxis.

Vaughan, R. A., & Wilson, S. T. (2001). Envisat – the mission. In *Remote Sensing and Climate Change: the Role of Earth Observation*, ed. A. P. Cracknell, pp. 241–252. Chichester: Springer-Praxis.

Wadhams, P. (2000). *Ice in the Ocean*. London: Taylor & Francis.

Wadhams, P., & Holt, B. (1991). Waves in frazil and pancake ice and their detection on Seasat SAR imagery. *J. Geophys. Res.*, **96**, 8835–8852.

Walton, C. C., Pichel, W. G., & Sapper, J. F. (1998). The development and operational application of nonlinear algorithms for the measurement of sea surface temperatures with the NOAA polar-orbiting environmental satellites. *J. Geophys. Res.*, **103**, 27 999–28 012.

Wang, M. (1999). Atmospheric correction of ocean color sensors: computing atmospheric diffuse transmittance. *Appl. Opt.*, **38**, 451–455.

Wang, M. (2000). The SeaWiFS atmospheric correction algorithm updates. In *SeaWiFS Postlaunch Technical Report Series*, Vol. 9, ed. S. B. Hooker & E. R. Firestone, NASA/TM-20000206892, pp. 57–68. Greenbelt, MD: NASA Goddard Space Flight Center.

Wang, M., Bailey, S., & McClain, C. R. (2000). SeaWiFS provides unique global aerosol optical property data. *EOS*, **81**(18), 197 ff.

Watts, A. B. (1979). On geoid heights derived from Geos-3 altimeter data along the Hawaiian-Emperor seamount chain. *J. Geophys. Res.*, **84**, 3817–3826.

Weast, R. C. (ed.) (1976). *Handbook of Chemistry and Physics*, 57th edn. Boca Raton, FL: CRC Press.

Wentz, F. J. (1975). A two-scale model for foam-free sea microwave brightness temperatures. *J. Geophys. Res.*, **80**, 3441–3446.

(1978). The forward scattering of microwave solar radiation from a water surface. *Radio Sci.*, **13**(1), 131–138.

(1981). *The effect of sea-surface sun glitter on microwave radiometer measurements.* RSS Tech. Report No. 110481. Santa Rosa, CA: Remote Sensing Systems.

(1983). A model function for ocean microwave brightness temperatures. *J. Geophys. Res.*, **102**, 1892–1908.

(1992). Measurement of oceanic wind vector using satellite microwave radiometers. *IEEE Trans. Geosci. Remote Sens.*, **30**, 960–972.

(1997). A well-calibrated ocean algorithm for Special Sensor Microwave/Imager. *J. Geophys. Res.*, **102**, 8703–8718.

Wentz, F. J., & Meissner, T. (1999). *AMSR Ocean Algorithm, Version 2*. RSS Tech. Report 121599A. Santa Rosa, CA: Remote Sensing Systems.

Wentz, F. J., & Smith, D. K. (1999). A model function for the ocean-normalized radar cross section at 14 GHz derived from NSCAT observations. *J. Geophys. Res.*, **104**, 11 499–11 514.

Wentz, F. J., & Spencer, R. W. (1998). SSM/I rain retrievals within a unified all-weather ocean algorithm. *J. Atmos. Sci.*, **55**, 1613–1627.

Wentz, F. J., Cardone, V. J., & Fedor, L. S. (1982). Intercomparison of wind speeds inferred by the SASS, Altimeter and SMMR. *J. Geophys. Res.*, **87**, 3378–3384.

Wentz, F. J., Mattox, L. A., & Peteherych, S. (1986). New algorithms for microwave measurements of ocean winds: applications to SEASAT and the Special Sensor Microwave Imager. *J. Geophys. Res.*, **91**, 2289–2307.

Wentz, F. J., Peteherych, S., & Thomas, L. A. (1984). A model function for ocean radar cross sections at 14.6 GHz. *J. Geophys. Res.*, **89**, 3689–3704.

Wilheit, T. T. (1978). A review of applications of microwave radiometry to oceanography. *Boundary-Layer Meteorol.*, **13**, 277–293.

Wilson, W. S. (2001). Oceanography from space in the U.S.A. *Backscatter*, **12**(2), 31–36.

Wilson, W. S., Apel, J. R., & Lindstrom, E. J. (2001). Satellite oceanography, history and introductory concepts. In *Encyclopedia of Ocean Sciences*, ed. J. H. Steele, K. K. Turekian & S. A. Thorpe, Vol. 5, pp. 2517–2530. London: Academic Press.

Wu, J., (1990). Mean square slopes of the wind-disturbed water surface, their magnitude, directionality, and composition. *Radio Sci.*, **25**, 37–48.

Wu, X., & Smith W. L. (1997). Emissivity of rough sea surface for 8–13μm: modeling and validation. *Appl. Opt.*, **36**, 1–11.

Wunsch, C. (2002). What is the thermohaline circulation? *Science*, **298**, 1179–1180.

Wunsch, C., & Stammer, D. (1998). Satellite altimetry, the marine geoid, and the oceanic general circulation. *Ann. Rev. Earth Planet. Sci.*, **26**, 219–253.

Wunsch, C., Anderle, R. J., Bryden, H., Douglas, B., Halpern, D., Haxbe, W., Marsh, J., Rapp, R., Reid, J., Schott, F., Stewart, R., Tapley, B., Thompson, D., Walsh, E., & Wyrtki, K. (1981). *Satellite Altimetric Measurements of the Ocean, Report of the TOPEX Science Working Group*. Pasadena, CA: NASA Jet Propulsion Laboratory, California Institute of Technology.

Yueh, S. H. (1997). Modeling of wind direction signals in polarimetric sea surface brightness temperatures. *IEEE Trans. Geosci. Remote Sens.*, **35**, 1400–1418.

Yueh, S. H., Wilson, W. J., Li, K., & Dinardo, S. J. (1999). Polarimetric microwave brightness signatures of ocean wind directions. *IEEE Trans. Geosci. Remote Sens.*, **37**, 949–959.

(2002). Polarimetric radar remote sensing of ocean surface wind. *IEEE Trans. Geosci. Remote Sens.*, **40**, 793–800.

Xie, S.-P., Liu, W. T., Liu, Q., & Nonaka, M. (2001). Far-reaching effects of the Hawaiian Islands on the Pacific ocean-atmosphere system. *Science*, **292**, 2057–2060.

Zaneveld, J. R. V. (1995). A theoretical derivation of the dependence of the remotely sensed reflection of the ocean on the inherent optical properties, *J. Geophys. Res.*, **100**, 13 135–13 142.

Zaneveld, J. R. V., & Kitchen, J. C. (1995). The variation in the inherent optical properties of phytoplankton near an absorption peak as determined by various models of cell structure. *J. Geophys. Res.*, **100**, 13 309–13 320.

Zeng, L., & Brown, R. A. (1998). Scatterometer observations at high wind speeds. *J. Applied Meteor.*, **37**, 1412–1420.

Zieger, A. R., Hancock, D. W., Hayne, G. S., & Purdy, C. L. (1991). NASA radar altimeter for the TOPEX/POSEIDON Project. *Proc. IEEE*, **79**, 810–826.

Zierden, D. F., Bourassa, M. A., & O'Brien, J. J. (2000). Cyclone surface pressure fields and frontogenesis from NASA Scatterometer (NSCAT) winds. *J. Geophys. Res.*, **105**, 23 967–23 981.

Zwally, H. J., Comiso, J. C., Parkinson, C. L., Campbell, W. J., Carsey, F. D., & Gloersen, P. (1983). *Antarctic Sea Ice, 1973–1976: Satellite Passive-Microwave Observations*. NASA Spec. Publ. 459. Washington, DC: NASA.

Zwally, H. J., Comiso, J. C., Parkinson, C. L., Cavalieri, D. J., & Gloersen, P. (2002). Variability of Antarctic sea ice 1979–1998. *J. Geophys. Res.*, **107**(C5), 3041, doi: 10.1029/2000JC00733.

Index

411